Lecture Notes in Mathematics

A collection of informal reports and seminars
Edited by A. Dold, Heidelberg and B. Eckmann, Zürich

T0183968

212

Bruno Scarpellini

Universität Basel, Basel/Schweiz

Proof Theory and Intuitionistic Systems

Springer-Verlag
Berlin · Heidelberg · New York 1971

AMS Subject Classifications (1970): Primary: 02D99, 02E05. Secondary: 02C15

ISBN 3-540-05541-X Springer-Verlag Berlin · Heidelberg · New York
ISBN 0-387-05541-X Springer-Verlag New York · Heidelberg · Berlin

Offsetdruck: Julius Beltz, Hemsbach/Bergstr.

PREFACE

The aim of this monograph is to show that the methods used by Gentzen in his second consistency proof for number theory can be extended and used in order to exhibit properties of mathematical interest of certain intuitionistic systems of analysis. The monograph has its root in a paper [8] in which familiar properties of number theory have been derived with the aid of Gentzen methods. An outline of the material contained in chapter IV has been presented at the Buffalo conference on intuitionism and proof theory (1968) [9], while other parts have been discussed in seminaries on mathematical logic at the university of Basel. A detailed introduction, containing a review of the content of the monograph, is given at the beginning of chapter I.

The author would like to express his gratitude to the Swiss national foundations whose financial support made this work possible. Thanks are also due to the Freiweillige Akademische Gesellschaft Basel which supplied the major part of the typing costs.

CONTENTS

CHAPTER V:

Transfinite induction with respect to recursive wellorderings without function parameters

CHAPTER VI:

A formally intuitionistic theory equivalent to classical transfinite induction with respect to primitive recursive wellfounded trees with function parameters

CHAPTER VII:

A system containing bar induction with respect to decidable predicates

CHAPTER VIII:

Harrop formulas

CONTENTS

CHAPTER V:

Transfinite induction with respect to recursive wellorderings without function parameters

CHAPTER VI:

A formally intuitionistic theory equivalent to classical transfinite induction with respect to primitive recursive wellfounded trees with function parameters

CHAPTER VII:

A system containing bar induction with respect to decidable predicates

CHAPTER VIII:

Harrop formulas

CHAPTER IX:

The Markov principle

CHAPTER X:

Relative consistency proof of ZT/V with respect to ZTi/I_N^*

CHAPTER I:

Introduction and preliminaries

1.1. Introductory remarks

A. The work presented in this monograph consists essentially of two components: 1) the results which are proved, 2) the techniques which are used in proving them. Let us begin with a quick review of the kind of results which we are going to prove. For certain intuitionistic theories T and certain families F of formulas we are going to prove a statement (to be denoted by (s) in the sequel) of the following kind:

(S) Let $A_1,\dots,A_s$ be closed formulas from F and A, B, $(E\xi)C(\xi)$, $(Ex)D(x)$ be arbitrary closed formulas.

a) If $A_1,\dots,A_s$, $T \vdash A \vee B$ then $A_1,\dots,A_s$, $T \vdash A$ or $A_1,\dots,A_s, T \vdash B$.

b) If $A_1,\dots,A_s$, $T \vdash (E\xi)C(\xi)$, then there is a functor F such that $A_1,\dots,A_s$, $T \vdash (C(F)$ holds.

c) If $A_1,\dots,A_s$, $T \vdash (Ex)D(x)$ then there is a numeral n such that $A_1,\dots,A_s$, $T \vdash D(n)$ holds.

The language from which the formulas of the theories to be considered are constructed, is that of second order arithmetic, that is essentially the language used in the book of Kleene-Vesley. The theories T for which we are going to prove statement (S) (and whose exact definition will be given in the course of the work) are theories which are obtained from intuitionistic number theory by addition of certain axiom schemas of transfinite induction. Among these we mention in particular: 1) an intuitionistic theory which has the same strength as classical analysis, 2) the intuitionistic theory of barinduction with respect to primitive recursive wellfounded trees, 3) the intuitionistic theory of barinduction with respect to decidable wellfounded trees. The families F which are admitted in statement (S) are: 1) a family F of formulas considered for the first time by R. Harrop in [2], and which we call for simplicity the family of Harrop formulas, 2) the subfamily of classically true Harrop formulas. Two major applications will be presented: 1) an application to questions connected with the Markov principle, 2) a relative consistency proof of the classical theory of barinduction with respect to wellfounded primitive recursive trees modulo a weak system of intuitionistic analysis. Many further applications of the methods developed here have been omitted in order to keep the size of the monograph within reasonable limits.

B. Before proceeding further we would like to stress the fact that our results are not contained in the results obtained by Kleene in [6] (which are perhaps more interesting from an intuitionistic point of view) who proves the statement (S) for the system treated in [5], but without the family F and the formulas $A_1,\dots,A_s$. On the other hand Kleene's result is not contained in ours and there are reasons which suggest that there is no straightforward extension of our technique in order to recover Kleene's result.

C. Now a few words about the techniques used in this work. A first application to intuitionistic systems of the methods used in Gentzen's second consistency proof has been presented in [8] , where several familiar properties of intuitionistic number theory have been derived by means of Gentzen's techniques; among others we mention in particular statement (S), a result which has for the first time been proved by R. Harrop in [2] (with F the family of Harrop formulas, of course). At about that time, the author discovered what he calls the basic lemma; he then recognized that the basic lemma permitted a proof theoretic treatment of certain intuitionistic systems of analysis, some of them as strong as classical analysis. The basic lemma really deserves its name as the reader will see; everything presented in this work depends completely on its validity.

At first sight one might believe that the restriction to intuitionistic systems of analysis in this work is due to a deficiency of the method and that more refined methods permit us to treat classical systems in a similar way. However, by using a result due to Kreisel (whose proof he sketched in the first volume of the Stanford report [12]) one can show that the techniques used by Gentzen in his second consistency proof cannot be applied to sufficiently strong systems of classical analysis if they are formulated in the language of second order arithmetic. Hence, one of the main reasons, why proof theoretic methods can successfully be applied to the systems considered in this monograph is that this systems are intuitionistic.

D. As mentioned, there are many results which the author did not present in this monograph. However, there are also many problems which came up in the course of the work, which the author could not solve. Among these we would like to mention just one: to recover Bachmann's ordinal $\varphi\varepsilon_{\Omega+1}(1)$ from the reduction technique presented in chapter V.

E. Next, some words about the organisation of the work. In chapter I we present preliminaries and list the formal systems which will find consideration later on; some elementary properties of this systems are discussed. In chapter II we present a short repetition of Gentzen's second consistency proof together with a mild generalization. In chapter III we discuss the application of Gentzen's methods to intuitionistic number theory; the basic lemma is proved. In chapter IV we consider an intuitionistic system (call it T_0) which is as strong as classical analysis and show that Gentzen's proof theoretic methods can be applied to this system. For this system we prove among others a weak version of statement (S), that is, statement (S) but without F and $A_1,\dots,A_s$. An outline of the material contained in chapter IV has been presented at the Buffalo conference of proof theory and intuitionism [9]. In chapters V, VI and VII we consider consecutively three systems of intuitionistic analysis; we denote them by T_1, T_2 and T_3 for the moment being. Theory T_1 is equivalent to the intuitionistic theory of barinduction over wellfounded, primitive recursive trees, with function parameters absent. In order to explain the strength of T_2 let T_2^* be the classical theory of barinduction over primitive recursive wellfounded trees, with function parameters admitted. Next, for any formula A, let A^0 be the result of replacing $\vee$ and E by $\neg$, $\wedge$ and $\forall$ in the well-known way described eg. in [4], p. 493. Now T_2 is a formally intuitionistic theory having the property : if the sequent $\longrightarrow$ A is provable in T_2^* then $\longrightarrow A^0$ is provable in T_2. The theory T_3 finally is essentially equivalent to the theory which one obtains if one omits from the system of Kleene-Vesley the axiom of choice and the axiom of continuity. For each of these systems we prove the weak form of statement (S) (that is without F and $A_1,\dots,A_s$) with the aid of a method which differs considerably from that one used in chapter IV. The advantage of this method becomes clear in chapter VIII, which is so to speak the main chapter of our monograph, in that it contains the most general results. In this chapter we prove three results: 1) as a preparation full statement (S) for intuitionistic number theory, with F the family of Harrop formulas, 2) statement (S) for the intuitionistic theory T_0 of chapter IV, with F the family of classically true Harrop formulas, 3) statement (S) for the intuitionistic theories T_1 and T_3, (considered in chapter V and VII) with F the full family of Harrop formulas. In order to prove 2) we use the methods of chapter IV combined with some new ideas involved in the proof of 1), in order to prove 3) we use the methods of chapter V and VII respectively, combined with the ideas used in the proof of 1). Chapter IX contains some applications of the results obtained in chapter VIII to questions centering around the Markov principle. Its main result is the following: (with F the family of Harrop formulas) if $A_1,\ldots,A_s \in F$ and if $A_1,\ldots,A_1$, T_3 is consistent then Markov's principle is not derivable from $A_1,\ldots,A_s$, T_3 . Chapter X finally contains a kind of consistency proof for the theory T_2 (and hence for T_2^*) considered in chapter VI. More precisely we show that the consistency of T_2 can be reduced to the consistency of a certain (seemingly) weak subtheory $\hat{T}_1$ of T_1 . The basic idea used is the following: one shows that the apparently unconstructive method used in chapter VI can be made constructive to such an extent that it can be formalized in $\hat{T}_1$.

F. Now some remarks about the presentation. The presentation is not polished and many similar things are presented in a different way at different places. The reason for this is that many results were found when the monograph was already under preparation (in particular the results in chapters VIII and IX). It would have been possible to condense chapters V, VII and VIII into one single chapter. The reason for not having done this is that it would have been difficult for the reader to grasp the simple mathematical ideas which lie behind the sometimes rather involved syntactical considerations. Most of the theorems stated in this work are proved in detail; however, if a proof is only a slight variant of an other, similar one, given earlier, then we content ourself with an outline or an indication. An exception is perhaps the consistency proof presented in chapter X. There, we did not present all the details, since this would have increased the size of the monograph considerably. However, we have worked out the consistency proof to such an extent that it will become clear to the reader that the details omitted can be supplied without difficulty.

G. The monograph is not selfcontained. The reader is supposed to have a good knowledge of Gentzen's second consistency proof [1] and at least a superficial knowledge of [8]. Concerning ordinal notations the reader is supposed to be familiar with the ordinal functions $\omega_n(\alpha)$, $\alpha \# \beta$, $\alpha + \beta$, $\alpha\beta$ and their properties, such as discussed in Schütte's book [10]. It is not absolutely necessary, but highly recommendable to have some further familiarity with Schütte's book. Finally, it is indispensable for the reader to be familiar with Kleene's "Introduction to Metamathematics" [4], at least with that parts which are concerned with sentential calculus and recursive functions.

1.2. Preliminaries and notations

A. In this section we collect some notions and notations which will be used throughout the rest of this work. We start with a few remarks on primitive recursive functions. By N we denote the set of natural numbers, if not otherwise stated. By N^N we denote the set of mappings from N into N, that is the set of one place number-theoretic functions (or sometimes simply numbertheoretic functions). If S is any set, then S^n denotes the n-fold cartesian product of S; if $S_1,\dots,S_m$ are sets then $S_1 x \dots x S_m$ denotes the cartesian product of $S_1,\dots,S_m$. A function of type (s,t) is a mapping from $(N^N)^s \times N^t$ into N; a functional of type (s,t) is a mapping from $(N^N)^s \times N^t$ into N^N. If $s=t=0$, then f will be identified with an element in N, while F will be identified with an element in N^N. Let f be a function of type $(s,t+1)$. With f we associate a functional F of type (s,t), which satisfies the following equation: for all $g_i \in N^N$, $i=1,\dots,s$ and all $n,n_1,\dots,n_t \in N$
$F(g_1,\dots,g_s,n_1,\dots,n_t)(n)=f(g_1,\dots,g_s,n_1,\dots,n_j,n,n_{j+1},\dots,n_t)$
(with $1 \leq j \leq t$). The uniquely determined F will be denoted by $\Lambda_j f$ or Λf in case $j=t$, or also by
$\Lambda y f(\alpha,\dots,\alpha_s,x_1,\dots,x_j,y,x_{j+1},\dots,x_t)$, where α_j, x_k indicate function and number arguments and where y is "bound" by the abstraction operator Λ.

In this work it is convenient to use a particular notion of primitive function and primitive recursive functional. Their inductive definition is given by the clauses listed below, where Greek letters

α_i, γ_k represent elements from N^N, while x_i, y_k run over N. a) The natural numbers are primitive recursive (p.r.) functions of type $(0,0)$. b) The successor function s (of type $(0,1)$) given by $s(x)=x+1$, is a p.r. function. c) The functions $f_i^{s,t}$ of type (s,t), given by $f_i^{s,t}(\alpha_1,\ldots,\alpha_s,x_1,\ldots,x_t)=x_i$ $(1\leq i\leq t)$, are p.r. functions. d) The functions $f_{i,k}^{s,t}$ of type (s,t), given by $f_{i,k}^{s,t}(\alpha_1,\ldots,\alpha_s,x_1,\ldots,x_t)=\alpha_i(x_k)$, are p.r. functions (with $1\leq i\leq s$, $1\leq k\leq t$). e) If f of type (s,t) is p.r. then $\Lambda_i f$ $(1\leq i\leq t)$ is a p.r. functional. f) Let f of type $(s,t+1)$ and g of type $(a+b,c+d)$, with $a\leq s$, $c\leq t$, be p.r. functions. Let $\vec{\alpha}$, $\vec{\gamma}$, $\vec{x}$ be short for $\alpha_1,\ldots,\alpha_s$ and $\gamma_1,\ldots,\gamma_b$ and $x_1,\ldots,x_{t+d}$ respectively; assume $1\leq i\leq t$.

The function

$$f(\vec{\alpha},x_1,\ldots,x_i,g(\alpha_1,\ldots,\alpha_a,\vec{\gamma},x_1,\ldots,x_c,x_{t+1},\ldots,x_{t+d}),x_{i+1},\ldots,x_t)$$

is a p.r. function of type $(s+b,t+d)$. g) Let f be a p.r. function of type $(s+1,t)$ and F a p.r. functional of type $(a+b,c+d)$ with $a\leq s$, $c\leq t$; assume $1\leq i\leq s$. Let $\vec{\alpha}$, $\vec{\gamma}$ be as before.

The function

$$f(\alpha_1,\ldots,\alpha_i,F(\alpha_1,\ldots,\alpha_a,\vec{\gamma},x_1,\ldots,x_c,x_{t+1},\ldots,x_{t+d}),$$
$$\alpha_{i+1},\ldots,\alpha_s,x_1,\ldots,x_t)$$

is a p.r. function of type $(s+b,t+d)$. h) Let f and g be p.r. functions of type (s,t) and $(a+b,c+d)$ respectively, with $a\leq s$, $c\leq t$. Assume $1\leq i\leq s+d+1$. Then we can define a function φ by means of the following inductive clauses:

1) $\varphi(\vec{\alpha},\vec{\gamma},\vec{x},0) = f(\vec{\alpha},x_1,\ldots,x_t)$,

2) $\varphi(\vec{\alpha},\vec{\gamma},\vec{x},n+1) = g(\vec{\alpha},\vec{\gamma},x_1,\ldots,x_i,n,x_{i+1},\ldots,x_{t+d},\varphi(\vec{\alpha},\vec{\gamma},\vec{x},n))$.

Then φ is a p.r. function of type $(s+b,t+d+1)$ (with $\vec{\alpha}$, $\vec{\gamma}$ and $\vec{x}$ as before). Clauses f), g) simply state, that the set of p.r. functions is closed under substitution; h) means that the set of p.r. functions is closed under primitive recursion. We note three facts: 1) the functions f given by $f(\vec{\alpha},\vec{x})=n=\text{constant}$ are p.r. functions in virtue of clauses a), c) and f), 2) if f is a p.r. function and $\vec{\beta}$ and $\vec{y}$ permutations of $\vec{\alpha}$ and $\vec{x}$ respectively, then f^*, given by $f^*(\vec{\beta},\vec{y})=f(\vec{\alpha},\vec{x})$, is a p.r. function, 3) if F is a p.r.

functional of type $(s,t+1)$, then f, given by $f(\vec{\alpha},\vec{x},y) = F(\vec{\alpha},\vec{x})(y)$, is a p.r. function.

B. Sequences of numbers are codified in the usual way: with $a_0,\ldots,a_{s-1}$ we associate the number $p_0^{a_0+1}\cdot\ldots\cdot p_{s-1}^{a_{s-1}+1}$, where $p_0,p_1,\ldots$ is the list of primes, starting with 2 and listed in increasing order. A number of the form $p_0^{a_0+1}\cdot\ldots\cdot p_{s-1}^{a_{s-1}+1}$ is called sequence number. Sequence numbers will usually be denoted by letters such as $u,v,w,u_1,v_1,w_1,\ldots$ etc.; the sequence number associated with $a_0,\ldots,a_{s-1}$ will also be denoted by $\langle a_0,\ldots,a_{s-1}\rangle$. The empty sequence is represented by 1 and often written as $\langle\ \rangle$. Concatenation of $u = \langle a_0,\ldots,a_{s-1}\rangle$ with $v = \langle b_0,\ldots,b_{t-1}\rangle$ is given by $\langle a_0,\ldots,a_{s-1}, b_0,\ldots,b_{t-1}\rangle$ and written as $u * v$. As length of $u=\langle a_0,\ldots,a_{s-1}\rangle$ we take s; we write length$(u)=s$ or simply $l(u)=s$. If $u=\langle a_0,\ldots,a_{s-1}\rangle$ and if f is a one place numbertheoretic function then $u*f$ denotes the one place numbertheoretic function g given by: 1) $g(i)=a$ for $i<s$, 2) $g(i)=f(i-s)$ for $i\geq s$. With $f\in N^N$ and $n\in N$ we can associate the sequence number $\langle f(0),\ldots,f(s-1)\rangle$, which will be denoted by $\bar{f}(s)$. Sequence numbers can always be represented in the form $\bar{f}(s)$. A partial ordering $\subseteq_K$ can be introduced as follows: 1) if $n\subseteq_K m$ then n and m are sequence numbers, 2) for $f,g\in N^N$, $\bar{f}(s)\subseteq_K\bar{g}(t)$ iff $s\geq t$ and $f(i)=g(i)$ for $i<t$. The Kleene-Brouwer partial ordering $\subset_K$ is given by: $n\subset_K m$ iff $n\neq m$ and $n\subseteq_K m$. There is a well known total linear ordering of sequence numbers, the so-called Kleene-Brouwer linear ordering. It is denoted by $<_K$ and its definition is as follows: 1) if $n<_K m$, then n and m are sequence numbers, 2) for $f,g\in N^N$, $\bar{f}(s)<_K\bar{g}(t)$ iff either $\bar{f}(s)\subset_K\bar{g}(t)$ or else $\bar{f}(i)=\bar{g}(i)$ and $f(i+1)<g(i+1)$ for some $i<\min(s,t)-1$. The sequence number $u=\langle a_0,\ldots,a_{s-1}$ is said to be an initial segment of $f\in N^N$ if $\bar{f}(s)=u$.

C. Another important notion is that of continuity function. An element $\tau \in (N^N)^s$ is said to be a continuity function if the following holds: 1) if $\tau(n_1,\dots,n_s) \neq 0$ then all n_i are sequence numbers and length (n_i) = length (n_{i+1}) for $i=1,\dots,s-1$,
2) if $\tau(\bar{f}_1(n),\dots,\bar{f}_s(n)) \neq 0$ and $n < m$, then $\tau(\bar{f}_1(n),\dots,\bar{f}_s(n)) = \tau(\bar{f}_1(m),\dots,\bar{f}_s(m))$ (with $f_1,\dots,f_s \in N^N$),
3) for every s-tupel $f_1,\dots,f_s$ of elements from N^N there is an n with $\tau(\bar{f}_1(n),\dots,\bar{f}_s(n)) \neq 0$. An element $\tau \in (N^N)^s \times N^t$ is said to be a generalized continuity function of type $[s,t]$ if for every t-tupel of natural numbers $n_1,\dots,n_t$ $\tau(x_1,\dots,x_s,n_1,\dots,n_t)$ is a continuity function with respect to the variables $x_1,\dots,x_s$. In order to exhibit the particular role of the first s arguments we sometimes write $\tau(x_1,\dots,x_s/y_o,\dots,y_t)$ instead of $\tau(x_1,\dots,x_s,y_1,\dots,y_t)$. Generalized continuity functions can be used in order to describe the behaviour of primitive recursive functions, as the following theorem shows:

Theorem: Let f be a p.r. function of type (s,t). Then we find effectively a generalized p.r. continuity function τ of type $[s,t]$ with the property: for all natural numbers $m,n_1,\dots,n_t$ and all numbertheoretic functions $f_1,\dots,f_s$, if $\tau(\bar{f}_1(m),\dots,\bar{f}_s(m),n_1,\dots,n_t)=k+1$, then $f(f_1,\dots,f_s,n_1,\dots,n_t)=k$.

There are many elementary proofs of this theorem (see section 1.4 for an indication); we omit the details of such a proof. A continuity function, having the properties described by the theorem will be called a continuity function related with f. The word "effective" could easily be made precise with the aid of partial recursive functions and Goedel numbers.

D. The main formalism used in this work is that of Gentzen's sequential calculus, also treated by Kleene in [4]. In connection with sequential calculus we adopt the notions and notations used by Kleene; as example we cite the notion of principal formula of a logical inference. An expression such as eg. $\longrightarrow \supset$ indicates an inference "introduction of an implication on the right"; similarly with $\supset \longrightarrow$, $\longrightarrow \wedge$ etc.. We also use capital Greek letters such as $\Gamma, \Pi, \Sigma, \Lambda$ in order to denote sequences of formulas. The following notation is very convenient:
a) if S_1, S_2 are premisses of a two-premiss inference and S its

conclusion then we express this by writing $S_1,S_2/S$, b) if S_1 is the premiss of a one-premiss inference and S its conclusion then we write S_1/S .

E. Proofs in sentential calculus are treated in an obvious way as finite trees (infinite at some places); we call them proof trees or simply proofs. We could characterize such proof trees in a precise way (see eg. [10]); however, we omit such a characterization and use the properties of proof trees without proving them explicitly whenever they are intuitively evident. With respect to formulas, sequents and proofs we have to be a bit careful in one respect: a formula can occur at several places in a proof and we should actually speak of an "occurence of a formula in a proof". However, in order to avoid lengthy formulations we mostly simply speak of "formula in a proof". It will always be clear from the context whether the formula itself or rather an occurence of the formula in the proof is meant. Similar remarks apply to formulas in sequents and to sequents in proofs. In most of the cases "formula in a proof", "formula in a sequent" and "sequent in a proof" mean "occurence of the formula in the proof" etc., Similarly we have to distinguish between a particular inference, say $S_1,S_2/S$, itself and its occurences in a given proof. Again we speak of an "inference $S_1,S_2/S$ in a proof P" meaning in most of the cases a particular occurence of $S_1,S_2/S$ in P . Some further notions are needed in connection with proof trees. In order to explain them we do not fix the formal system, to which the notion of proof refers. All we have to know about this formal system is that all its inferences have the form $S_1,S_2/S$ or S_1/S . Consider a proof P and two occurences S and S' of sequents in P . We call S the successor of S' if there is either (an occurence of) a one premiss inference S_1/S^* in P or else (an occurence of) a two premiss inference $S_1,S_2/S^*$ in P such that S' is S_1 or S_2 and such that S is S^* ; we call S' a predecessor of S (the predecessor in the first case). A path in P is a list $S_1,\ldots,S_m$ of (occurences of) sequents in P such that S_{i+1} is the successor of S . (An occurence of) a sequent in a proof P , say S , is called an axiom, if S has no predecessors, or in other words, if S is an uppermost sequent in P ; the lowest sequent of P (the only one without successor) is called the endsequent of P . A sequent S in P is said to be situated below the sequent S' in P if there is a path $S_o,\ldots,S_m$ in P such that $m > 0$ and $S'=S_o$

and $S=S_m$. We express this by writing $S \prec S'$ and use $S \preceq S'$ as abbreviation for $S \prec S' \vee S=S'$. If S is a sequent in P then we can consider the set of those occurences of sequents S' in P for which $S \preceq S'$ holds. If we restrict the tree relation to the set $\{S'/S \preceq S'\}$ then we obtain a subtree of P , called the subproof of S in P and denoted by P_S . The occurences of sequents in a proof P are sometimes also called the nodes of P .

F. We also need a small portion of ordinal arithmetics in our work. All that has to be known are essentially the ordinal functions $\omega_n(\alpha)$, $\alpha+\beta$, $\alpha\beta$ and $\alpha \# \beta$ (natural sum) and their properties. The reader will find everything needed about these functions in Schütte's book.

1.3. Languages, Syntax

In this section we introduce the languages on which the systems considered in this work are mainly based.

A.1. The most important of the languages to be used is (apart from minor differences) that one used in [5] . We denote it by L . The alphabet of L consists of the following symbols: 1) the logical signs $\wedge$, $\vee$, $\neg$, $\supset$, $\forall$, E which in this order denote conjunction, disjunction, negation, implication, all-quantifier and existential quantifier; 2) number variables x, y, z, x_i $(i<\omega)$ etc.; 3) variables for one place number theoretic functions α , β , γ , α_i $(i<\omega)$ etc.; 4) an individual constant 0 ; 5) a denumerable list of constants f_o, f_1,... for primitive recursive functions among which the first three f_o, f_1, f_2 play a particular role and are denoted by ' , + and x respectively; 6) for every finite sequence $\vec{u} = \langle u_o, \ldots, u_{x-1} \rangle$ of natural numbers a denumerable list $\alpha^i_{\vec{u}}$ $(i<\omega)$ of so called special function constants; 7) commas and parentheses; 8) the two-place predicate constant =, called equality; 9) the abstraction symbol λ ; 10) the sequential arrow $\longrightarrow$. With every constant f_i we associate in a fixed way an ordered pair of natural numbers $\langle n_i, m_i \rangle$, called the type of f_i . For i=0,1,2 these pairs are in particular $\langle 0,1 \rangle$, $\langle 0,2 \rangle$ and $\langle 0,3 \rangle$ respectively. Now we define the notions "term" and "functor" in the same way as in [5] , namely: 1) number variables and constants are terms; 2) the function variables are functors; 3) the constants for special functions and the constants of type $\langle 0,1 \rangle$ are functors; 4) if $F_1,\dots,F_{n_i}$ are functors and $t_1,\dots,t_{m_i}$ are terms then $f_i(F_1,\dots,F_{n_i}, t_1,\dots,t_{m_i})$ is a term; 5) if F^1 is a functor and t a term then $F(t)$ is a term; 6) if t is a term then (λxt) is a functor. The particular terms $0, 0', (0')'$ etc. are called numerals.

2. The inductive definition of formulas is given as follows: 1) if t_1, t_2 are terms then $t_1 = t_2$ is a prime formula, 2) if A, B are formulas then $(A \wedge B)$, $(A \vee B)$, $(\neg A)$, $(A \supset B)$, $(\forall x)A$, $(Ex)A$, $(\forall \alpha)A$ and $(E\alpha)A$. If no confusion arises we omit brackets and use current abbreviations such as $A_1 \vee A_2 \vee A_3$ for $((A_1 \vee A_2) \vee A_3)$ etc.; universal quantification is often written more simply $(x)A$, $(\alpha)A$. The notions "free occurence of a number variable in a term" (λ binds variables!), "free occurence of a (number or function) variable in a formula", "bound occurence of a (number or function) variable in a formula (term)" are introduced as in [4], § 18, but now taking into account the symbol λ. A closed formula is a formula without free variables (but special function constants may occur in it); a constant functor (term) is a functor (term) which does not contain free variables (but it may contain special function constants).

Let $t, q_1,\dots,q_n$ be terms, $F, G_1,\dots,G_m$ functors, A a formula, $x_1,\dots,x_n$ pairwise distinct number variables and $\alpha_1,\dots,\alpha_m$ pairwise distinct function variables. By $S^{G_1,\dots,G_m,q_1,\dots,q_n}_{1,\dots,\ m,x_1,\dots,x_n}A$ we denote the expression which we obtain if we replace for each i every free occurence of α_i by G_i and for each k every free occurence of x_k by q_k; if no α_i and no x_k occurs free in A then this expression is simply A. The expressions $S^{G_1,\dots,G_m,q_1,\dots,q_n}_{1,\dots,\ m,x_1,\dots,x_n}F$ and $S^{G_1,\dots,G_m,q_1,\dots,q_n}_{1,\dots,\ m,x_1,\dots,x_n}t$ are defined analogously. Clearly, the result of this substitution is again a formula, a functor and a term respectively. Frequently we use more suggestive notations such as $A(G_1,\dots,G_m,q_1,\dots,q_n)$ etc. in order to denote the result of replacing $\alpha_1,\dots,\alpha_m,x_1,\dots,x_n$ wherever they occur free. Of course, we can also replace special function constants by functors: if e.g. $\xi_1,\dots,\xi_s$ are special function constants which occur in a formula A, if $F_1,\dots,F_s$ are functors then $S^{F_1,\dots,F_s}_{\xi_1,\dots,\xi_s}A$ is the expression which we obtain when we replace each ξ_i by F_i wherever ξ_i occurs in A. Similarly with a term t or a functor G in place of A. In this connection we use the notions "t is free for x in A", "G is free for α in A" etc. which are defined in the same way as in [4], § 18. We note: for every term t (functor F, formula A) there is an other term t' (functor F', formula A') without special function constants, pairwise distinct function variables $\alpha_1,\dots,\alpha_s$ and special function constants $\xi_1,\dots,\xi_s$ such that $t = S^{\xi_1,\dots,\xi_s}_{1,\dots,x}t'$ (or $F = S^{\xi_1,\dots,\xi_s}_{1,\dots,s}F'$ or $A = S^{\xi_1,\dots,\xi_s}_{1,\dots,s}A'$). One can easily prove that t' is essentially determined by t (that is up to the function variables $\alpha_1,\dots,\alpha_q$ which one is going to replace by $\xi_1,\dots,\xi_s$ respectively). However we do not need this. We merely assume that the term t' has been associated in a fixed and well determined way: we call t' <u>the term associated with t</u>. If t contains no special function constants, then clearly $t=t'$. The variables $\alpha_1,\dots,\alpha_s$ in t' which we are going to replace by $\xi_1,\dots,\xi_s$ are called the <u>substitution variables</u> of t' (with respect to t).

<u>3.</u> We now make a convenient assumption which is supposed to be satisfied throughout the whole work.

<u>Assumption A:</u> With every constant f_i we associate (in an effective way) once and for all a fixed primitive recursive function φ_i of type $\langle n_i,m_i\rangle$. Moreover this assignment is such that every primitive recursive function φ is associated with at least one f_i. In particular φ_0 is the successor function, φ_1 is addition and φ_2 is multiplication.

From now on we will work with primitive recursive functions in a liberal way and introduce special notations for particular ones whenever we find it convenient. Let M be the set of terms and functors containing no special function constants. Taking assumption A as basis and making use of the remarks on primitive recursive functions and functionals stated in "Preliminaries and Notations" we can associate with every term $t(\alpha_1,\ldots,\alpha_s,x_1,\ldots,x_q)$ and every functor $F(\alpha_1,\ldots,\alpha_s,x_1,\ldots,x_q)$ belonging to M a primitive recursive function $\varphi(\alpha_1,\ldots,\alpha_s,x_1,\ldots,x_q)$ and a primitive recursive functional $\widetilde{F}(\alpha_1,\ldots,\alpha_s,x_1,\ldots,x_q)$ respectively in an obvious and well determined way. Of course this assignment is defined in such a way as to be compatible with the inductive definition of terms and functors: if e.g. φ is associated with t then $\Delta x \varphi$ is associated with (λxt), if in turn F is associated with F and φ with t, then $\widetilde{F}(\varphi)$ is associated with F(t) etc. We call φ the primitive recursive function associated with t or more briefly the primitive recursive function of t and $\widetilde{F}$ the primitive recursive functional associated with F (of F). As pointed out in "Preliminaries and Notations" one can relate with every primitive recursive function $\varphi(\alpha_1,\ldots,\alpha_s,x_1,\ldots,x_t)$ a generalized continuity function $\tau(y_1,\ldots,y_s/x_1,\ldots,x_t)$ which "describes" the behaviour of τ for its arguments in the way explained in "Preliminaries and Notations".

<u>Definition 0:</u> Let t be a term in M , φ its primitive recursive function and τ a (generalized) continuity function related with φ. Then we call τ <u>a</u> continuity function of t related with t.

<u>Assumption B:</u> With every term t from the set M we associate in an effective way once and for all a fixed continuity function τ related with t , called <u>the</u> continuity function of t.

There are many possibilities of associating with a term t a continuity function τ related with t . A particular way of doing this will be described at the end of the next section; this particular assignment will find application in chapter IX.

<u>Definition 1:</u> Let t be a term containing no free variables, let $\alpha^{1}_{\vec{u}_1},\ldots,\alpha^{s}_{\vec{u}_s}$ be the list of special function constants occuring in t , let t' be the term associated with t and $\alpha_1,\ldots,\alpha_s$ its substitution variables (with respect to t). Let

finally $\tau(y_1,\ldots,y_s)$ be the continuity function of t' . We say that t is <u>saturated</u> if $\tau(\vec{u}_1,\ldots,\vec{u}_s)$ is greater than zero. In this case we denote the number $\tau(\vec{u}_1,\ldots,\vec{u}_s)-1$ by $/t/$.

In other words, if eg. $a(\alpha,\beta)$ is a term from M , whose only free variables are the function variables α, β, if $\varphi(\alpha,\beta)$ is its primitive recursive function, if $\xi_{\vec{u}}$, $\eta_{\vec{v}}$ are two constants for special functions, then $q(\xi_{\vec{u}}, \eta_{\vec{v}})$ is saturated if we are able to calculate the value of $\varphi(f,g)$ under the sole assumption that $\vec{u}$ is an initial segment of f and $\vec{v}$ an initial segment of g . At this point we can briefly explain the role of the special function constants $\alpha^1_{\vec{u}}$, $\alpha^2_{\vec{u}}$,... . In the formal systems which we are going to consider, this symbols are treated like constants for functions. Their semantical meaning, however, is rather the same as that of function variables: $\alpha^1_{\vec{u}}$ represents so to speak a function f , about which we only know that $f(i)=u_i$ for $i<n$ where $\vec{u} = \langle u_0,\ldots,u_{x-1}\rangle$ and which is undetermined otherwise. In principle one could avoid the use of special function constants; their introduction, however, turns out to be very convenient.

5. Next we need

<u>Definition 2:</u> Two formulas A,B are called isomorphic (with each other) if there is a formula $C(x_1,\ldots,x_s)$ containing the free individual variables $x_1,\ldots,x_s$ and two lists of saturated terms $t_1,\ldots,t_s$ and $q_1,\ldots,q_t$ such that: a) $|q_i| = t_i$ for $i=1,\ldots,s$, b) $C(t_1,\ldots,t_s)$ is A , c) $C(q_1,\ldots,q_s)$ is B . Similarly for terms p,q and functors F,G .

6. Sequents are expressions of the form $A_1,\ldots,A_s \longrightarrow B_1,\ldots,B_t$, where the formulas A_i or the formulas B_k or both may be absent. The list $A_1,\ldots,A_s$ is called the antecedent, the list $B_1,\ldots,B_t$

the succedent. Prime formulas are those of the form $t_1=t_2$. A sequent which contains only prime formulas is called a prime sequent. A saturated prime sequent is one which contains only prime formulas $t_1=t_2$ with t_1,t_2 saturated.

7. It remains to explain what a true prime sequent is. To this end, let $t_1,\ldots,t_s$ be a list of terms, let $\alpha_1,\ldots,\alpha_m,x_1,\ldots,x_q,\beta^{1}_{\vec{u}_1},\ldots,\beta^{p}_{\vec{u}_p}$ be an enumeration (without repetition) of the free function variables, free number variables and special function constants which occur in at least one t_i . A list $\xi^{1}_{\vec{v}_1},\ldots,\xi^{m}_{\vec{v}_m},\eta^{1}_{\vec{w}_1},\ldots,\eta^{p}_{\vec{w}_p}$ of special function constants and a list of numerals $n_1,\ldots,n_q$ is called a saturating list for $t_1,\ldots,t_s$ (with respect to the given enumeration) if the following holds: a) every $\vec{w}_i$ is a (proper or improper) extension of $\vec{u}_i$, b) replacement of α_i by $\xi^{i}_{\vec{v}_i}$, of $\beta^{j}_{\vec{u}_j}$ by $\eta^{j}_{\vec{w}_j}$ and of x_k by n_k $(i=1,\ldots,m\ ,\ j=1,\ldots,p\ ,\ k=1,\ldots,n)$ transforms every term t_q into a saturated term t'_q . We express the relation between $t_1,\ldots,t_s,t'_1,\ldots,t'_s$, the enumeration $\alpha_1,\ldots,\alpha_m,x_1,\ldots,x_q,\beta^{1}_{\vec{u}_1},\ldots,\beta^{p}_{\vec{u}_p}$ and the saturating list $\xi^{1}_{\vec{v}_1},\ldots,\xi^{m}_{\vec{v}_m},\eta^{1}_{\vec{w}_1},\ldots,\eta^{p}_{\vec{w}_p},n_1,\ldots,n_q$ briefly by saying that the given saturating list transforms $t_1,\ldots,t_s$ into $t'_1,\ldots,t'_s$ without mentioning the enumeration $\alpha_1,\ldots,\alpha_m,x_1,\ldots,x_q,\beta^{1}_{\vec{u}_1},\ldots,\beta^{p}_{\vec{u}_p}$ explicitly. Now to the truth definition for prime sequents. If the sequent S , which we assume to be given explicitly by

$t_1=p_1,\ldots,t_s=p_s \longrightarrow q_1=r_1,\ldots,q_t=r_t$, is saturated, then S is of course true if either $|t_i| \neq |p_i|$ or $|q_k| = |r_k|$ for at least one i or k . Now assume that S is not saturated. Then S is called true in the first sense if every saturating list for

$t_1,\dots,t_s,p_1,\dots,p_s,q_1,\dots,q_t,r_1,\dots,r_t$ transforms this list into $t_1',\dots,t_s',p_1',\dots,p_s',q_1',\dots,q_t',r_1',\dots,r_t'$ such that the (necessarily saturated) sequent S' : $t_1'=p_1' ,\dots,t_s'=p_s' \longrightarrow q_1'=r_1' ,\dots,q_t'=r_t'$ is true. There is of course another more natural definition of truth for prime sequents. Let S be as above and let φ_i^t , φ_j^p , φ_k^q, φ_h^r be the primitive recursive functions associated with t_i , p_j , q_k , r_h respectively. Then S is true in the second sense if the following holds : in whatever way we put functions and numbers at the respective argument places of φ_i^t , φ_j^p , φ_k^q , φ_h^r , the resulting intuitive implication "if $\varphi_i^t = \varphi_i^p$ for all $i \leqq s$, then $\varphi_k^q = \varphi_k^r$ for at least one k " is true. For us it is useful to note the following, easily provable fact: a prime sequent is true in the first sense if and only if it is true in the second sense. This closes our discussion of the language L and the concepts immediately related with it. The discussion of the notion of truth for arbitrary formulas and sequents will be postponed to a later section.

B. On many occasions we have to consider formulas and sequents which are constructed with respect to a certain restricted language L* . This language L* is obtained from L merely by deleting the constants for special functions. Then all definitions and statements made in part A of this section specialize immediately to the case of the language L* , by omitting all references to special function constants. The resulting notions then essentially coincide (apart from minor differences) with the corresponding notions in 1.3.,A.

1.4. Some basic systems

The aim of this section is to introduce some formal systems which will serve as basis for all later considerations. One of these systems is essentially number theory, formalized in terms of sentential calculus. All these systems have L and L* respectively as their basic languages.

A. Let f_i, f_j, f_k be three different constants for primitive recursive functions and let $\varphi_i, \varphi_j, \varphi_k$ be the primitive recursive functions associated with f_i, f_j, f_k respectively. The types of f_i, f_j and f_k are for simplicity assumed to be $\langle 1,1 \rangle$, $\langle 1,3 \rangle$ and $\langle 1,2 \rangle$ respectively. Now let us assume that φ_k is defined from φ_i and φ_j by means of the following recursion scheme:

1) $\varphi_k(\alpha, 0, x) = \varphi_i(\alpha, x)$,

2) $\varphi_k(\alpha, y+1, x) = \varphi_j(\alpha, y, \varphi_k(\alpha, y, x), x)$. Then we call the following two sequents the defining sequents of f_k :

$\longrightarrow f_k(\alpha, 0, x) = f_i(\alpha, x)$, $\longrightarrow f_k(\alpha, y', x) = f_j(\alpha, y, f_k(\alpha, y, x) x)$.

Similarly let g_1, g_2, h_1, h_2, f and f^* be a list of different constants for primitive recursive functions. For simplicity we assume that the types of this constants are $\langle 1,2 \rangle$, $\langle 1,2 \rangle$, $\langle 1,1 \rangle$, $\langle 1,1 \rangle$, $\langle 2,2 \rangle$ and $\langle 1,1 \rangle$ respectively. Let $\varphi_1(\alpha, x, y)$, $\varphi_2(\alpha, x, y)$, $\varphi_1(\alpha, x)$, $\varphi_2(\alpha, x)$, $\varphi(\alpha, x)$ and $\theta(\beta_1, \beta_2, y_1, y_2)$ be the primitive recursive functions associated with g_1, g_2, h_1, h_2, f and f^* respectively. Now let us assume that $\varphi(\alpha, x)$ is defined from $\varphi_1, \varphi_2, \varphi_1, \varphi_2$ and θ by means of substitution as follows:

$$\varphi(\alpha, x) = \theta(\Lambda y\, \varphi_1(\alpha, x, y), \Lambda y\, \varphi_2(\alpha, x, y), \phi_1(\alpha, x), \phi_2(\alpha, x)) .$$

Then we call the following sequent the defining sequent of f :

$$\longrightarrow f(\alpha, x) = f^*(\lambda y g_1(\alpha, x, y), \lambda y g_2(\alpha, x, y), h_1(\alpha, x), h_2(\alpha, x)) .$$

If constants of more general types are involved then the corresponding definitions are of course completely analogous. Next, let f_i and f_k be two constants, whose associated primitive recursive functions $\varphi_i(\alpha_1, \ldots, \alpha_{n_i}, x_1, \ldots, x_{m_i})$ and $\varphi_k(\alpha_1, \ldots, \alpha_{n_k}, x_1, \ldots, x_{m_k})$ satisfy the equations $\varphi_i(\alpha_1, \ldots, \alpha_{n_i}, x_1, \ldots, x_{m_i}) = \alpha_j(x_r)$ $(j \leq n_i,\ r \leq m_i)$ and $\varphi_k(\alpha_1, \ldots, \alpha_{n_k}, x_1, \ldots, x_{m_k}) = x_p$ $(p \leq m_k)$ respectively. In this case we call $\longrightarrow f_i(\alpha_1, \ldots, \alpha_{n_i}, x_1, \ldots, x_{m_i}) = \alpha_j(x_r)$ the defining sequent of f_i and $\longrightarrow f_k(\alpha_1, \ldots, \alpha_{n_k}, x_1, \ldots, x_{m_k}) = x_r$ the defining sequent of f_k respectively. Finally, if f_i has as associated primitive recursive function the successor function $\varphi_i(x) = x+1$, then we take as defining sequents of f_i the following ones: $f_i(x) = f_i(y) \longrightarrow x = y$ and $f_i(x) = 0 \longrightarrow$. Thus the defining sequents of $'$ (that is f_o) are $x'=y' \longrightarrow x=y$ and $x'=0 \longrightarrow$, the defining sequents of $+$ (that is f_1) are $\longrightarrow x+0=x$ and $\longrightarrow x+y'=(x+y)'$ and the defining sequents of x finally are $\longrightarrow x \times 0 = 0$ and $\longrightarrow x \times y' = x \times y + x$.

<u>Notation:</u> from now on we write $a \cdot b$ or sometimes even more simpler ab in place of $a \times b$. <u>Remark:</u> Up to now the assignment of φ_i with f_i has been arbitrary except that both have to be of the same type and that assumption A has to be satisfied. One can always choose this assignment in such a way that the following assumption is satisfied.

<u>Assumption C:</u> Every primitive recursive function φ occurs exactly once in the list $\varphi_o, \varphi_1, \varphi_2, \dots$. Each φ_i is either a basic function or defined in terms of previous ones by means of substitution or the schema of primitive recursion.

Actually, we never make use of assumption C; however the reader who likes can always assume C to be satisfied.

<u>B.</u> A sequent which contains at most one formula in the succedent will be called <u>normal</u>. Next let S be a sequent without constants for special functions, whose list of free variables is given by $\alpha_1, \dots, \alpha_s$, $x_1, \dots, x_t$. Then S' is called a substitution instance of S if there is a list of functors $F_1, \dots, F_s$ and terms $q_1, \dots, q_t$ (with F_i free for α_i and q_k free for x_k) such that S' is obtained from S by replacing for each i every free occurence of α_i by F_i and for each k every free occurence of x_k by q_k . Of course, S is a substitution instance of itself. Now we define some sets of sequents. M_o is the set of all sequents of the form $\longrightarrow (\lambda x t(x))=t(q)$ where q is free for x in t . M_1 is the set of all true, saturated normal prime sequents. M_2^i is the set of all defining sequents of f_i ; hence M_2^i contains one or two sequents according to which of the cases, which have been listed under A , applies to f_i . M_2^* is $\bigcup_i M_2^i$. Finally, a sequent S' is in M_2 , if and only if it is a substitution instance of some sequent S in M_2^* . By M_3 we understand the set which contains precisely those sequents having one of the following forms:

$t=p,\ p=q \longrightarrow t=q$, $t=p \longrightarrow p=t$, $\longrightarrow t=t$,

$t=p \longrightarrow S_x^t q = S_x^p q$ (with t,p free for x in q). Next,

M_4 is the set of sequents of the form $D \longrightarrow D'$ where D and D' are isomorphic. As M_5 we take the set of all the sequents $\longrightarrow \alpha_{\vec{u}}^{\geq}(j)=k$, where $\vec{u} = \langle u_o,\ldots,u_{n-1}\rangle$, $j<n$ and $u_j=k$. Finally, let M_6^* be any set of true normal prime sequents not containing special function constants and let S' be in M_6 if and only if it is a substitution instance of some S in M_6^*. The set M_6^* (and hence M_6) is allowed to be void.

Remark: If we agree to associate with $\alpha(x)$ the continuity function $\tau(\vec{u}/x)$ given by $\tau(\vec{u}/x)=0$ for $x \geq n$ and $\tau(\vec{u}/x)=u_x+1$ for $x<n$ where $\vec{u} = \langle u_o,\ldots,u_{n-1}\rangle$, then M_5 is of course a subset of M_1 according to def. 1. In order to exhibit the particular role of the special function constants we have preferred to consider them separately.

We note the trivial

Lemma 0: If $S \in \bigcup_o^6 M_i$ and if S' is a substitution instance of S then $S' \in \bigcup_o^6 M_i$ too.

Clearly, all sequents in $\bigcup_o^6 M_i$ are normal.

C. Now we introduce a formal system ZT whose structure is essentially that of number theory except that it may contain additional true normal prime sequents as axioms (namely those in M_6). The set M of axioms of ZT is $\bigcup_o^6 M_i$. The rules of ZT are the following ones: 1) the structural rules of sentential calculus such as thinning, interchange, contraction and cut; 2) the propositional rules of sentential calculus; 3) the four quantifier rules for number quantification, namely

a) $\dfrac{A(t), \Gamma \longrightarrow \Delta}{(\forall x)A(x), \Gamma \longrightarrow \Delta}$ b) $\dfrac{\Gamma \longrightarrow \Delta, A(y)}{\Gamma \longrightarrow \Delta, (\forall x)A(x)}$

c) $\dfrac{A(y), \Gamma \longrightarrow \Delta}{(Ex)A(x), \Gamma \longrightarrow \Delta}$ d) $\dfrac{\Gamma \longrightarrow \Delta, A(t)}{\Gamma \longrightarrow \Delta, (Ex)A(x)}$

where t is a number term free for x in $A(x)$, where y does not occur free in the conclusions of b) and c), and where y is free for x in $A(x)$;

4) four quantifier rules for function quantification, namely

a') $\dfrac{A(F), \Gamma \longrightarrow \Delta}{(\forall\alpha)A(\alpha), \Gamma \longrightarrow \Delta}$ b') $\dfrac{\Gamma \longrightarrow \Delta, A(\beta)}{\Gamma \longrightarrow \Delta, (\forall\alpha)A(\alpha)}$

c') $\dfrac{A(\beta), \Gamma \longrightarrow \Delta}{(E\alpha)A(\alpha), \Gamma \longrightarrow \Delta}$ d') $\dfrac{\Gamma \longrightarrow \Delta, A(F)}{\Gamma \longrightarrow \Delta, (E\alpha)A(\alpha)}$

where F is a functor free for α in $A(\alpha)$, where β does not occur free in the conclusions of b') and c'), and where β is free for α in $A(\alpha)$;

5) a so-called conversion rule (or more briefly conversion)

$$\frac{A_1,\dots,A_s \longrightarrow B_1,\dots,B_t}{A_1',\dots,A_s' \longrightarrow B_1',\dots,B_t'}$$

where A and B are isomorphic with A' and B' respectively;

6) the induction rule

$$\frac{A(x), \Gamma \longrightarrow \Delta, A(x')}{A(0), \Gamma \longrightarrow \Delta, A(t)}$$

with t a term free for x .

Rule 5) is just another version of Schütte's "Umsetzungsregel" [10], also used in [8] . What we understand by a <u>proof</u> in ZT is clear; we will always consider proofs as certain finite trees of sequents (at many places however we will have to consider infinite trees!). In particular, there is the notion of <u>pure variable proof</u>, introduced in [4] , § 78.

<u>D.</u> Let ZT' be the system which differs from ZT only in that it contains no conversion rule. Let ZT* be that subsystem of ZT' which we obtain by dropping special function constants; that is, a proof in ZT' is a proof in ZT* if it contains only formulas built up from the symbols of the language L* . The following lemma is easily provable:

Lemma 1: a) If a sequent S is provable in ZT then in ZT'.
b) If a sequent S which does not contain special function constants is provable in ZT', then it is provable in ZT*.

Hint: If S is as in pt. b) of the lemma and P a proof in ZT' of S, then we obtain a proof P* in ZT* of S by replacing every special function constant $\alpha_{\vec{u}}$ in P by a constant functor $F_{\vec{u}}$ whose associated primitive recursive function(al) φ has $\vec{u}$ as initial segment. Concerning pt. a) it is sufficient to note that we can derive the conclusion $A_1', \ldots, A_s' \longrightarrow B_1', \ldots, B_t'$ of a conversion from its premiss $A_1, \ldots, A_s \longrightarrow B_1, \ldots, B_t$ by means of structural rules with the aid of the axioms $A_i' \longrightarrow A_i$ and $B_k \longrightarrow B_k'$.

E. A proof P in ZT (in ZT*, ZT') is said to be intuitionistic if it contains only normal sequents, that is sequents which contain at most one formula in the succedent. By restricting attention to intuitionistic proofs we obtain the subsystem ZTi of ZT, called the intuitionistic version of ZT* and ZT', to be denoted by ZTi* and ZTi' respectively. Of course, we have the

Lemma 2: a) If S is provable in ZTi then in ZTi'.
b) If S does not contain special function constants and is provable in ZTi', then it is provable in ZTi*.
The justification of the term "intuitionistic" will be given below.

F. With each of the systems ZT* and ZT' we associate a corresponding Hilbert-type system ZH* and ZH^o, respectively. We give only the description of ZH^o; the description of ZH* is completely analogous. The formulas of ZH^o are the same as those of ZT. The set MA of mathematical axioms of ZH* is given as follows:
a) if $S \in \bigcup_0 M_i$ has the form $A_1, \ldots, A_s \longrightarrow B$ with antecedent and succedent both nonempty, then
$A_1 \supset (A_2 \supset \ldots \supset (A_s \supset B)\ldots)$ is in MA; b) if
$S \in \bigcup_0^6 M_i$ has the form $\longrightarrow B$ then $B \in MA$; c) if $S \in \bigcup_a^6 M_i$ has the form $A_1, \ldots, A_s \longrightarrow$ then
$A_1 \supset (A_2 \supset \ldots\ldots \supset (A_s \supset 0=1)\ldots)$ is in MA;
d) $F \in MA$ only in virtue of a), b), c). The so-called logical axioms listed in [4], p. 82 (such as $A \supset (B \supset A)$, $A \supset A \vee B$ etc. for all formulas A,B); b) two groups of axioms for number quantification, namely $(x)A(x) \supset A(t)$ and

$A(t) \supset (Ex)A(x)$ with t free for x in A ; c) two groups of axioms for function quantification, namely $(\alpha)A(\ ) \supset A(F)$ and $A(F) \supset (E\alpha)A(\alpha)$ where the functor F is free for α in A . Finally, we have the group of induction axioms: $A(0) \wedge (x)(A(x) \supset A(x')) . \supset . A(t)$ with t free for x in $A(x)$ .The rules of ZH are: a) modus ponens $A; A \supset B \ / \ B$; b) two rules for number quantification $C \supset A(x) \ / \ C \supset (x)A(x)$ and $A(x) \supset C \ /(Ex)A(x) \supset C$ with x not free in C ; c) two rules for function quantification $C \supset A(\alpha) \ / \ C \supset (\alpha)A(\alpha)$ and $A(\alpha) \supset C \ / \ (E\alpha)A(\alpha) \supset C$ with α not free in C .

The corresponding intuitionistic version of ZH^o , to be denoted by ZHi^o , is obtained by omitting all propositional axioms of the form $\neg\neg A \supset A$ and by adding in their place all propositional axioms of the form $\neg A \supset (A \supset B)$ ([4] , pp. 82, 101) . The systems ZH* and ZHi* are related to ZT* in the same way as ZH^o and ZHi^o to ZT' .

G. Further systems which will find consideration are the following ones: ZT^*_o , ZTi^*_o , ZT'_o , ZTi'_o and ZH^*_o , ZHi^*_o , ZH^o_o , ZHi^o_o . Each of the systems with index 0 follows from the corresponding one without index by omitting the induction rule (in case of a Gentzen type system) or the group of induction axioms in case of a Hilbert type system.

H. In order to explain the connection between these different systems we recall the notion of a "derivation from given assumptions with all variables held constant", [4] , § 22. In the theorem below and throughout the work, we indicate e.g. the fact that a formula A is derivable from assumptions $A_1,\dots,A_s$ on the basis of ZH* by ZH*: $A_1,\dots,A_s \vdash A$; similarly, if by adding sequents $S_1,\dots,S_n$ to the axioms of ZT* we can derive (by means of the rules of ZT*) the sequent S, then we denote this fact by ZT*: $S_1,\dots,S_n \vdash S$. Analogous notations are used in connection with other systems.

Theorem 0: a) If ZH: $A_1,\dots,A_s \vdash A$ with all variables held constant, then ZT': $\vdash A_1,\dots,A_s \longrightarrow A$. On the other hand, if $F_1,\dots,F_s$ are closed formulas from the language L*, and if ZT': $\longrightarrow F_1,\dots, \longrightarrow F_n \vdash A_1,\dots,A_s \longrightarrow B_1,\dots,B_t,C$, then

ZH: $F_1,\dots,F_n,A_1,\dots,A_s, \neg B_1,\dots, \neg B_t \vdash C$ with variables held constant. b) Likewise in the case of ZHi and ZTi' but with $B_1,\dots,B_t$ absent. c) Likewise in the case of ZH* and ZT* . d) Likewise in the case of ZHi* and ZTi* but with $B_1,\dots,B_t$ absent. The proof of th. 0 is up to a few minor modifications the same as the proofs of theorems 46 and 47 in [4] and will be omitted.

I. In order to study the connection between classical and intuitionistic number theory, Kleene introduces in [4] § 8 two mappings ^{o} and $^+$ of formulas, whose definition is given as follows: 1) A^+ is obtained from A by replacing each prime part P in A by $\neg\neg P$; 2) if P is prime, then P^o is P ; 3) $(A \supset B)^o$, $(A \wedge B)^o$ and $(\neg A)^o$ are $A^o \supset B^o$, $A^o \wedge B^o$ and $\neg A^o$ respectively ; 4) $((x)A)^o$ and $((\alpha)A)^o$ are $(x)A^o$ and $(\alpha)A^o$ respectively ; 5) $(A \vee B)^o$ is $\neg(\neg A^o \wedge \neg B^o)$; 6) $((Ex)A(x))^o$ and $((E\alpha)A(\alpha))^o$ are $\neg(x)\neg A(x)^o$ and $\neg(\alpha)\neg A(\alpha)^o$ respectively.

The connection between ZH* and ZHi* and also between ZH* and ZHi* is described by the following theorem whose proof parallels that one of theorem 60 in [4] :

Theorem 1: If ZH*: $A_1,\dots,A_s \vdash A$, then ZHi*: $A_1^o,\dots,A_s^o \vdash A^o$. Similarly, if ZH_o^*: $A_1,\dots,A_s \vdash A$ then ZHi_o^*: $A_1^{o+},\dots,A_s^{o+} \vdash A^{o+}$. The connection between ZT* and ZTi* and also between ZT* and ZTi* now follows immediately via theorems 0 and 1 :

Corollary: If ZT*: $\longrightarrow F_1,\dots,\longrightarrow F_s \vdash A_1,\dots,A_n \longrightarrow B$, then ZTi*: $\longrightarrow F_1^o,\dots,\longrightarrow F_s^o \vdash A_1^o,\dots,A_n^o \longrightarrow B$, where $F_1,\dots,F_s$ are closed formulas. Similarly, if
ZT*: $\longrightarrow F_1,\dots,\longrightarrow F_s \vdash A_1,\dots,A_n \longrightarrow B$, then
ZTi*: $\longrightarrow F_1^{o+},\dots,\longrightarrow F_s^{o+} \vdash A_1^{o+},\dots,A_n^{o+} \longrightarrow B^{o+}$.

K. The set PR of bounded formulas is defined as follows: 1) a prime formula p=q is in PR ; 2) if A, B are in PR, then so are $A \supset B$, $A \wedge B$, $A \vee B$ and $\neg A$; 3) if A is in PR , if t is a term not containing y free, then $(Ey)(y<t \wedge A)$ and $(y)(y<t \supset A)$ are in PR. By PR* we denote the set of all formulas in PR which do not contain special function constants. We note the following trivial fact: for every formula $A \in PR$ there is

a formula $B(\alpha_1,\ldots,\alpha_s)$ PR* and pairwise distinct special function constants $\zeta^1_{\bar{n}_1},\ldots,\zeta^s_{\bar{n}_s}$ such that A is $B(\zeta^1_{\bar{n}_1},\ldots,\zeta^s_{\bar{n}_s})$.
The following theorem is easily proved by induction with respect to the number of logical symbols in the formula A; its proof is omitted.

<u>Theorem 2:</u> For every formula $A \in PR^*$ one finds effectively a term t containing exactly the same free variables as A and containing no special function constants for which the following holds:

a) $ZTi^* \vdash t=0 \longrightarrow A$,

b) $ZTi^* \vdash A \longrightarrow t=0$, c) $TZi^* \vdash \quad \longrightarrow t=0 \vee t=1$.

Theorem 2 is not indispensable, but its use is convenient in many places.

Notation: the term t associated with A in virtue of theorem 2 will be denoted in the sequel by t_A.

<u>L.</u> As promised in the last section we will briefly describe a particular assignment which associates with every term t a continuity function τ related with t. To this end we will use a result which will not be proved and which has already been mentioned (in a somewhat different form) in the "Preliminaries". Let ZTi_c be obtained from ZTi by omission of the conversion rule. Let $t(\alpha_1,\ldots,\alpha_s)$ be a term without free number variables and special function constants whose free function variables are precisely $\alpha_1,\ldots,\alpha_s$. Then we can prove the following statement ST_o : for given numbertheoretic functions $f_1,\ldots,f_s$ there exist numbers n and m such that $ZTi_c \vdash t(\alpha^1_{u_1},\ldots,\alpha^s_{u_s})=n$ holds, where $u_i = \bar{f}_i(m)$, $i = 1,\ldots,s$. The proof of this statement does not make use of the full force of ZTi_c but depends merely on the fact that the whole calculus of primitive functions is formalized within ZTi_c. The statement then follows by means of arguments which are very

similar to those presented in [11], 8.4. Now let $\tau(x_1,\ldots,x_s)$ be a numbertheoretic function defined as follows: if $u_1,\ldots,u_s$ are sequence numbers of equal length, then $\tau(u_1,\ldots,u_s)=n+1$ if and only if there exists a Goedel number $e \leq \text{length}(u_1)$ of the proof in ZTi_c of $t(\alpha^1_{u_1},\ldots,\alpha^s_{u_s})=n$. Now ZTi_c has a primitive recursive proof predicate (" e is (Goedel number of) a proof of the formula (with Goedel number) b"), as is easy to show. Therefore τ is primitive recursive. Moreover, τ is a continuity function in virtue of the statement ST_0 . Finally, by showing that every formula provable in ZTi_c is "true" in the usual sense, it follows that τ is indeed related with t . Furthermore, it is clear that as soon as we are given t we are given τ . If we use this particular assignment as basis for the definition of "saturated", then one can easily prove with the aid of statement ST_0 the statement ST_1: if a sequent S is provable in ZTi then it is provable in ZTi_c . The advantage of this particular assignment is that the syntax of ZTi becomes primitive recursive. It will not be until chapter IX that we will make use of this advantage.

1.5. Some systems of analysis

In this section we introduce those systems of analysis which will be considered most of the time in this work.

A. Below we consider some particular primitive recursive functions and relations. With respect to them we adopt a particular convention which is useful for typographical reasons: we use one and the same sign in order to denote both the intuitively given object and its formal counterpart in the theory.

1. Intuitively we have the natural numbers at our disposal; they are represented formally in ZT by the list 0,0',0",... of terms, called numerals. By symbols such as n,m,a,b etc. we denote both certain particular numbers as well as their corresponding numerals.

2. As is evident from the axioms, the symbols $', +, ^{\circ}$ represent in our formal systems successor function, addition and multiplication. By the very same symbols we denote also the intuitively given functions successor, addition and multiplication.

3. The function $f(x,y)=\frac{1}{2}((x+y)^2+3x+y)$ maps the ordered pairs (a,b) of natural numbers in a one way into the set of natural numbers. There are, of course, infinitely many terms in L* whose associated primitive recursive function is $f(x,y)$. Among these we choose in a welldetermined way a particular one t and call t the term representing f in ZT. Both the term and the function will be denoted by $\langle x,y \rangle$.

4. There is a primitive recursive function $\phi(\alpha, x)$ (of type $\langle 1,1 \rangle$) which associates with every function f and every number n the sequence number $\langle f(0),\ldots,f(n-1)\rangle = p_0^{f(0)+1} \ldots p_{s-1}^{f(s-1)+1}$ if $n \neq 0$ and 1 otherwise. Again there is a welldetermined term t in L* whose associated primitve recursive function is ϕ. Both ϕ and t will be denoted by $\bar{\alpha}(x)$ as in [5].

5. There is another primitive recursive function seq(x), which has the property: seq(n)=0 iff n is a sequence number, that is, a number of the form $\langle f(0),\ldots,f(s-1)\rangle$ for some f and s (s=0 included). The function seq(x) has a formal counterpart in the theory (a term $t \in L^*$ having only x free); we denote this counterpart also by seq(x).

6. The primitive recursive function $\varphi(x,y)$ which associates with two sequence numbers $u = \langle u_0,\ldots,u_{s-1}\rangle$, $v = \langle v_0,\ldots,v_{t-1}\rangle$ its concatenation $u*v$ will be denoted by $x*y$; as above, we denote also the formal counterpart of $x*y$ in ZT by $x*y$.

7. Let R(x,y) be the Kleene-Brouwer partial ordering. There is a welldetermined term $t(x,y) \in L^*$ whose associated primitive recursive function f(x,y) has the property: a) R(n,m) iff f(n,m)=0, b) f(n,m)=0 or f(n,m)=1 for all n,m. Both R(x,y) and t(x,y) will be denoted by $x \subset_k y$. We recall that the definition of $x \subset_k y$ is such that $n \subset_k m$ always implies that both n and m are sequence numbers. The sequents $x \subset_k y \longrightarrow \mathrm{seq}(x)$, $x \subset_k y \longrightarrow \mathrm{seq}(y)$ are both provable in ZTi and we can even

assume that they occur among the axioms (in the set M_6).

8. By $x \prec_k y$ we denote the Kleene-Brouwer linear ordering of sequence numbers and at the same time a certain prime formula $q(x,y)=0$ which is related to $x \prec_k y$ in the same way as $t(x,y)$ to $x \subset_k y$ before. We use $x \subseteq_k y$ and $x \preceq_k y$ as abbreviations for $x \subset_k y \vee x=y$ and $x \prec_k y \vee x=y$ respectively.

<u>B.</u> Next we introduce some particular types of formulas. Let $R(x)$ be an arbitrary formula. We use $x \subset_R y$ as abbreviation for $x \subset_k y \wedge R(x) \wedge R(y)$ and $x \prec_R y$ as abbreviation for $x \prec_k y \wedge R(x) \wedge R(y)$. We use $x \subseteq_R y$ and $x \preceq_R y$ as abbreviations for $x \subset_R y \vee x=y$ and $x \prec_R y \vee x=y$ respectively. By $W(\subset_R)$ we denote the formula $(\alpha)(Ex)(\neg\, \alpha(x+1) \subset_R \alpha(x))$; by $\mathfrak{W}(\subset_R)$ we denote the formula $(\alpha)\neg(x)(\alpha(x+1) \subset_R \alpha(x))$. $W(\prec_R)$ and $\mathfrak{W}(\prec_R)$ are defined similarly but with $\prec_R$ in place of $\subset_R$. The meaning of $\subset_R$ and $\prec_R$ is clear: $x \subset_R y$, e.g. represents the restriction of $x \subset_k y$ to the set of those sequence numbers which belong to $\{x/R(x)\}$. The formulas $W(\subset_R)$ and $\mathfrak{W}(\subset_R)$ express classically both that $\subset_R$ is wellfounded. The expression $(x) \subset_R y A(x)$ serves as abbreviation for the formula $(x)(x \subset_R y \supset A(x))$. An important class of formulas are those which do not contain function parameters. A formula A is said <u>to contain no function parameters</u> if the following holds: there is a formula $B(x_1,\ldots,x_s) \in L^*$ (that is, without special function constants) which does not contain free function variables and there are terms $t_1,\ldots,t_s$ free for $x_1,\ldots,x_s$ in B such that A is $B(t_1,\ldots,t_s)$. Eg. $(Ey)(\alpha(x)=y+1)$ is such a formula while $(x)(x \leq y \supset \alpha(x)=0)$ is not. In other words: a formula without function parameters may contain free function variables and special function constants, however, only in an "inessential" way.

Another important class of formulas is that one described by
<u>Definition 3:</u> a) A Π^1_1-formula is a formula of the form $(\alpha)(Ex)R(\bar{\alpha}(x))$ where $R \in PR$. b) The set W of formulas is determined as follows: α) Π^1_1-formulas are in W, β) if A does not contain bounded function variables, then $A \in W$, γ) if $A,B \in W$ then $A \supset B$, $A \wedge B$, $A \vee B$, $\neg A$, $(Ex)A$, $(x)A$ are all in W.
c) $A \in W_N$ iff $A \in W$ and iff A does not contain function parameters.

Finally we note that, in view of theorem 2 and the remarks preceding it, we can associate with every $R \in PR$ effectively a term t containing exactly the same free variables and the same special function constants as $\subset_R$ such that $ZTi \vdash t(x,y)=0 \longrightarrow x \subset_R y$, $ZTi \vdash x \subset_R y \longrightarrow t(x,y)=0$ and $ZTi \vdash \longrightarrow t=0 \vee t=1$. We abbreviate $t(x,y)=0$ by $x <_R y$. Similarly, there is another term g containing exactly the same free variables and the same special function constants as $\subset_R$ such that $ZTi \vdash \longrightarrow g=0 \vee g=1$, $ZTi \vdash g(x,y)=0 \longrightarrow \neg x \subset_R y$, and $ZTi \vdash \neg x \subset_R y \longrightarrow g(x,y)=0$ hold. In view of theorem 2 we can choose $t(x,y)$ and $g(x,y)$ both in such a way that if R (and hence $\subset_R$) does not contain function parameters, then $t(x,y)$ and $g(x,y)$ do not contain function parameters. We use $(x) <_R y A(x)$ as abbreviation for $(x)(x <_R y \supset A(x))$, $x \nless_R y$ as abbreviation for $g(x,y)=0$ and $W'(<_R)$ as abbreviation for $(\alpha)(Ex)(\alpha(x+1) \nless_R \alpha(x))$. Finally we need the notion of <u>standard formula</u>. A formula $R(y)$ is called a <u>standard formula</u> if it has the form $Q(y) \wedge seq(y)$ where $Q(y)$ is an arbitrary formula. The only purpose of standard formulas is to secure the following implication: if $R(q)$ holds, then q is a sequence number.

<u>C.</u> In order to define the systems of analysis needed, we have to introduce a number of rules, all representing essentially transfinite induction with respect to $\subset_R$. The formula $R(y)$ which occurs in all these rules <u>is by definition a standard formula</u>. These rules are

I. $$\frac{R(y), (x) \subset_R y A(x), \Gamma \longrightarrow \Delta, A(y)}{R(q), W(\subset_R), \Gamma \longrightarrow \Delta, A(q)}$$

II. $$\frac{R(y), (x) \subset_R y A(x), \Gamma \longrightarrow \Delta, A(y)}{R(q), W(\subset_R), \Gamma \longrightarrow \Delta, A(q)}$$

III. $$\frac{t_R(y)=0, (x) <_R y A(x), \Gamma \longrightarrow \Delta, A(y)}{t_R(q)=0, W(<_R), \Gamma \longrightarrow \Delta, A(q)}$$

IV. $$\frac{t_R(y)=0,\ (x) <_R y\, A(x),\ \Gamma \longrightarrow \Delta,\ A(y)}{t_R(q)=0,\ W'(<_R),\ \Gamma \longrightarrow \Delta,\ A(q)}$$

V. $$\frac{t_R(x)=0,\ (x) <_R y\, A(x),\ \Gamma \longrightarrow \Delta,\ A(y)}{t_R(q)=0,\ \mathring{W}(<_R),\ \Gamma \longrightarrow \Delta,\ A(q)}$$

In all these rules y does not occur free in the conclusion and q is free for x in A. Of importance are some rules which are obtained by imposing certain restrictions concerning A and R on the above rules. The rules thus obtained are as follows: 1) the rules I_N, II_N,... are obtained from I, II, ... respectively by admitting only such formulas R which do not contain function parameters, 2) the rules I', II', ... are obtained from I, II, ... respectively by admitting only formulas R from PR (this is automatically satisfied for III, IV, V), 3) the rules I'_N, II'_N, ... are obtained from I, II, ... by admitting only formulas $R \in PR$ which do not contain function parameters, 4) the rules I*, II*, ... are obtained from I, II, ... by requiring $R \in PR$ and $A \in W$, 5) the rules I^*_N, II^*_N,... finally are obtained from I, II, ... by admitting only such formulas R and A which are in PR and in W respectively and which do not contain function parameters.

<u>Notation:</u> From now on we will abbreviate $t_R(x)=0$ by $d_R(x)$ or sometimes more simpler by $d(x)$.

<u>D.</u> In sect. 1.4 we have defined a set M of sequents which serves as axiom set for the systems ZT, ZTi*,... . M is the union of seven sets M $(0 \leq i \leq 6)$. With exception of M_6, every other set M_i is a well defined set of sequents; M_6, however, plays the role of a parameter set and has remained undetermined up to now. From now on however we make the following assumption:

<u>Assumption D:</u> The set M_6 contains for all terms p,q,t the following sequents: a) $p \subset_K q,\ q \subset_K t \longrightarrow p \subset_k t$, b) $p <_R q \longrightarrow t_R(q)=0$ and $p <_R q \longrightarrow t_R(p)=0$ for all $R \in PR$.

Actually, assumption D is redundant: using only axioms from $\bigcup_{o}^{5} M_i$

we can prove $x \subset_K y,\ y \subset_K z \longrightarrow x \subset_K z$, $x <_R y \longrightarrow t_R(y)=0$ and $x <_R y \longrightarrow t_R(x)=0$ in ZTi (in ZTi' if $R \in PR^*$) . We assume D merely for technical convenience.

E. By adding one of the new rules to any of the systems ZT, ZTi, ZT* etc. we obtain quite a series of more or less similar systems. Consider eg. the system ZT . By adding to ZT the new rule I we obtain a new system, to be denoted by ZT/I. The system ZT/I differs from ZT in that we can now use the new rule I in addition to the old ones in order to generate proofs: whenever P is a proof of a sequent of the form $R(y), (x) \subset_R y A(x), \Gamma \longrightarrow \Delta, A(y)$ for some R , then we can apply rule I^R to the endsequent of P in order to obtain a proof P' of $W(\subset_R), R(q), \Gamma \longrightarrow \Delta, A(q)$, provided that y does not occur free in $W(\subset_R), R(q), \Gamma \longrightarrow \Delta, A(q)$ (and where q is free for x in A). Proofs are of course identified with certain finite trees of sequents. A proof P (with respect to ZT/I) is again called intuitionistic if there is no sequent in P containing more than one formula in the succedent. If we restrict our attention to intuitionistic proofs only, then we obtain a subsystem which will be denoted by ZTi/I. The system ZT*/I is obtained from ZT/I by considering only such proofs which do not contain special function constants; the system ZTi*/I is obtained from ZT/I by restricting attention to intuitionistic proofs not containing special function constants. Quite similarly, if we combine any of the systems of sect. 1.4 with any of the above rules we obtain a whole list of new systems, to be denoted in a selfexplanatory way by ZT/I, ZTi/I, ZT/I*, ZTi/I*, ... ZT^*/I_N, ZTi^*/I'_N etc. A first superficial insight into the strength of some of these systems is given by

Theorem 3: a) ZT/I' has the same strength as the theory TI_{QF} in [3] ; b) ZT/I' has the same strength as ZT/III ; c) ZT/I has the same strength as ZT*/I ; d) ZT/I' has the same strength as ZT*/I'; e) ZT/I has the same strength as ZT*/I ; f) ZT*/I and ZTi*/II have the same strength; g) ZT*/I is as strong as classical analysis.

Proof: Most of these relationships are trivial. We just consider a), f) and g).

a) We merely show that ZT^*/I' is at least as strong as TI_{QF} . The proof of the converse makes use of th.2 and is almost trivial. First we show that for each A and each $R \in PR^*$ we can derive

I. $\longrightarrow W(\subset_R) \supset .(y)(R(y) \supset .(x) \subset_R{}^y A(x) \supset A(y)) \supset (z)(R(z) \supset A(z))$.

Let us denote to this end

$(y)(R(y) \supset .(x) \subset_R{}^y A(x) \supset A(y))$ by $Progr(R,A)$ and consider the sequent $Progr(R,A) \longrightarrow Progr(R,A)$ which is an axiom of ZT^*/I'. By a bit of intuitionistic predicate calculus we can derive $R(y),(x) \subset_R{}^y A(x), Progr(R,A) \longrightarrow A(y)$. Application of rule I' to this sequent yields the conclusion $W(\subset_R), R(z), Progr(R,A) \longrightarrow A(z)$ with suitably chosen free z . By intuitionistic predicate calculus we immediately derive the sequent I. That is, ZT^*/I' is at least as strong as the theory T which we obtain by adding to ZT^* all sequents of the form

I. (for $R \in PR^*$) as axioms. In virtue of theorem 0, this theory has the same strength as the theory T^* which we obtain from ZH^* by adding to it as axioms all formulas of the form

II. $W(\subset_R) \supset .Progr(R,A) \supset (z)(R(z) \supset A(z))$ for all $R \in PR^*$ and all formulas A . The only thing which remains to be done is to show that in T^* one can derive all formulas of the form

III. $W(\subset_R) \supset .(y)((x) \subset_R{}^y A(x) \supset A(y)) \supset (z)A(z)$. But this is an easy task if one notes the provability of the formulas

IV. $\neg R(y) \supset (x) \subset_R{}^y A(x)$ and

V. $(y)((x) \subset_R{}^y A(x) \supset A(y)) \supset Progr(R,A)$. Since PR^* contains in particular all quantifierfree formulas without special function constants, we conclude $TI_{QF} \subseteq T^*$. We note that all derivations are entirely intuitionistic; the rule of excluded middle is only used in the form $R(y) \vee \neg R(y)$ and this is intuitionistically correct in virtue of $R \in PR^*$.

Next, to g). We content ourself to show that ZT^*/I is at least as strong as classical analysis; the converse is more routine work. By proceeding as in the proof of a) we conclude that ZT^*/I is at least as strong as a theory T which is obtained from ZH^* by adding to it all formulas of the form III., but now for all formulas

A and R and not merely for formulas R in PR* ; now, of course, we use the law of excluded middle in a nontrivial way, namely in the form $R(y) \vee \neg R(y)$ for arbitrary R . It remains to show that T has indeed the strength of whole classical analysis. But this has essentially been proved by W. Howard in chapter II , p. 2.8 of the Stanford report, vol. I ([12]) . More precisely, one first shows that the axiom of bar induction

VI. $(\alpha)(Ex)P(\bar{\alpha}(x)) \wedge (\alpha)(x)(P(\bar{\alpha}(x)) \supset A(\bar{\alpha}(x))) \wedge.$
$\wedge (\alpha)(x)((y)A(\bar{\alpha}(x)*y) \supset A(\bar{\alpha}(x))). \supset (\alpha)(x)A(\bar{\alpha}(x))$

can be derived in T for all formulas A and R . This task is easily achieved by transforming the bar induction into a transfinite induction over $\subset_p$. Thus the theory T is at least as strong as the theory BI which is obtained from ZH* by adding all formulas of the form VI. as axioms. But according to Howard's result, BI has the same strength as classical analysis what proves one half of the statement g).

In order to prove f), one shows that whenever a sequent $\longrightarrow$ G has been proved in ZT*/I , then $\longrightarrow \mathfrak{G}$ is provable in ZTi*/II . To this end let Tr be the set of all formulas of the form II. above (for all R and all A not containing special function constants) and let Tr^o be the set of all formulas of the form $\mathfrak{W}(\subset_R) \supset .Progr(R,A) \supset (z)(R(z) \supset A(z))$. Let finally STr be the set of sequents of the form $\longrightarrow$ F with $F \in Tr$; let STr^o be the set of sequents of the form $\longrightarrow$ F with $F \in Tr^o$. Denote by ZT** the theory obtained by adding to ZT* all the sequents from STr as axioms. By the same reasoning used in the proof of a) one shows that ZT** $\vdash$ S iff ZT*/I $\vdash$ S . Now assume ZT*/I $\vdash \longrightarrow$ G. Then ZT** $\vdash \longrightarrow$ G , that is ZT*: $\longrightarrow F_1, \dots, \longrightarrow F_n \vdash \longrightarrow G$ for some F_i's from Tr . In virtue of theorem 2 this implies ZH*: $F_1, \dots, F_n \vdash G$ and therefore ZHi*: $\mathfrak{F}_1, \dots, \mathfrak{F}_n \vdash \mathfrak{G}$ again by theorem 2. A third application of theorem 2 finally yields ZTi* $\vdash \mathfrak{F}_1, \dots, \mathfrak{F}_n \longrightarrow \mathfrak{G}$. On the other hand $\longrightarrow \mathfrak{F}_i$ $(i \leq n)$ are all provable sequents in ZTi*/II , as a repetition of the argument used in the proof of a) shows. Hence we obtain ZTi*/II $\vdash \longrightarrow \mathfrak{G}$.

F. The theories on which we will concentrate mainly are ZTi/I , ZTi/II , ZTi/IV , ZTi/V and ZTi/IV^*_N , but other theories from our list will be considered from time to time. The theories ZTi/I , ZTi/II etc. have not yet the form suitable for a proof theoretic treatment. This will be achieved by considering certain conservative extensions of the above theories. Thus eg. we will consider in place of ZTi/IV a certain conservative extension, to be denoted by T for the moment, which is obtained from ZTi/IV by adding to ZTi/IV a set of rules, all of which are derivable in ZTi/IV; that is T and ZTi/IV have the same theorems. This conservative extensions serve only technical purposes and have no interest in their own; we will therefore define these extensions at the places where they are needed.

CHAPTER II:

A review of Gentzen's second consistency proof

In this chapter we present a brief repetition of Gentzen's second consistency proof and a mild generalisation of it, to be of use later on. This chapter can, of course, not replace a detailed study of [1], with which the reader is assumed to be familiar. In this and the next chapter we include some material contained in [8]. We will base our discussion on the system ZT and a system $ZT(\subset_D)$ (to be defined below) which contains a principle of transfinite induction with respect to a fixed given primitive recursive wellordering.

2.1. Some preliminary notions

From now on a proof (in ZT or any other system) will always be a finite tree (a proof tree) with sequents as nodes, which satisfies the following requirements: a) uppermost sequents are axioms; b) if S is a node of the tree which is not an uppermost one, then S has either one or two predecessors; c) if S is a node and S' its only predecessor, then S'/S is a one-premiss inference; d) if S is a node and S_1, S_2 its predecessors from left to right, then $S_1, S_2/S$ is a two-premiss inference; e) the tree has exactly one lowest node, which is called the endsequent of the proof. Let S be (an occurence of) a sequent in a proof P; let N_S be the set of nodes which contains precisely S together with all sequents S' in P which are situated above S. By restricting P to N_S we obtain a subtree P_S of P which is obviously a proof of S. We call P_S the subproof of S in P. An important notion connected with a proof tree is that of its <u>final part</u>: 1) the endsequent belongs to the final part; 2) if S'/S is a conversion or a one-premiss structural rule and if S belongs to the final part of P, then S' belongs to the final part of P; 3) if $S_1, S_2/S$ is a cut in P and if S belongs to the final part of P, then both S_1 and S_2 belong to the final part of P; 4) S belongs to the final part of P only in virtue of 1), 2), 3). Clearly, an uppermost sequent of the part is either an axiom or the conclusion of a logical inference or an induction.

Definition 4: Let P be a proof. An inference S'/S or $S_1,S_2/S$ in P is called critical if it is neither a conversion nor a structural inference and if its conclusion S belongs to the final part of P .

In the following definition Γ denotes the list $A_1,\ldots,A_s$ of formulas, Δ denotes $B_1,\ldots,B_t$, Σ denotes $C_1,\ldots,C_p$ and Π denotes $D_1,\ldots,D_q$; the formulas $A'_1,\ldots,A'_s$ and $B'_1,\ldots,B'_t$ are isomorphic with $A_1,\ldots,A_s$ and $B_1,\ldots,B_t$, respectively, and the two lists are denoted by Γ' and Δ', respectively.

Definition 5: Let A be a formula (more precisely an occurence of a formula) in the final part of P . A formula B in P is called successor of A if one of the following clauses is satisfied:
1) there is a right interchange
$\Gamma \longrightarrow \Delta, F_1, F_2, \Sigma / \Gamma \longrightarrow \Delta, F_2, F_1, \Sigma$ in the final part of P and A is A_i, B_j, F_k or C_m in the premiss, while B is A_i, B_j, F_k or C_m respectively in the conclusion; 2) similarly in case of a left interchange; 3) there is a conversion Γ —— Δ / Γ' —— Δ' and A is A_i or B_j in the premiss, while B is A'_i or B'_j respectively in the conclusion; 4) there is a right contraction $\Gamma \longrightarrow \Delta, F, F / \Gamma \longrightarrow \Delta, F$ in the final part, and A is A_i or B_j or one of the F's in the premiss, while B is A_i, B_j or F , respectively, in the conclusion; 5) similarly in case of a left contraction; 6) there is a right thinning $\Gamma \longrightarrow \Delta / \Gamma \longrightarrow \Delta, D$ in the final part and A is A_i or B_j in the premiss, while B is A_i or B_j , respectively, in the conclusion; 7) similarly in case of a left thinning; 8) there is a cut $\Gamma \longrightarrow \Delta, F$; $F, \Sigma \longrightarrow \Pi / \Gamma, \Sigma \longrightarrow \Delta, \Pi$ in the final part, and A is A_i, B_j, C_k or D_m in the premiss, while B is A_i, B_j, C_k or D_m , respectively, in the conclusion.

Since the final part of a proof is also a finite tree, all notions introduced in connection with finite trees retain their meaning for the final part.

Definition 6: Let $S_1,\ldots,S_n$ be a path in the final part of P , let $A_i,\ldots,A_{i+k}$ be a list of formulas in $S_i,\ldots,S_{i+k}$ respectively such that A_{n+1} is the successor of A_n for $i \leq n < i+k$ according to definition 5. Then A_{i+k} is called the image of A_i in S_{i+k}.

We note that in connection with logical inferences we use the notions "principal formula" and "side formula(s)" of the inference in the same sense as Kleene in [4] , p . 443.

Definition 7: Let P be a proof and $\Gamma \longrightarrow \Delta, A(\alpha) / \Gamma \longrightarrow \Delta, (\forall \alpha) A(\alpha)$ a quantifier inference where α is subject to the usual restriction on variables. We call α the quantified variable of this inference. Similarly, in case of a quantifier inference $A(\alpha), \Gamma \longrightarrow \Delta / (E\alpha) A(\alpha), \Gamma \longrightarrow \Delta$ and similarly with x in case of the quantifier inferences $\Gamma \longrightarrow \Delta, A(x) / \Gamma \longrightarrow \Delta, (\forall x) A(x)$ and $A(x), \Gamma \longrightarrow \Delta / (Ex) A(x), \Gamma \longrightarrow \Delta$, respectively. If $A(x), \Gamma \longrightarrow \Delta, A(x') / A(0), \Gamma \longrightarrow \Delta, (q)$ is an induction inference in P, then x is called the induction variable of this inference.

Remark: If eg. we say that α is the quantified variable of a quantifier inference, then we tacitly assume that this inference is an $E \longrightarrow$ or an $\longrightarrow \forall$ with α as the quantified variable.

Definition 8: A proof P is called normal if it has the following properties: 1) no variable occurs both free and bound in it; 2) if α is the quantified variable of a quantifier inference S/S' in P, then α does not occur free in any sequent S" below S ; 3) if x is the quantified variable of a quantifier inference S/S' or the induction variable of an induction S/S', then x does not occur free in any sequent S" below S; 4) if α occurs free in a sequent S in P but not in the endsequent, then there is a quantifier inference $S_1/_2$ in P with α as quantified variable and such that S_2 is below S ; 5) if x occurs free in a sequent S in P but not in the endsequent of P, then there is either a quantifier inference S_1/S_2 with x as quantified variable and S_2 below S, or an induction inference S'/S" with x as induction variable and S" below S .

Remark: A pure variable proof always satisfies 1), 2), 3) of def. 8 . On the other hand, if P is a proof which satisfies 1), 2), 3) of def. 8, then we can always transform P into a normal proof P' by replacing certain free variables in P by appropriately choosen constant functors and terms. If P satisfies 1), 2), 3), if S is a sequent in P and P_S the subproof of S in P, then P_S satisfies 1), 2), 3).

Definition 9: A proof is called saturated if every constant term (that is, term without free variables of both kinds) occuring in the final part is saturated.

Definition 10: Let P be a proof and I_1, I_2 two logical inferences in it. We call I_1, I_2 dual to each other if one of the following clauses is satisfied: 1) I_1 is an $\longrightarrow \supset$ and I_2 is an $\supset \longrightarrow$ inference; 2) I_1 is an $\longrightarrow \neg$ and I_2 is an $\neg \longrightarrow$ inference; 3) similarly with $\wedge$ or $\vee$ in place of $\supset$; 4) I_1 and I_2 are both number quantifications $\longrightarrow \forall$ and $\forall \longrightarrow$, respectively; 5) I_1 and I_2 are both function quantifications $\longrightarrow \forall$ and $\forall \longrightarrow$, respectively; 6) similarly with E in place of $\forall$.

The next few definitions are intimately connected with Gentzen's second consistency proof. In this connection we use the very convenient notion of "fork" which has been introduced by D. Isles in an as yet unpublished work on proof theory.

Definition 11: Let P be a proof. Let there be three inferences in P which we denote symbolically by I_1, I_2, I_3 ; let S_1 be the conclusion of I_1 and S_2 the conclusion of I_2 . The ordered triple I_1, I_2, I_3 is called a fork in P if the following conditions are satisfied: 1) I_1 and I_2 are critical logical inferences with principal formulas A_1 and A_2 respectively; 2) I_3 is a cut $S', S''/S$ where S' and S'' are $\Gamma \longrightarrow \Delta, F$ and $F, \Sigma \longrightarrow \Pi$, respectively, while S is, of course, $\Gamma, \Sigma \longrightarrow \Delta, \Pi$; 3) S and hence S' and S'' belong to the final part of P ; 4) A_1 has the cut formula F as image in S' ; 5) A_2 has the cut formula F as image in S'' .

Remark: Retain the notation of def. 11 and assume that I_1, I_2, I_3 is a fork. Then we can draw immediately the following conclusions: 1) S' is equal to S_1 or situated below S_1 ; 2) S'' is equal to S_2 or situated below S_2 ; 3) F is isomorphic with A_1 and A_2 , respectively; 4) hence A_1 is isomorphic with A_2;

5) I_1 and I_2 are dual to each other. The clauses 1) - 5) follow immediately from our preceding definitions. With respect to forks we adopt the following expressions: 1) if the inference I_1 (and hence I_2) is a propositional inference and the symbol introduced a $\supset$, $\wedge$, $\vee$ or $\neg$, then we say that I_1, I_2, I_3 is an $\supset$-, $\wedge$-, $\vee$- or $\neg$-fork, respectively; 2) if I_1 is a number quantification and the symbol introduced an $\forall$, then we call I_1, I_2, I_3 a numerical $\forall$-fork; 3) if I_1 is a function quantification and the symbol introduced an $\forall$, then we call I_1, I_2, I_3 a functional $\forall$-fork; 4) similarly with E in place of $\forall$.

In [1] Gentzen associates with every cut $\Gamma \longrightarrow \Delta, F\ ;\ F, \Sigma \longrightarrow \Pi \,/\, \Gamma, \Sigma \longrightarrow \Delta, \Pi$ and every induction $A(x), \Gamma \longrightarrow \Delta, A(x') / A(0), \Sigma \longrightarrow \Pi, A(t)$ a natural number called complexity of the cut and the induction, respectively: in case of the cut this number is equal to the number of logical symbols contained in F, in case of the induction this number is equal to the number of logical symbols in $A(x)$. Next, Gentzen associates with every sequent S in a proof P another number, called its height and denoted by $h(S)$, according to the following

Definition 12: Let S be a sequent in P. If S is the endsequent then $h(S)=0$. Now let S be a premiss of an inference I with conclusion S'. If I is a cut, then $h(S)$ is $\max(h(S'),d)$, where d is the complexity of the cut. If I is an induction then $h(S)$ is $\max(h(S'),d)$, where d is the complexity of the induction. In all other cases $h(S)=h(S')$.

Remark: If $S', S''/S$ is a cut in P, then by definition $h(S')=h(S'')$. If $S_0, \dots, S_n$ is a path in P, then clearly $h(S_i) \geqq h(S_{i+1})$. If, in particular, $S', S''/S$ is a cut in P such that $h(S') > h(S)$, then we say that $S', S''/S$ is a cut with jump ("Höhensprung" in [1]).

Lemma 3: Let I_1, I_2, I_3 be a fork in P according to def. 11 and let $S', S''/S$ be the cut I_3; assume that I has complexity $d > 0$. Then there is exactly one cut $S_0', S_0''/S_0$ in P having the following properties: 1) S_0 is equal to S or situated below S, 2) $h(S_0')=h(S')$, 3) $h(S_0) < h(S_0')$.

Proof: Trivial

Definition 12*: The cut $S_o', S_o''/S_o$ in lemma 3, which is unequally determined by the fork I_1, I_2, I_3, is called the cut associated with the fork I_1, I_2, I_3.

2.2. The reduction steps

We are now ready to give a short account of Gentzen's second consistency proof. In this respect we explain a few essential points and refer the reader otherwise to [1]. In the sequel we will always observe the following convention: by a logical axiom in the final part of a proof P we will always understand an uppermost sequent S of the final part which has the form $D \longrightarrow D'$, where D and D' are isomorphic with each other.

A. In [1], Gentzen introduces certain syntactical transformations of proofs which he calls reduction steps. We can distinguish three kinds of reduction steps: 1) removing all thinnings and logical axioms from the final part; 2) removing critical inductions from the final part; 3) removing forks from the final part. Reduction steps of type 1) will be called preliminary reduction steps, those of type 2) and 3) essential reduction steps. We start with a brief discussion of the preliminary reduction steps. We omit a precise definition of the preliminary reduction steps and content ourselves by discussing some typical cases. Assume e.g. that in the final part of a proof P there is a left thinning whose conclusion is the right premiss of a cut:

$$\cfrac{\Gamma \longrightarrow \Delta, F \qquad \cfrac{\Sigma \longrightarrow \Pi}{F, \Sigma \longrightarrow \Pi}}{\Gamma, \Sigma \longrightarrow \Delta, \Pi}$$

in this case we can obviously omit the cut and derive the conclusion by a series of thinnings and interchanges from $\Sigma \longrightarrow \Pi$:

$$\frac{\Sigma \longrightarrow \Pi}{\Gamma, \Sigma \longrightarrow \Delta, \Pi} \quad \text{thinnings, interchanges}$$

The proof P' which results from this alteration is said to follow from P by means of a preliminary reduction step.

In order to consider a similar but more general situation, let us for the moment introduce the so-called identity rule which permits us to derive S' from S . Now assume that in the final part of P there is a path $S_o,\ldots,S_n$,with S_o an uppermost sequent of the final part and S_n the endsequent of P , such that there is an i with the property: S_{i+1} follows from S_i by means of a left thinning, that is, S_i is $\Gamma \longrightarrow \Delta$ and S_{i+1} is $A, \Gamma \longrightarrow \Delta$. We distinguish two cases: 1) there is an S_m with $i<m<n$ such that S_m is the right premiss of a cut $S', S_m/S_{m+1}$ whose cut formula F in S_m is the image of A ; 2) the endsequent S_n contains an image A' of A (in S_{i+1}) . In the first case we proceed similarly as in our example above, that is, we cancel A in S_{i+1} together with all its images in P, obtaining thus a new path $S_o,\ldots,S_i, S'_{i+1},\ldots,S'_m, S_{m+1},\ldots,S_n$, then we cancel the subproof $P_{S'}$ and derive S_{m+1} by thinning and interchange from S'_m in the same way as in the example above. This operation transforms P into a tree P* which is a proof tree in a slightly more general sense: it contains in addition to the ordinary inferences also some identity inferences (they all occur in the part $S_i,S'_{i+1},\ldots,S'_m$ of the altered path). By cancelling these identity inferences in P* we finally obtain a proof P' in the ordinary sense; P' is said to follow from P by means of a preliminary reduction step. In case 2) we proceed as follows: we cancel A together with all its images in $S_{i+1},\ldots,S_n$. This operation transforms P into a generalized proof tree P* in the above sense, containing among others some identity inferences. By cancelling in P* all identity inferences we obtain again an ordinary proof P', whose endsequent S' is related to the endsequent S of P in the following way: S is derivable from S' by means of a thinning and some interchanges. Here too we say that P' follows from P by means of a preliminary reduction step.

Another situation to be treated is the following: assume that in P there is a left premiss $\Gamma, D_1, \Sigma \longrightarrow \Pi, D'_1, \Delta$ (to be denoted by S) of a cut, whose subproof P_S in P has the particular form

$$\frac{D \longrightarrow D'}{\Gamma, D_1, \Sigma \longrightarrow \Pi, D_1', \Delta} \quad \text{one premiss structural rules and conversions}$$

where $D \longrightarrow D'$ is an axiom, D_1 image of D and D_1' image of D'. Of course D_1 and D_1' are isomorphic with each other. We distinguish two subcases: 1) Δ is empty and the cut in question is

$\Gamma, D_1, \Sigma \longrightarrow \Pi, D_1'$;

$D_1', \Gamma' \longrightarrow \Pi' / \Gamma, D_1, \Sigma, \Gamma' \longrightarrow \Pi, \Pi'$,

2) D_1' is not the cut formula, Δ has the form Δ', F and the cut in question is

$\Gamma, D_1, \Sigma \longrightarrow \Pi, D', \Delta', F$;

$F, \Gamma' \longrightarrow \Pi' / \Gamma, D_1, \Sigma, \Gamma' \longrightarrow \Pi, D_1', \Delta', \Pi'$.

In both cases we can derive the conclusion of the cut by canceling in P the subproof of its right premiss and by deriving its conclusion by means of thinnings and interchanges from the axiom $D \longrightarrow D'$. Here, too, we say that the resulting proof P' is obtained from P by application of a preliminary reduction step.

The three cases presented are typical; all other cases can be obtained from them by interchanging the roles of left and right.

The properties of the preliminary reduction steps are summarized by

Theorem 4: There is a primitive recursive relation PR(X,Y) and two primitive recursive functions $\varphi(X)$, $\tau(X)$ such that for all proofs P,P' the following holds: 1) PR(P,P') iff P' is obtained from P by means of a preliminary reduction step; 2) if PR(P,P') then P' has less than $\varphi(P)$ symbols; 3) every sequence $P_o, \ldots, P_N$ (with $P_o = P$) such that $PR(P_i, P_{i+1})$ for $i < N$ has length $< \tau(P)$, that is $N < \tau(P)$; 4) if PR(P,P') then either P and P' have the same endsequent or we can derive the endsequent S of P by thinnings and interchanges from the endsequent S' of P' . The proof of theorem 4 is completely elementary and hence omitted.

B. In order to describe a reduction step of type 2), also called induction reduction, let

$$\text{I:}\quad \frac{A(x), \Gamma \longrightarrow \Delta, A(x')}{A(0), \Gamma \longrightarrow \Delta, A(q)}$$

be a critical induction in a proof P and q a saturated term with $|q| = n$. Let P_w be the subproof of the premiss and P_i the result of replacing every free occurence of x in P_w by i; let $\hat{P}$ be the subproof of $A(0), \Gamma \longrightarrow \Delta, A(q)$ in P. Denote $A(i), \Gamma \longrightarrow \Delta, A(i+1)$ by S_i and $A(0), \Gamma \longrightarrow \Delta, A(k)$ by S^*_k. We distinguish three cases.

a) $|q| = 0$. Then we replace the subproof of $A(0), \Gamma \longrightarrow \Delta, A(q)$ in P by the following derivation:

$$\frac{A(0) \longrightarrow A(0)}{\overline{A(0), \Gamma \longrightarrow \Delta, A(0)}}\quad \text{thinnings, interchanges}$$

b) $q = 1$. Then we replace $\hat{P}$ by the following derivation:

$$\begin{array}{c} P_0 \\ \vdots \\ A(0), \Gamma \longrightarrow \Delta, A(1) \\ \hline A(0), \Gamma \longrightarrow \Delta, A(q) \end{array}\quad \text{conversion}$$

c) $|q| = m+1$ and $m \geqq 1$. Now S^*_{i+1} can be derived from S^*_i and S_i by means of a cut and some interchanges and contractions as follows:

$$\frac{S^*_i \quad S_i}{\overline{S^*_{i+1}}}\quad \text{cut, interchanges, contractions.}$$

Hence we can replace $\hat{P}$ by the following derivation:

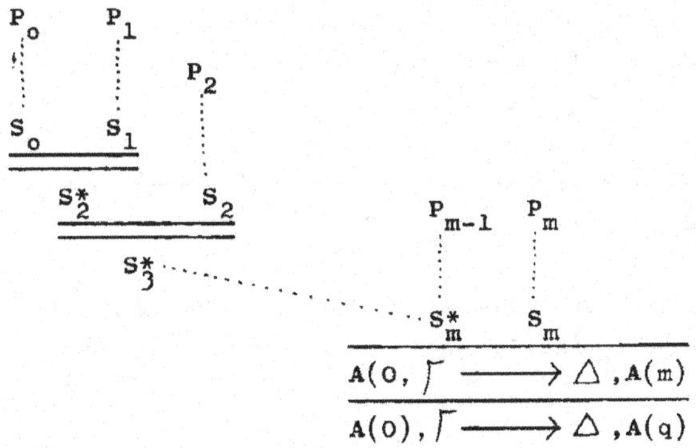

$$\frac{A(0, \Gamma \longrightarrow \triangle, A(m)}{A(0), \Gamma \longrightarrow \triangle, A(q)} \quad \text{conversion}$$

In each case we say that the resulting new proof P' is obtained
from P by application of an induction reduction.

<u>C</u>. The most sophisticated among the reduction steps are those of
type 3). We explain two of them, namely the case of an $\supset$ -fork and
the case of a functional $\forall$ -fork. All other cases are treated in an
analoguous way; for further details the reader may consult [1] .
In order to discuss the elimination of a <u>functional $\forall$-fork</u> from the
final part, we note the following

<u>Lemma 4</u>: A. Let P be a proof which satisfies 1), 2) and 3) of
definition 8 . Let F be a constant functor whose bound variables
do not occur free in P . Let α be a function variable which occurs
free in the endsequent E of P . If we replace every (free) occurence of α in P by F , then we obtain a proof which still satisfies 1), 2) and 3) .

B. Similarly in case of a number variable x and a number term t
in place of α and F .

<u>Proof</u>: The statement follows immediately by "finite bar induction"
over P.

<u>Corollary</u>: A. Let P be a normal proof whose endsequent does not
contain free variables. Let E/E_1 be a quantifier inference in P
with α as quantified variable. Let P contain a critical function
quantification
$$\vee \longrightarrow \quad \text{or} \longrightarrow \quad E \text{ , say } \quad A(F), \Gamma \longrightarrow \triangle /(\forall \beta)A(\beta), \Gamma \longrightarrow \triangle .$$

If we replace α wherever it occurs (free) in the subproof P_E of E (in P) by F, then we obtain a proof of $S_\alpha^F E$, which satisfies properties 1), 2), 3) of definition 8.

B. Similarly in case of number quantification $\forall \longrightarrow$ or $\longrightarrow E$ with x and t in place of α and F.

<u>Proof:</u> Since P is normal, it follows that F is constant. The conditions of lemma 4 are obviously satisfied; hence the statement follows.

The first case to be treated is that of a function $\forall$-fork. The treatment of this case is precisely the same as that of a numerical $\forall$-fork considered in [1], but for illustration we treat this case in some detail. To this end let P be a normal proof whose endsequent does not contain free variables and I_1, I_2, I_3 a functional $\forall$-fork in P. Let I_1, I_2, I_3 be as follows:

$$I_1: \frac{\Gamma_o \longrightarrow \Delta_o, A_1(\alpha)}{{}_o \longrightarrow \quad {}_o, (\quad) A_1(\quad)} \qquad I_2: \frac{A_2(F), \Sigma \longrightarrow \Pi}{(\forall\beta) A_2(\beta), \Sigma \longrightarrow \Pi}$$

$$I_3: \frac{\Gamma_1 \longrightarrow \Delta_1, (\forall\beta) A(\beta) \qquad (\forall\beta) A(\beta), \Sigma_1 \longrightarrow \Pi_1}{\Gamma_1 \; \Sigma_1 \longrightarrow \Delta_1 \; \Pi_1}$$

As noted earlier, $(\forall\beta) A_i(\beta)$ $(i=1,2)$ and $(\forall\beta) A(\beta)$ are all isomorphic with each other. The inferences I_1, I_2 and I_3 will also be written more briefly as S_1/S_1', S_2/S_2' and $S_3, S_4/S_5$ respectively. Let furthermore P_i $(i \leq 5)$ and P_k' be the subproofs in P of S_i $(i \leq 5)$ and S_k' $(k=1,2)$, respectively. Let in addition $I: S', S''/S$ be the cut associated with I_1, I_2, I_3 and assume that S', S'' and S are $\Gamma_2 \longrightarrow \Delta_2, F$ and $F, \Sigma_2 \longrightarrow \Pi_2$ and $\Gamma_2, \Sigma_2 \longrightarrow \Delta_2, \Pi_2$ respectively; S will also be written more briefly as $\Gamma_3 \longrightarrow \Delta_3$. Let finally P', P'' and P_o be the subproofs of S', S'' and S respectively.

On P we perform a syntactical transformation to be described in the sequel. First we replace every free occurence of α in P_1 by F; in view of lemma 4 and its corollary this transforms P_1 into a proof P_1^F of $\Gamma_o \longrightarrow \Delta_o, A_1(F)$. Then we replace in P_o the

subproof P_1' by the following derivation of
$\Gamma_0 \longrightarrow A(F), \Delta_0, (\forall\alpha)A_1(\alpha)$:

$$\begin{array}{c} P_1^F \\ \vdots \\ \underline{\underline{\Gamma_0 \longrightarrow \Delta_0, A_1(F)}} \\ \Gamma_0 \longrightarrow A(F), \Delta_0, (\forall\alpha)A_1(\alpha) \end{array} \quad \text{conversion, interchanges, thinning}$$

This transforms P_0 into a proof P^* of $\Gamma_3 \longrightarrow A(F), \Delta_3$. By adding some interchanges to P^*, we obtain a proof $\bar{P}_1$ which can symbolically be written as follows :

$$\begin{array}{c} P^* \\ \vdots \\ \underline{\underline{\Gamma_3 \longrightarrow A(F), \Delta_3}} \\ \Gamma_3 \longrightarrow \Delta_3, A(F) \end{array} \quad \text{interchanges}$$

Next we perform another, similar transformation on P_0 . First we replace in P_0 the subproof P_2' by the following derivation of $(\forall\alpha)A_2(\alpha), \Sigma, A(F) \longrightarrow \Pi$:

$$\begin{array}{c} P_2 \\ \vdots \\ \underline{\underline{A_2(F), \Sigma \longrightarrow \Pi}} \\ (\forall\alpha)A_2(\alpha), \Sigma, A(F) \longrightarrow \Pi \end{array} \quad \text{conversion, interchanges, thinning}$$

This transforms P into a proof P^{**} of $\Gamma_3, A(F) \longrightarrow \Delta_3$. By adding some interchanges to P^{**}, we obtain a proof $\bar{P}_2$ which can symbolically be written as follows:

$$\begin{array}{c} P^{**} \\ \vdots \\ \underline{\underline{\Gamma_3, A(F) \longrightarrow \Delta_3}} \\ A(F), \Gamma_3 \longrightarrow \Delta_3 \end{array} \quad \text{interchanges}$$

Finally we replace the subproof P_o of S in P by the following derivation, to be denoted by $\widetilde{P}$:

$$\dfrac{\dfrac{\begin{matrix}\bar{P}_1\\ \vdots\\ \Gamma_3 \longrightarrow \Delta_3, A(F)\end{matrix} \qquad \begin{matrix}\bar{P}_2\\ \vdots\\ A(F), \Gamma_3 \longrightarrow \Delta_3\end{matrix}}{\Gamma_3, \Gamma_3 \longrightarrow \Delta_3, \Delta_3}}{\Gamma_3 \longrightarrow \Delta_3} \text{ interchanges, contractions}$$

The final result of this transformation, call it $\hat{P}$, is again a normal proof, having the same endsequent as P . We say that $\hat{P}$ follows from P by means of a functional $\forall$-reduction step. The second case to be treated is that of an <u>implicational fork</u>. Let again P be a normal proof and I_1, I_2, I_3 an implicational fork in P . Let I_1, I_2, I_3 be

$$I_1: \quad \dfrac{A_1, \Gamma_o \longrightarrow \Delta_o . B}{\Gamma_o \longrightarrow \Delta_o, A_1 \supset B_1}$$

$$I_2: \quad \dfrac{\Sigma \longrightarrow \Pi, A_2 \qquad B_2, \Sigma' \longrightarrow \Pi'}{A_2 \supset B_2, \Sigma, \Sigma' \longrightarrow \Pi, \Pi'}$$

$$I_3: \quad \dfrac{\Gamma_1 \longrightarrow \Delta_1, A \supset B \qquad A \supset B, \Sigma_1 \longrightarrow \Pi_1}{\Gamma_1, \Sigma_1 \longrightarrow \Delta_1, \Pi_1}$$

Of course, A is isomorphic with A_1 and A_2 and B is isomorphic with B_1 and B_2 . Let us write the inferences I_1, I_2, I_3 more symbolically as follows: 1) S_1'/S_1 in the case of I_1, 2) S_2'/S_2 in the case of I_2 , 3) $S_3, S_4/S_5$ in the case of I_3 . Let $S', S''/S$ be the cut associated with the fork in question and let S', S'' and S be $\Gamma_2 \longrightarrow \Delta_2, F$ and $F, \Sigma_2 \longrightarrow \Pi_2$ and $\Gamma_2, \Sigma_2 \longrightarrow \Delta_2, \Pi_2$ respectively; S will also be written more briefly as $\Gamma_3 \longrightarrow \Delta_3$. Finally let us denote the subproofs of $S_1, S_2, S_1', S_2', S_2'', S', S''$ and S by $P_1, P_2, P_1', P_2', P_2'', P', P''$ and P_o, respectively. First we describe a syntactical transformation to be performed on P . We replace P_1 in P_o by the following derivation:

$$\begin{array}{c} P_1' \\ \vdots \\ A_1, \Gamma_o \longrightarrow \Delta_o, B_1 \\ \hline\hline \Gamma_o, A \longrightarrow B, \Delta_o, A_1 \supset B_1 \end{array} \qquad \text{interchanges, conversions, thinnings.}$$

This transforms P_o into a proof P* of $\Gamma_3, A \longrightarrow B, \Delta_3$. By adding some interchanges we obtain a proof $\bar{P}_1$ of $A, \Gamma_3 \longrightarrow \Delta_3, B$:

$$\begin{array}{c} P^* \\ \vdots \\ \Gamma_3, A \longrightarrow B, \Delta_3 \\ \hline\hline A, \Gamma_3 \longrightarrow \Delta_3, B \end{array} \qquad \text{interchanges}$$

Next, we perform another transformation on P_o . We replace P_2 in P_o by the following derivation:

$$\begin{array}{c} P_2' \\ \vdots \\ \Sigma \longrightarrow \Pi, A \\ \hline\hline A_2 \supset B_2, \Sigma, \Sigma' \longrightarrow A, \Pi, \Pi' \end{array} \qquad \text{interchanges, conversion thinnings.}$$

This transforms P into a proof P** of $\Gamma_3 \longrightarrow A, \Delta_3$. By adding some interchanges we obtain a proof $\bar{P}_2$ of $\Gamma_3 \longrightarrow \Delta_3, A$ as follows:

$$\begin{array}{c} P^{**} \\ \vdots \\ \Gamma_3 \longrightarrow A, \Delta_3 \\ \hline\hline \Gamma_3 \longrightarrow \Delta_3, A \end{array} \qquad \text{interchanges.}$$

Finally, we perform a third transformation on P_o . We replace P_2 in P_o by the following derivation:

$$\dfrac{\begin{array}{c} P''_2 \\ \vdots \\ B, \Sigma' \longrightarrow \Pi' \end{array}}{A_2 \supset B_2, \Sigma, \Sigma', B \longrightarrow \Pi, \Pi'} \;\text{(double line) interchanges, conversion, thinning}$$

This transforms P into a proof P*** of $\Gamma_3, B \longrightarrow \Delta_3$. By adding some interchanges, we obtain similarly as above a proof $\bar{P}_3$ of B , $\Gamma_3 \longrightarrow \Delta_3$. The proofs $\bar{P}_1, \bar{P}_2$ and $\bar{P}_3$ can now be composed by means of cuts, interchanges and contractions in order to yield a new proof $\tilde{P}$ of $\Gamma_3 \longrightarrow \Delta_3$ as follows:

$$\dfrac{\dfrac{\dfrac{\begin{array}{c}\bar{P}_2\\ \vdots\end{array}}{\Gamma_3 \longrightarrow \Delta_3, A} \quad \dfrac{\begin{array}{c}\bar{P}_1\\ \vdots\end{array}}{A, \Gamma_3 \longrightarrow \Delta_3, B}}{\Gamma_3, \Gamma_3 \longrightarrow \Delta_3, \Delta_3, B}\text{cut} \quad \dfrac{\begin{array}{c}\bar{P}_3\\ \vdots\end{array}}{B, \Gamma_3 \longrightarrow \Delta_3}}{\dfrac{\Gamma_3, \Gamma_3, \Gamma_3 \longrightarrow \Delta_3, \Delta_3, \Delta_3}{\Gamma_3 \longrightarrow \Delta_3}\text{(double line) interchanges, contractions}}\text{cut}$$

Now we replace P_o in P by $\tilde{P}$. This transforms P into a new proof $\hat{P}$. Clearly $\hat{P}$ has the same endsequent as P and is again normal. We say that $\hat{P}$ follows from P by means of an implication reduction (or an $\supset$-reduction) .

2.3. Properties of reduction steps

A. In order to discuss some properties of reduction steps we need

Definition 12**: The two-place relation W applies to proofs P and P' (in symbols W(P,P')) if and only if P and P' are normal, have endsequents without free variables and satisfy the following conditions: 1) there is a list $P_o, \dots, P_N$ of proofs such that $P_o = P$ and $PR(P_i, P_{i+1})$ (see th. 4) for all $i < N$; 2) P' follows from P_N by exactly one application of an essential reduction step; 3) no preliminary reduction step is applicable to P_N .

The properties of W are described by

Theorem 5: a) W is recursive. b) There is a recursive function θ having the following property: if $W(P,P')$ then P' has at most $\theta(P)$ symbols.

In connection with W we introduce some notations. Let P be a normal proof whose endsequent does not contain free variables. By D_P we denote the set of those proofs P' which satisfy one of the following two conditions: 1) P' is P, 2) there is a list $P_o,\dots,P_N$ such that $P_o = P$, $P_N = P'$ and $W(P_i,P_{i+1})$ for $i < N$. By W_P we denote the restriction of W to D_P. By W* we denote the two-place relation which is induced by W in the following way: $W^*(P,P')$ iff there is a list $P_o,\dots,P_N$ with $P_o = P$, $P_N = P'$ such that $W(P_i,P_{i+1})$ for $i<N$. By W^*_P we denote the restriction of W* to D_P .

The reduction steps have an elementary but fundamental property, which is described by

Theorem 6: Let P be a normal saturated proof whose endsequent does not contain free variables. Assume that P does not admit reduction steps (neither preliminary nor essential ones) and that P is different from its final part. Then there is a critical logical inference, whose principal formula has an image in the endsequent.

A proof of this theorem can be found in [1] or in [2]. Before mentioning the main application of th. 6, we note

Lemma 5: Let P be a saturated proof and E its endsequent. Let P have the following properties: a) it contains no logical axioms, b) it contains only conversions, cuts, interchanges and contractions. Then E is a true saturated prime sequent.

The proof is trivial and hence omitted. The main conclusion which can be drawn from th. 6 is

Corollary: Let P be a normal saturated proof of $\longrightarrow m = n$ which does not contain special function constants. If W is well-founded (that is, does not allow strictly descending sequences) then $\longrightarrow m = n$ is true.

<u>Proof:</u> Let us call a proof P' "good" if it has the same properties as P, except that its endsequent may be $\longrightarrow$ m = n or $\longrightarrow$. Then one easily shows: if P' is "good" and if W(P',P") holds, then P" is also "good". Next we take an arbitrary but fixed strictly descending sequence $P_o,\ldots,P_N$ such that $P_o = P$, $W(P_i,P_{i+1})$ and $(\forall X)\neg W(P_N,X)$ (such a sequence exists in view of our assumptions). Since the endsequent of P_N does not contain any logical symbol, one concludes from th. 6 that P_N is identical with its final part. The statement then follows via lemma 5.

In view of the above corollary, Gentzen directed his main effort toward a proof of the wellfoundedness of W. How he achieved this with the aid of ordinal numbers will be outlined in the next section.

<u>Notation:</u> Since from now on we will almost always be concerned with normal proofs whose endsequent does not contain free variables, we will introduce a new name for them and call them <u>strictly normal</u>. Strictly normal proofs which are also saturated will also be called <u>strongly normal</u> proofs. We note

<u>Lemma 6:</u> Let P be strictly normal. If P' is obtained from P by means of a reduction step (preliminary or essential) then P' is strictly normal (but not necessarily strongly normal).

2.4. Assignment of ordinals to proofs

As mentioned above, we present an outline of Gentzen's proof that the relation W is wellfounded.

<u>A.</u> Let P be an arbitrary proof. With every sequent S in P we associate an ordinal to be denoted by o(S), inductively as follows: 1) if S is an axiom, then $o(S)=1$; 2) if S is the conclusion of a one-premiss structural inference or a conversion S_o/S, then $o(S)=o(S_o)$; 3) if S is the conclusion of a one-premiss logical inference S_o/S, then $o(S)=o(S_o)+1$; 4) if S is the conclusion of a two-premiss logical inference $S_1,S_2/S$, then $o(S)=o(S_1)\ o(S_2)\#1$; 5) if S is the conclusion of a cut S_1S_2/S, then $o(S)=\omega_d(o(S_1)\#o(S_2))$ where $d=h(S_1)-h(S)$ (with $h(S_1)$ and $h(S)$ the heights according to def. 12); 6) if S is the conclusion of an induction S_o/S, then

$O(S) = \omega_d(O(S_o)\omega)$ with $d=h(S_o)-h(S)$. As ordinal of the proof P , sometimes denoted by $O(P)$, we take the ordinal $O(E)$ of the end-sequent E of P .

B. The essential step is to prove the following

Theorem 7: Let P be a strictly normal proof. Let P' be obtained from P by means of an essential reduction step. Then $O(P') < O(P)$.

We also need

Theorem 8: Let P be an arbitrary proof. Let P' be obtained from P by means of a preliminary reduction step. Then $O(P') \leqq O(P)$.

Before discussing theorem 7, let us comment briefly theorem 8. In [1] Gentzen sketched a proof of theorem 8. For the time beeing (that is in this and the next section) we assume theorem 8 to be true. However, in view of the importance of preliminary steps for intuitionistic systems, we will look more closer at theorem 8 in the last section of this chapter. Concerning theorem 7, we are content to prove the statement for the case of an implication reduction. The treatment of the other cases is similar but simpler; we refer to [1].

Let P be a strictly normal proof, I_1, I_2, I_3 an implicational fork in P and $S', S''/Q$ the cut associated with I_1, I_2, I_3 ; let Q be more explicitly $\Gamma \longrightarrow \Delta$. Let $A_k \supset B_k$ be the principal formulas of I_k for k=1,2 , let $A \supset B$ be the cutformula of I_3 and let F be the cutformula of $S', S''/Q$. Let finally $h=h(S')$ be the height of S' (and S") and h_o the height of $\Gamma \longrightarrow \Delta$ (in P) . From the definition of cut associated with the fork I_1, I_2, I_3 one immediately deduces the following inequalities:
1) $h=N(F)$, 2) $N(A \supset B) \leqq h$, 3) $h_o < N(F)$. Here $N(F)$, $N(A \supset B)$, $N(A)$ and $N(B)$ denote the number of logical symbols in F, A, B and $A \supset B$, respectively. The proof P can symbolically be written as follows:

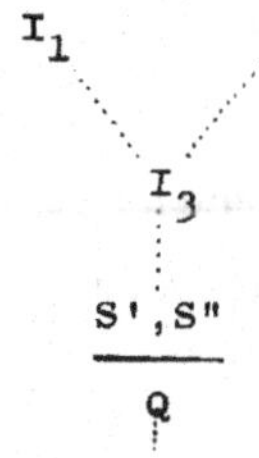

In view of our definition of implicational reduction step, we can write the altered proof symbolically as follows:

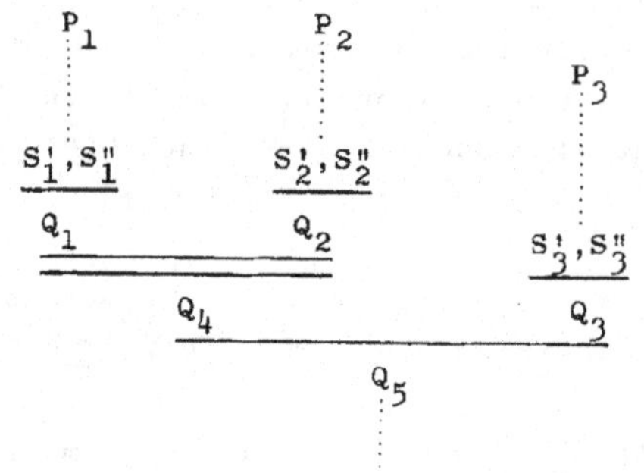

Here Q_1 is $\Gamma \longrightarrow \Delta, A$, Q_2 is $A, \Gamma \longrightarrow \Delta, B$, Q_4 is $\Gamma \longrightarrow \Delta, B$, Q_3 is $B, \Gamma \longrightarrow \Delta$ and Q_5 is $\Gamma \longrightarrow \Delta$. A double line indicates a cut followed by some interchanges and contractions. The cuts $S_i', S_i''/Q_i$ all have the same cut formula, namely F. The heights of Q_i, $i=1,\ldots,5$ and S_k', $k=1,2,3$ are given as follows: a) $h(Q_5)$ is h_o; b) $h(Q_3)$ and $h(Q_4)$ are $\max(h_o, N(B))$ and will be denoted by h_2; c) $h(Q_1)$ and $h(Q_2)$ are $\max(h_o, N(A), N(B))$ and will be denoted by h_1; d) $h(S_1')$ and $h(S_2')$ are $\max(h_o, N(A), N(B), F)$; e) $h(S_3')$ is $\max(h_o, N(B), F)$. From our inequalities 1)-3) listed above, one immediately deduces that $h(S')$ are all equal to h and that the following inequalities are satisfied: $h_o \leq h_2 \leq h_1 < h$. We note the following easily established fact: if the ordinal of Q in P' is smaller than the ordinal of Q in P, then P' has smaller ordinal than P. In order to calculate the ordinals of Q in P and of Q' in P', let us introduce the following notations:
1) by $O(S)$ we denote the ordinal of a sequent S in P, by $O'(S)$ we denote the ordinal of a sequent S in P'; 2) we put $O(S') = \alpha'$, $O(S'') = \alpha''$, $O'(S_i') = \alpha_i'$ and $O'(S_i'') = \alpha_i''$;
3) we put $O(Q) = \xi$

and $O'(Q_i)=\beta_i$; 4) $\alpha' \# \alpha''$ is denoted by α and $\alpha_i' \# \alpha_i''$ is denoted by α_i . Clearly, the following inequalities are satisfied: $\alpha_i < \alpha$ for i=1,2,3 . Now we have by definition: a) $\xi = \omega_{h-h_o}(\alpha)$, b) $\beta_1=\omega_{h-h_1}(\alpha_1)$, c) $\beta_2=\omega_{h-h_1}(\alpha_2)$, d) $\beta_4=\omega_{h_1-h_2}(\beta_1 \# \beta_2)$, e) $\beta_3=\omega_{h-h_2}(\alpha_3)$, f) $\beta_5=\omega_{h_2-h_o}(\beta_3 \# \beta_4)$. We distinguish two cases : $h_1 > h_2$ and $h_1=h_2$.

<u>Case 1:</u> $h_1=h_2$. Then β_5 is given as follows:
$\beta_5 = \omega_{h_2-h_o}(\omega_{h-h_2}(\alpha_1) \# \omega_{h-h_2}(\alpha_2) \# \omega_{h-h_2}(\alpha_3))$. Since $\alpha_1, \alpha_2, \alpha_3 < \alpha$ and $h_2 < h$, we have in view of the properties of ω_d (see preliminaries) the following relation:

$$\sum_1^3 \omega_{h-h_2}(\alpha_i) < \omega_{h-h_2}(\alpha)$$. Therefore
$\beta_5 < \omega_{h_2-h_o}(\omega_{h-h_2}(\alpha))$ that is $\beta_5 < \omega_{h-h_o}(\alpha) = \xi$.

<u>Case 2:</u> $h_2 < h_1$. Then β_5 is given as follows:
$\beta_5 = \omega_{h_2h_o}(\omega_{h_1-h_2}(\omega_{h-h_1}(\alpha_1) \# \omega_{h-h_1}(\alpha_2)) \# \omega_{h_1-h_2}(\omega_{h-h_1}(\alpha_3)))$
where use has been made of $\beta_3 = \omega_{h_1h_2}(\omega_{h-h_1}(\alpha_3))$. Put $\sum_1^3 \omega_{h-h_1}(\alpha_i) = \lambda$. Again in view of $\alpha_1, \alpha_2, \alpha_3 < \alpha$ and $h_1 < h$, one finds $\lambda < \omega_{h-h_1}(\alpha)$. On the other hand, $\beta_5 \leq \omega_{h_2-h_o}(\omega_{h_1-h_2}(\lambda) \# \omega_{h_1-h_2}(\lambda))$ and, since $h_2 < h_1$ by assumption and $\lambda < \omega_{h-h_1}(\alpha)$, as noted, we conclude $\beta_5 < \omega_{h_2-h_o}(\omega_{h_1-h_2}(\omega_{h-h_1}(\alpha))) = \xi$, what proves the statement also in this case.

<u>2.5. A generalization</u>

In this section we discuss a simple and straightforward generalization of Gentzen's procedure which will play an important role in the sequel.

<u>A.</u> Let D be an arbitrary standard formula containing the number variable x free; let $\subset_D$ be the partial ordering associated

with $D(x)$ acording to chapter 1, section 1.5., part B. We note

Lemma 7: a) For terms t,p,q we can prove the sequents $t \subset_d p \longrightarrow D(t)$ and $t \subset_D p, p \subset_D q \longrightarrow t \subset_D q$ in ZTi without cuts and inductions. b) Let $D(x)$ be $u(x)=v(x)$. Let t,q be terms such that $u(t), v(t), u(q)$ and $v(q)$ all are saturated. If $t \subset_D q \longrightarrow$ is true, then it is provable in ZTi without cut and induction.

Proof: a_1) From $D(t) \longrightarrow D(t)$ we can derive by means of two applications of $\wedge \longrightarrow$ the sequent $t \subset_K p \wedge D(t) \wedge D(q) \longrightarrow D(t)$, that is, $t \subset_D q \longrightarrow D(t)$.
a_2) From $D(t) \longrightarrow D(t)$ and $D(q) \longrightarrow D(q)$ we can derive by means of two applications of $\wedge \longrightarrow$, a thinning on the left and an interchange, the sequent $t \subset_D p, p \subset_D q \longrightarrow D(t)$ and $t \subset_D p, p \subset_D q \longrightarrow D(q)$. These two sequents can be combined by means of an $\longrightarrow \wedge$ in order to yield $t \subset_D p, p \subset_D q \longrightarrow D(t) \wedge D(q)$. On the other hand $t \subset_K p, p \subset_K q \longrightarrow t \subset_K q$ is an axiom. By means of several applications of $\wedge \longrightarrow$ and some interchanges to this sequent we can derive $t \subset_D p, p \subset_D q \longrightarrow t \subset_K q$. Combining this sequent with that one proved under a_1) by means of an $\longrightarrow \wedge$, we finally obtain a derivation of $t \subset_D p, p \subset_D q \longrightarrow t \subset_D q$.
b) Since $t \subset_D q$ is false, one of $t \subset_K q$, $D(t)$, $D(q)$ is false. Assume eg. $t \subset_K q$ to be false; then $t \subset_K q \longrightarrow$ is an axiom from which $t \subset_D q \longrightarrow$ can be derived by means of two applications of $\Delta \longrightarrow$. We proceed similarly in the other cases.

For the rest of this section let D be a prime formula, which for simplicity is assumed to contain no function variables or special function constants. Let us assume that for one reason or the other (eg. by means of a proof in Zermelo-Fränkel set theory) we know that $\subset_D$ is a wellordering. We construct a new formal system by adding to ZT suitably formulated rules which express transfinite induction with respect to $\subset_D$. The system so obtained and denoted by $ZT(\subset_D)$ is more precisely defined as follows:
a) its axioms are the same as those of ZT ; b) it contains all the rules of ZT ; c) it contains in addition the following rules

$$\text{TI:}\quad \frac{D(y),\ (x) \subset_{Dy} A(x),\ \Gamma \longrightarrow \Delta, A(y)}{D(q),\ \Gamma \longrightarrow \Delta, A(q)}$$

and for all saturated terms t <u>such that $D(t)$ is true</u>

$$\text{TI}_a\text{:}\quad \frac{y \subset_D t,\ (x) \subset_{Dy} A(x),\ \Gamma \longrightarrow \Delta, A(y)}{q \subset_D t,\ \Gamma \longrightarrow \Delta, A(q)}$$

where $/t/$ is assumed to be a. In both cases y does not occur free in the conclusion and q is supposed to be free for y in $A(y)$.

The rules TI_a are of course superfluous; they are derivable from TI, as can easily be seen. We have introduced them for technical purposes, as will be seen below. The system $ZT(\subset_D)$ thus introduced has the same strength as the Hilbert-type system which we obtain by adding to ZH all axioms of the following form:

$$(y)(D(y) \wedge (x) \subset_{Dy} A(x) \ .\supset.\ A(y)) \ .\supset. (z)(D(z) \supset A(z)).$$

We omit the easy proof.

Proofs are again considered as finite trees. Those proofs which contain only sequents with at most one formula in the succedent are called intuitionistic proofs; they give rise to the intuitionistic version of $ZT(\subset_D)$, to be denoted by $ZTi(\subset_D)$.

<u>B.</u> With the exception of definition 12, which will be modified slightly, we can carry over the whole content of section 2.1. to the present situation. That is, the notions such as final part, image, normal proof etc. can be defined for proofs in $ZT(\subset_D)$ in exactly the same way as in section 2.1. In order to modify definition 12 we associate natural numbers, called complexities, with cuts, inductions, TI- and TI_a-inferences. The complexity of a cut or an induction is the same as before, namely the number of logical symbols contained in the cut formula or the induction formula, respectively. If the premiss of the TI-inference in question is
$D(y), (x) \subset_{Dy} A(x), \Gamma \longrightarrow \Delta, A(y)$, then we take as complexity of this inference the number of logical symbols contained in

$(x) \subset_{D^y} A(x)$. Similarly, if the premiss of the TI-inference in question is $y \subset_D t, (x) \subset_{D^y} A(x), \Gamma \longrightarrow \Delta, A(y)$, then we take again the number of logical symbols contained in $(x) \subset_{D^y} A(x)$ as complexity of this inference.

Definition 12_1: With every sequent S in a proof P we associate a natural number h(S) , its height, inductively as follows: 1) if S is the endsequent, then h(S)=0; 2) if S is premiss of a logical inference, of a conversion, or a one-premiss structural rule with conclusion S' , then h(S)=h(S') ; 3) if S is a premiss of a cut with conclusion S', then h(S)=max(d,h(S')) where d is the complexity of the cut in question; 4) if S is premiss of an induction with conclusion S', then h(S)=max(d,h(S')) where d is the complexity of the induction in question; 5) if S is premiss of a TI- or TI_a-inference with conclusion S', then h(S)=max(d,h(S')) where d is the complexity of the TI- or TI_a-inference in question.

A cut with jump is, of course, the same as before, namely a cut $S_1,S_2/S$ such that $h(S_1) > h(S)$. It is clear that the height of a sequent in the final part is unaffected by this change of definition, and the same is true for the notion of cut associated with a given fork I_1,I_2,I_3 . A TI- or TI_a-inference will, of course, be called critical if its conclusion belongs to the final part; for logical and induction inferences the notion "critical" has the same meaning as before.

Next, we can carry over the whole body of section 2.2. to the present situation. That is, we can introduce preliminary reduction steps, induction reductions and elimination of forks from the final part in exactly the same way as in section 2.2. All the lemmas and theorems stated there remain invariably true in the present situation. In order to obtain a counterpart of theorem 6 in section 2.3., however, we have to introduce two new types of reduction steps, connected with the new rules TI and TI_a; they are called TI- and TI_a-reduction steps.

Let us first explain the TI-reduction step. To this end let P be a normal proof and assume that there is a critical TI-inference in P, say

$$\frac{D(y),\ (x) \subset_{D} y\, A(x),\ \Gamma \longrightarrow \Delta, A(y)}{D(q), \Gamma \longrightarrow \Delta, A(q)}$$

for which q is saturated; assume $|q| = a$. We denote this inference more symbolically by S/S' ; by P_S and $P_{S'}$ we denote the subproofs of S and S' in P, respectively. Now we distinguish two cases: 1) $D(q)$ is true; 2) $D(q)$ is false. We start with case 1). If we replace every (free) occurence of y in P_S by q then we obtain according to lemma 4 a new proof P_S^q of $D(q),\ (x) \subset_{D} q\, A(x),\ \Gamma \longrightarrow \Delta, A(q)$. On the other hand (lemma 7), there is a proof P_o not containing any cuts, inductions, TI- and TI_a-inferences, whose endsequent is $y \subset_D q \longrightarrow D(y)$. A new derivation P' of S' can now be obtained in the following way:

$$
\begin{array}{ll}
 & P_o \qquad\qquad\qquad\qquad P_s \\
 & \vdots \qquad\qquad\qquad\qquad\quad \vdots \\
\text{cut} & \dfrac{y \subset_D q \longrightarrow D(y) \qquad D(y),(x) \subset_D y\, A(x), \Gamma \longrightarrow \Delta, A(y)}{y \subset_D q,\ (x) \subset_D y\, A(x), \Gamma \longrightarrow \Delta, A(y)} \\
TI_a & \dfrac{}{s \subset_D q, \Gamma \longrightarrow \Delta, A(s)} \\
\longrightarrow \supset & \dfrac{}{\Gamma \longrightarrow \Delta, s \subset_D q\ .\supset. A(s)} \\
 & \dfrac{}{\Gamma \longrightarrow \Delta, (x) \subset_D q\, A(x)} \qquad \begin{array}{c} P_S^q \\ \vdots \\ D(q),(x) \subset_D q\, A(x), \Gamma \longrightarrow \Delta, A(q) \end{array} \\
 & \overline{\overline{\qquad\qquad D(q), \Gamma \longrightarrow \Delta, A(q) \qquad\qquad}}
\end{array}
$$

Here the double line indicates a cut followed by some interchanges and contractions. Now we replace $P_{S'}$ in P by P' , obtaining thus a new proof P^* having the same endsequent as P . Thereby we can always choose the variable s in such a way that the new proof P^* is again normal; eg. by taking for s the first individual variable which does not occur in P at all.

Now to case 2): $D(q)$ is false. Since $D(q)$ is prime and false, $D(q) \longrightarrow$ is an axiom. Hence we can derive S' from $D(q) \longrightarrow$ by means of thinnings and interchanges alone. Let $\hat{P}$ be such a derivation. By replacing $P_{S'}$ in P by $\hat{P}$ we obtain a new proof P^* , having the same endsequent as P , which is also normal.

Both in case 1) and case 2) we say that P^* is obtained from P by means of a TI-reduction step.

Now to the TI_a-reduction step. Let P be a normal proof and assume that there is a critical TI_a-inference in P, say

$$\frac{y \subset_D t,\ (x) \subset_D y\, A(x),\ \Gamma \longrightarrow \Delta, A(y)}{q \subset_D t,\ \Gamma \longrightarrow \Delta, A(q)}$$

for which both t and q are saturated; let $|t|$ and $|q|$ be a and b respectively. Of course, $D(t)$ is true by assumption. We denote this inference more briefly by S/S' ; by P_S and $P_{S'}$ we denote the subproofs of S and S' in P, respectively. Again we have two cases to distinguish: 1) $t \subset_D q$ is true , 2) $t \subset_D q$ is false.

Let us start with case 1); note that $D(q)$ is true. Replacing every (free) occurence of y in P_S by q gives a proof P_S^q of

$q \subset_D t, (x) \subset_D q\, A(x), \Gamma \longrightarrow \Delta, A(q)$. According to lemma 7 there is a proof in ZTi not containing cuts and inductions of $y \subset_D t, q \subset_D t \longrightarrow y \subset_D t$; call it P_o. A new deduction P' of S' can now be obtained in the following way:

$$\begin{array}{ll}
P_o & \qquad\qquad\qquad\qquad P_S \\
\vdots & \qquad\qquad\qquad\qquad \vdots \\
\multicolumn{2}{l}{y \subset_D q,\ q \subset_D t \longrightarrow y \subset_D t \qquad y \subset_D t, (x) \subset_D y\, A(x), \Gamma \longrightarrow \Delta, A(y)} \\
\hline\hline
\multicolumn{2}{l}{\quad y \subset_D q, (x) \subset_D q\, A(x), q \subset_D t, \Gamma \longrightarrow \Delta, A(y)} \\
\hline
TI_b & \quad s \subset_D q, q \subset_D t, \Gamma \longrightarrow \Delta, A(s) \qquad\qquad P_S^q \\
\hline
\multicolumn{2}{l}{\quad q \subset_D t, \Gamma \longrightarrow \Delta, s \subset_D q \supset A(s)} \\
\hline
\multicolumn{2}{l}{\quad q \subset_D t, \Gamma \longrightarrow \Delta, (x) \subset_D q\, A(x) \qquad q \subset_D t, (x) \subset_D q\, A(x), \Gamma \longrightarrow \Delta, A(q)} \\
\hline\hline
\multicolumn{2}{l}{\qquad\qquad q \subset_D t, \Gamma \qquad \Delta, A(q)}
\end{array}$$

A double line indicates again a cut followed by interchanges and contractions. Now we replace $P_{S'}$ in P by P' ; this gives a new proof P* , having the same endsequent as P . By choosing for s the first number variable which does not occur in P , we can achieve that P* is again normal.

Now to the second case: $q \subset_D t$ is false. Then there is a proof P_o in ZTi of $q \subset_D t \longrightarrow$ which does not contain cuts and inductions. By adding some thinnings and interchanges, we obtain a proof $\hat{P}$ in ZTi of S' which does not contain cuts and inductions. By replacing $P_{S'}$ in P by $\hat{P}$, we obtain a new normal proof P* which has the same endsequent as P . In both cases we say that P* is obtained from P by means of a TI_a-reduction step.

C. Now we can divide the set of reduction steps again in two classes: 1) preliminary reduction steps (elimination of logical axioms and thinnings from the final part) ; 2) essential reduction steps (elimination of forks, induction reductions, TI- and TI_a-reduction steps). For this enlarged set of reduction steps we can introduce a relation W in the same way as in definition 12**, sect. 2.3. ; with this W we can associate the sets D_P and the relations W_P, W* and W^*_P precisely as in section 2.3., pt. A. It is an easy matter to verify that theorems 5,6 and its corollaries also hold in the present case (with the new set of reduction steps, of course). Hence a formal consistency proof for $ZT(\subset_D)$ is obtained if we can show that the relation W is wellfounded. We prove this by associating ordinals with proofs in such a way that an essential reduction step applied to a proof P lowers its ordinal. More precisely, given a proof P , we associate inductively from above with every sequent S in P an ordinal, to be denoted by O(S). The inductive definition of O(S) goes as follows: 1) if S is an axiom then O(S)=1 ; 2) if S is the conclusion of a structural inference, a conversion, a logical inference or an induction then we proceed as in pt. A of sect. 2.4. ; 3) if S is the conclusion of a TI-inference with premiss S_1, then we put $O(S)=\omega_d((\alpha \# \omega^{\xi+1})\omega^{\xi+1})$, where $\alpha = O(S_1)$, $d=h(S_1)-h(S)$ and where ξ is the ordinal associated with the wellfounded relation $\subset_D$; 4) if S is the conclusion of a TI_a-inference S_1/S then we put $O(S)=\omega_d((\alpha \# \omega^{\lambda+1})\omega^{\lambda+1})$ where $\alpha = O(S_1)$, $d=h(S_1)-h(S)$ and where λ is the ordinal associated with the partial ordering $\{\langle x,y\rangle \,/\, x \subset_D a \wedge y \subset_D a \wedge x \subset_D y\}$. The ordinal of a proof P is now by definition the ordinal of its endsequent; we denote it by O(P) .

It remains to show that a preliminary reduction step does not increase the ordinal of a proof, and that an essential reduction step lowers the ordinal of a proof. Again we postpone the discussion of the first half of this statement (corresponding to th. 7) to the next section and look at the second half (corresponding to theorem 8). So we have to prove that $O(P^*) < O(P)$ holds whenever P^* is obtained from P by means of an essential reduction step. The proof is by cases according to the kind of reduction step which transforms P into P^*.

<u>Case 1:</u> P^* follows from P by means of an induction reduction. The verification of $O(P^*) < O(P)$ is achieved in exactly the same way as in [1].

<u>Case 2:</u> P^* follows from P by means of a fork elimination. Here too, the verification is word by word the same as in [1], or as in section 2.4. in case of an $\supset$-fork.

<u>Case 3:</u> P^* follows from P by means of a TI-reduction step. In order to verify $O(P^*) < O(P)$, we refer to the notation and the diagram which were introduced in connection with the definition of TI-reduction step. First we consider the <u>subcase 1</u>: $D(q)$ is true. Let us rewrite the diagram presented there in a shorter way, as follows:

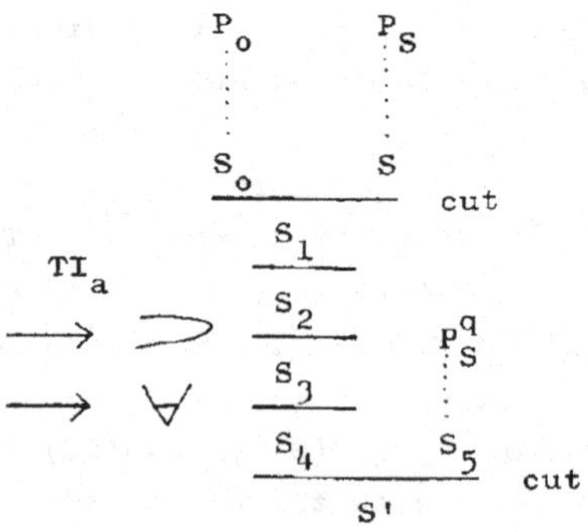

where S_5 is the endsequent of P_S, that is, $D(q), (x) \subset_{D^q} A(x), \Gamma \longrightarrow \Delta, A(q)$. Let us denote the ordinals of S and S' in P by α and γ respectively, the ordinals of S and S' in P^* by α' and γ' respectively, the ordinals of

S_i (in P^*) by α_i. In addition, let us denote by $h(S)$, $h(S')$ the heights of S,S' in P, and by $h'(S)$, $h'(S')$, $h'(S_i)$ $(i \leq 5)$ the heights of S,S',S_i' $(i\leq 5)$ respectively in P^*. A quick inspection shows: 1) $h'(S_1)=h'(S)=h(S)$; 2) $h'(S')=h(S')$; 3) $\alpha'=\alpha$; 4) $\alpha_5=\alpha$; 5) $\alpha_0=m<\omega$. By definition, $\zeta = \omega_d((\alpha \# \omega^{\xi+1})\omega^{\xi+1})$ where $d=h(S)-h(S')$. Now let us calculate ζ'. First we note that the ordinal λ associated with $\{\langle x,y\rangle / x \subset_D a \wedge y \subset_D a \wedge x \subset_D y\}$ is smaller than ξ, (the ordinal associated with $\subset_D$). Next, we obtain for $\alpha_1,\dots,\alpha_4$ and ζ' successively the following values: 1) $\alpha_1=\alpha \# m$; 2) $\alpha_2=(\alpha_1 \# \omega^{\lambda+1})\omega^{\lambda+1}$; 3) $\alpha_3=\alpha_2 \# 1$; 4) $\alpha_4=\alpha_3 \# 1$; 5) $\zeta'=\omega_d(\alpha_4 \# \alpha)$. We want to prove $\zeta'<\zeta$. Since $\lambda<\xi$, we have $\lambda+1\leq\xi$ and therefore $\omega_d((\alpha \# \omega^{\lambda+2})\omega^{\lambda+2}) \leq \omega_d((\alpha \# \omega^{\xi+1})\omega^{\xi+1})$. Hence we are through if we have proved $\omega_d((\alpha \# m \# \omega^{\lambda+1})\omega^{\lambda+1} \# \alpha \# 2) < \omega_d((\alpha \# \omega^{\lambda+2})\omega^{\lambda+2})$. This in turn is a special case of the following inequality:

E. $\omega_d((\alpha \# m \# \omega^{\gamma})\omega^{\gamma} \# \alpha \# n) < \omega_d((\alpha \# \omega^{\gamma+1})\omega^{\gamma+1})$

(with $n,m<\omega$). Let us turn to the proof of E. For convenience, we use the shorthand writing $\sum_n \eta$ for $\eta \# \eta \# \dots \# \eta$, n times. Since $\sum_n \zeta < \zeta\omega$ (see preliminaries), we obtain successively the following inequalities: 1) $\alpha \# m \# \omega^{\gamma} \leq \alpha \# \sum_{m+1}\omega^{\gamma}$;
2) $\alpha \# \sum_{m+1}\omega^{\gamma} < \alpha \# \omega^{\gamma+1}$;
3) $(\alpha \# m \# \omega^{\gamma})\omega^{\gamma+1} < (\alpha \# \omega^{\gamma+1})\omega^{\gamma+1}$;
4) $\sum_{n+2}(\alpha \# m \# \omega^{\gamma})\omega^{\gamma} < (\alpha \# m \# \omega^{\gamma})\omega^{\gamma+1}$;
5) $(\alpha \# m \# \omega^{\gamma})\omega^{\gamma} \# \alpha \# n \leq \sum_{n+2}(\alpha \# m \# \omega^{\gamma})\omega^{\gamma}$;
6) $(\alpha \# m \# \omega^{\gamma})\omega^{\gamma} \# \alpha \# n < (\alpha \# \omega^{\gamma+1})\omega^{\gamma+1}$. From 6) one immediately derives inequality E. Hence, by putting $n=2$ in E, we obtain $\zeta'<\zeta$. The inequality $O(P^*) < O(P)$ is now an easy consequence of $\zeta'<\zeta$.

Now to subcase 2: $D(q)$ is false. Then we get P^* from P by replacing $P_{S'}$ in P by a derivation $\hat{P}$ of S' which does not contain cuts, inductions, TI- or TI_a-inferences. That is, the ordinal ζ' of S' in P^* is a natural number m which clearly satisfies the inequality $m<\zeta$, where $\zeta = \omega_d((\alpha \# \omega^{\xi+1})\omega^{\xi+1})$ is again the ordinal of S' in P. From $\zeta'<\zeta$ the inequality $O(P^*) < O(P)$ immediately follows.

Case 4: P^* follows from P by means of a TI_a-reduction step. Again we use the notation and the diagram introduced in connection with the definition of TI_a-reduction step. First to subcase 1: $q \subset_D t$ is true. The diagram used in the definition of TI_a-reduction step may be presented more symbolically, as follows:

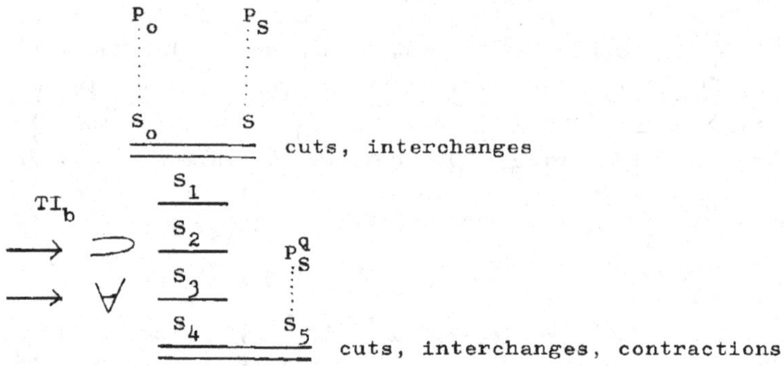
P_o
P_S
S_o
S
cuts, interchanges
S_1
TI_b
$\rightarrow \supset$
S_2
P^q_S
$\rightarrow \forall$
S_3
S_4
S_5
cuts, interchanges, contractions

where S_5 is $(x) \subset_{D^q} A(x),\ q \subset_D t, \Gamma \longrightarrow \Delta, A(q)$. By α and ζ we denote the ordinals of S and S' in P, by α' and ζ' the ordinals of S and S' in P*. In addition, h(S) and h(S') are the heights of S and S' in P, while h'(S) and h'(S') are the heights of S and S' in P*. Furthermore, λ is the ordinal of $\{\langle x,y\rangle / x \subset_D a \wedge y \subset_D a \wedge x \subset_D y\}$ and ν is the ordinal of $\{\langle x,y\rangle / x \subset_D b \wedge y \subset_D b \wedge x \subset_D y\}$. Since $q \subset_D t$ is true and $|t| = a$, $|q| = b$, it is clear that $\nu < \lambda$. The calculation of ζ and ζ' ensues in the same way as in case 3 and yields the same kind of expressions as there; that is, we obtain

$\zeta = \omega_d((\alpha \# \omega^{\lambda+1})\omega^{\lambda+1})$ and $\zeta' = \omega_d((\alpha \# m \# \omega^{\nu+1})\omega^{\nu+1} \# \alpha \# 2)$

where $d = h(S) - h(S')$. But the statement $\zeta' < \zeta$ is again a special case of the inequality E. which has been proved above under case 3. Finally, $O(P^*) < O(P)$ follows easily from $\zeta' < \zeta$.

Now to subcase 2: $q \subset_D t$ is false. We proceed in the same way as under subcase 2 of case 3.

<u>D.</u> The formal consistency proof for $ZT(\subset_D)$ thus obtained has, of course, not much interest in itself. The most which can be said is that all results proved in [1] (for ZT and ZTi essentially) can be proved also for $ZT(\subset_D)$ and $ZTi(\subset_D)$, as a straightforward analysis shows. However the technique used in this formal consistency proof will play an important role in the later chapters.

2.6. The preliminary reduction steps

A. As basis of our discussion we take the theory $ZT(\subset_D)$. Below P is an arbitrary proof in $ZT(\subset_D)$; the inferences in P are denoted symbolically by $I, I', I_1, I_2, \ldots$ etc. By $N(A)$ we denote the number of logical symbols in the formula A.

Definition 13: a) An inference I in a proof P is called strong if it is either a cut, an induction, a TI- or a TI_a-inference. All other inferences are called weak.

b) A function f which associates with every strong inference I in P, a natural number; $f(I)$ is called a complexity assignement for P.

c) Let f be a complexity assignement for P having the following properties: 1) if I is a cut with cut formula A then $f(I)=N(A)$; 2) if I is an induction with premiss $A(x), \Gamma \longrightarrow \Delta, A(x')$, then $f(I)=N(A)$; 3) if I is a TI- or a TI_a-inference with premiss $D(y), (x)_{\subset_D y} A, \Gamma \longrightarrow \Delta, A(y)$ or $y \subset_D t, (x)_{\subset_D y} a, \Gamma \longrightarrow \Delta, A(y)$ respectively, then $f(I)=N((x)_{\subset_D y} A)$. Then f^n is called the normal complexity assignement for P.

With such a complexity assignement f we may associate a notion of height in precisely the same way as in definition 12 or 12_1. That is, we have

Definition 14: Let f be a complexity assignement for P. A height $h(S)$ is associated with every sequent S in P as follows: 1) if S is the endsequent, then $h(S)=0$; 2) if S is the premiss of a weak inference I whose conclusion is S', then $h(S)=h(S')$; 3) if S is the premiss of a strong inference I whose conclusion is S', then $h(S)=\max(h(S'),f(I))$. With this notion of height we can associate ordinals with sequents in exactly the same way as before.

Definition 15: Let f be a complexity assignement for P, and h the height function associated with f according to def. 14 . Then an ordinal $O(S)$ can be associated with every S in P, as follows:

1) if S is an axiom, then $O(S)=1$; 2) if S is the conclusion of a one-premiss structural inference or a conversion S'/S, then $O(S)=O(S')$; 3) if S is the conclusion of a one-premiss logical inference S'/S, then $O(S)=O(S')\#1$; 4) if S is the conclusion of a two-premiss logical inference $S_1,S_2/S$, then $O(S)=O(S_1)\#O(S_2)\#1$; 5) if S is the conclusion of a cut $S_1,S_2/S$, then $O(S)=\omega_d(O(S_1)\#O(S_2))$ where $d=h(S_1)-h(S)$; 6) if S is the conclusion of an induction S'/S, then $O(S)=\omega_d(O(S')\omega)$ with $d=h(S')-h(S)$; 7) if S is the conclusion of a TI-inference S'/S, then $O(S)=\omega_d((O(S')\#\omega^{\xi+1})\omega^{\xi+1})$ with $d=h(S')-h(S)$ and where ξ is the ordinal associated with $\subset_D$; 8) if S is the conclusion of a TI_a-inference S'/S, then $O(S)=\omega_d((O(S')\#\omega^{\lambda+1})\omega^{\lambda+1})$ where $d=h(S')-h(S)$ and where λ is the ordinal associated with a with respect to $\subset_D$.

As ordinal of P, denoted by $O(P)$, we take the ordinal $O(S_E)$ of the endsequent S_E of P. In order to indicate the dependence of h and O on f and P, we write more explicitely $h(P,f/S)$ and $OP,f/S)$, respectively. Our main tool in treating preliminary reduction steps is

<u>Lemma 8:</u> Let P be a proof, I_o: $S_1,S_2/S^*$ a cut in P and f,g two complexity assignements for P having the following properties: 1) if I is a strong inference different from I, then $f(I)=g(I)$; 2) $g(I_o)+1 = f(I_o)$. Then the following holds: a) if S is a sequent in P which is different from S* and is neither above nor below S*, then $O(P,g/S)=O(P,f/S)$; b) if S is either S* or below S*, then $O(P,g/S)\leqq O(P,f/S)$. In particular, $O(P,g/S_E)\leqq O(P,f/S_E)$ where S_E is the endsequent of P.

<u>Proof:</u> Part a) of the statement is rather trivial to verify; we omit its proof. Part b) is essentially proved if we can show $O(P,g/S^*)\leqq O(P,f/S^*)$: if $S_o,\ldots,S_n$ (with $S_o=S^*$ and S_n the endsequent) is the path which leads from S* to the endsequent, one shows with an easy induction with respect to i (using part a)) that $O(P,g/S_i)\leqq O(P,f/S_i)$ holds. Hence, let us prove $O(P,g/S^*\leqq O(P,f/S^*)$. Here two subcases arise: 1) $h(P,f/S_1)=h(P,f/S^*)$; 2) $h(P,f/S^*)<h(P,f/S_1)$. In the first case, one easily verifies that $h(P,g/S)=h(P,f/S)$ holds for all S in P, and obtains as an immediate consequence that $O(P,f/S)=O(P,g/S)$ holds for all S in P. Hence, let us assume

$h(P,f/S^*) < h(P,f/S_1)$. Then the following relations hold, as is easily verified: 1) $h(P,f/S^* < f(I_o)$, 2) $h(P,f/S_1)=f(I_o)$, 3) $h(P,g/S_1)=f(I_o)-1$, 4) $h(P,f/S^*)=h(P,g/S^*)$. Now let us introduce the notion of "good" sequent with respect to S_1 inductively, as follows: 1) S_1 is good; 2) if S is a premiss of a weak inference whose conclusion is good, then S is good; 3) if S is a premiss of a strong inference I , whose conclusion is good, then S is good, provided that $f(I) < f(I_o)$ holds; 4) S is good only in virtue of 1)-3) . The set of good sequents (with respect to S_1) gives rise to a subtree P_1 of P: it is that subtree of P which contains precisely those sequents of P which are good with respect to S_1 . The following properties of good sequents are immediate consequences of their definition: α) $h(P,f/S)=h(P,f/S_1)=f(I_o)$; β) $h(P,g/S)=h(P,g/S_1)$; γ) if S is an uppermost element of P_1, then it is either an axiom or the conclusion of a strong inference I for which $f(I_o)\leqq f(I)$ holds; δ) if S is an uppermost element of P_1 and not an axiom, if furthermore S' is situated above S, then $O(P,g/S')=O(P,f/S')$. Now we will prove that the following inequality holds for every good sequent: A) $O(P,g/S)\leqq\omega_1(O(P,f/S))$ (where $\omega_1(\alpha)$ is, of course, only another way of writing ω^α) . We prove A) by induction over P_1 and proceed by cases.

Case 1: S is an axiom. Then $O(P,g/S)=O(P,f/S)=1$ and A) holds, since $1\leqq\omega^1$.

Case 2: S is the conclusion of a strong inference I such that $f(I_o)\leqq f(I)$. Let I be eg. a TI-inference S'/S and put $O(P,f/S')=\alpha$. In virtue of δ) above, $O(P,g/S')=\alpha$, too . In addition, $h(P,f/S')=f(I)$ and $h(P,g/S')=f(I)$, as is easily verified. On the other hand, $h(P,f/S)=f(I_o)=h(P,g/S)+1$. Putting $d=h(P,f/S')-h(P,f/S)$, we obtain $O(P,f/S)=\omega_d(\varrho)$ and $O(P,g/S)=\omega_{d+1}(\eta)$ where $\eta=(\alpha\#\omega^{\xi+1})\omega^{\xi+1}$ with ξ, as before the ordinal of $\subset_D$. Since $\omega_{d+1}(\eta)=\omega_1(\omega_d(\varrho))$, the inequality A) is clearly satisfied. The cases where I is a cut, an induction or a TI_a-inference are treated alike.

Case 3: S is the conclusion of an induction S'/S and S' is also a good sequent. Put $O(P,f/S')=\alpha$ and $O(P,g/S')=\alpha'$ and assume $\alpha'\leqq\omega^\alpha$ to be proved. Since $h(P,f/S)=h(P,f/S')=f(I_o)$ in virtue of property α) listed above, we find $O(P,f/S)=\alpha\cdot\omega$. On the other hand, we conclude from properties β) and 3) listed above that $h(P,g/S)=h(P,g/S')=f(I_o)-1$ holds. Hence, $O(P,g/S)=\alpha'\omega$. But $\alpha'\omega \leqq \omega^\alpha$. $\omega \leqq \omega^{\alpha+1} \leqq \omega^{\alpha\cdot\omega}$, that is, inequality A) is satisfied.

Case 4: S is the conclusion of a cut $S', S''/S$, and both S', S'' are good sequents. Put $O(P,f/S')=\alpha'$, $O(P,f/S'')=\alpha''$, $O(P,g/S')=\beta'$, $O(P,g/S'')=\beta''$, $O(P,f/S)=\alpha$ and $O(P,g/S)=\beta$. The inductive assumption is $\beta' \leq \omega^{\alpha'}$, $\beta'' \leq \omega^{\alpha''}$. As in case 3, we find $h(P,g/S')=h(P,g/S)$ and $h(P,f/S')=h(P,f/S)$. Therefore $\beta = \beta' \# \beta''$ and $\alpha = \alpha' \# \alpha''$. Since $\lambda = \max(\alpha', \alpha'') < \alpha' \# \alpha''$, we have $\omega^{\alpha'} \# \omega^{\alpha''} < \omega^\lambda \# \omega^\lambda < \omega^\alpha$. On the other hand, $\beta \leq \omega^{\alpha'} \# \omega^{\alpha''}$; hence, $\beta \leq \omega^\alpha$, that is, A) holds.

Case 5: S is the conclusion of a TI-inference S'/S and S' is a good sequent. Put $O(P,f/S')=\alpha$, $O(P,f/S)=\beta$, $O(P,g/S')=\alpha'$ and $O(P,g/S)=\beta'$. Again $h(P,g/S')=h(P,g/S)$ and $h(P,f/S')=h(P,f/S)$. Hence $\beta = (\alpha \# \omega^{\xi+1})\omega^{\xi+1}$ and $\beta' = (\alpha' \# \omega^{\xi+1})\omega^{\xi+1}$. Furthermore, by assumption with respect to δ one immediately proves $\alpha + \delta \leqq \alpha\omega^\delta$ for $\alpha \geqq 1$. Using this, we obtain the following list of inequalities, in which each is a consequence of the previous one or of the assumption $\alpha' \leq \omega^\alpha$: 1) $\alpha' \# \omega^{\xi+1} \leq \omega^\alpha \# \omega^{\xi+1}$;
2) $\alpha' \# \omega^{\xi+1} \leqq \omega^{\alpha \# (\xi+1)}$ (since $\max(\alpha, \xi+1) \leq \alpha \# (\xi+1)$);
3) $(\alpha' \# \omega^{\xi+1})\omega^{\xi+1} \leq \omega_1(\alpha \# \omega^{\xi+1}) \cdot \omega^{\xi+1}$;
4) $(\alpha' \# \omega^{\xi+1})\omega^{\xi+1} \leqq \omega_1((\alpha \# \omega^{\xi+1}) + \xi+1)$
(since $\omega_1(\mu)\omega^\nu = \omega^\mu \omega^\nu$);
5) $(\alpha' \# \omega^{\xi+1})\omega^{\xi+1} \leqq \omega_1((\alpha \# \omega^{\xi+1})\omega^{\xi+1})$
(since $\mu + \delta \leqq \mu\omega^\delta$ if $\mu \neq 0$). But 5) is nothing else than $\beta' \leqq \omega^{\alpha'}$, that is inequality A).

Case 6: S is the conclusion of a TI_a-inference S'/S and S' is good. The treatment is exactly the same as in case 5.

Hence, if we specialize to the case where S is S_1, we find $O(P,g/S_1) \leqq \omega_1(O(P,f/S_1))$. What has been done for S can be done in exactly the same way for S_2, and we find $O(P,g/S_2) \leqq \omega_1(O(P,f/S_2))$. Now let us put $h(P,f/S_1)-h(P,f/S^*)=d_o$, $h(P,g/S_1)-h(P,g/S^*)=d$ and $O(P,f/S_i)=\alpha_i$, $O(P,g/S_i)=\beta_i$ $(i=1,2)$. Then, obviously, $d_o=d+1$, $O(P,f/S^*)=\omega_{d+1}(\alpha_1 \# \alpha_2)$ and $O(P,g/S^*)=\omega_d(\beta_1 \# \beta_2)$. But $\beta_1 \leq \omega^{\delta_1}$, $\beta_2 \leq \omega^{\delta_2}$ in view of inequality A). Therefore $\omega_d(\beta_1 \# \beta_2) \leq \omega_d(\omega^{\alpha_1} \# \omega^{\alpha_2})$ and since

$\omega_d(\omega^{\alpha_1} \# \alpha_2) \leq \omega_{d+1}(\alpha_1 \# \alpha_2)$, we obtain the desired inequality $\omega_d(\beta_1 \# \beta_2) \leq \omega_{d+1}(\alpha_1 \# \alpha_2)$, that is, $O(P,g/S^*) \leq O(P,f/S^*)$.
From the preceding lemma we now obtain immediately the following

Theorem 9: Let P be a proof in $ZT(\subset_D)$ and f,g two complexity assignements for P which satisfy the following condition: for every strong inference I, we have $g(I) \leq f(I)$. Then $O(P,g/S_E) \leq O(P,f/S_E)$ where S_E is the endsequent of P.

Proof: One constructs a list of complexity assignements $g_o,\dots,g_{n+1}$ with the following properties: 1) g_o is f ; 2) g_{n+1} is g ; 3) for every $i \leq n$ there is a strong inference I_i in P such that $g_i(I_i) = g_{i+1}(I_i)+1$, while $g_i(I) = g_{i+1}(I)$ for all other strong inferences. The theorem then follows by some successive applications of the previous lemma.

B. We are now ready to discuss preliminary reduction steps. Among the operations involved in preliminary reduction steps, there is just one for which it is not evident that it does not increase the ordinal of the proof to which it is applied. This operation applies in case there is a cut $S_1,S_2/S$ in the final part of a proof P which has the property: S is derivable from S_1 (or S_2) by means of thinnings and interchanges. The operation then consists in the following: one replaces the subproof P_S in P by the following derivation

$$\begin{array}{l} P_{S_1} \\ \vdots \\ S_1 \\ \hline\hline S \end{array} \quad \text{thinnings, interchanges}$$

obtaining thus a new proof P* having the same endsequent as P .
If the roles of S_1 and S_2 are interchanged, then one replaces P_S , of course, by

$$\begin{array}{l} P_{S_2} \\ \vdots \\ S_2 \\ \hline\hline S \end{array} \quad \text{thinnings, interchanges}$$

In order to have a name for it, let us call the operation just described omission of a cut; we say that P^* follows from P by omission of a cut. The main property of this operation is described by

Theorem 10: Let P^* follow from P by omission of a cut. Then $O(P^*)$ is smaller than $O(P)$: $O(P^*) < O(P)$.

Proof: Let $I_o : S_1, S_2/S$ be a cut in P which eg. has the property: S can be derived from S_1 by means of thinnings and interchanges. Let P^* be obtained from P by replacing the subproof P_S of S in P by the following derivation:

$$\begin{array}{l} P_{S_1} \\ \vdots \\ \vdots \\ S_1 \\ \hline\hline S \end{array} \quad \text{thinning, interchanges}$$

(P_{S_1} is the subproof of S_1 in P) . Let finally f and f^* be the normal complexity assignements for P and P^* respectively. The theorem is proved if we can show $O(P^*,f^*/S) < O(P,f/S)$. In order to prove this, let us first consider the proof P but provided with a complexity assignement g having the following properties: 1) $g(I_o)=0$; 2) if I is a strong inference different from I_o, then $g(I)=f(I)$. From lemma 9 we obtain $O(P,g/S) \leqq O(P,f/S)$. On the other hand, one easily verifies that, if S' is a sequent in P_{S_1} , then $O(P,g/S')=O(P^*,f^*/S')$. Now put $O(P,g/S_1)=\alpha_1$, $O(P,g/S_2)=\alpha_2$; in view of the last remark we have $O(P^*,f^*/S_1)=\alpha_1$. Then $O(P,g/S)=\alpha_1 \# \alpha_2$, while $O(P^*,f^*/S)=\alpha_1$. Since $0<\alpha_2$, we obtain $O(P^*,f^*/S)<O(P,g/S)$, that is, $O(P^*,f^*/S)<O(P,f/S)$, what proves the statement.

With the aid of theorem 10 it is now almost trivial to verify

Theorem 11: If P^*,P are two proofs in $ZT(\subset_D)$ such that P^* is obtained from P by means of a series of preliminary reduction steps, then $O(P^*) \leqq O(P)$.

We omit the proof.

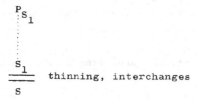

C. In this section, we have presented in some detail a generalization of Gentzen's second consistency proof to systems of the type ZT($\subset_D$) . Now, as noted, theories of this type have no real interest in themselves. Our main objects of investigation will be the theories ZTi/I , ZTi/II etc., which where introduced in the preceding chapter. However, it turns out that these theories are amenable to a Gentzen-like treatment which behaves with respect to reduction steps and ordinal assignements in essentially the same way as the treatment of ZT($\subset_D$) presented in this chapter, and we will see that most of the results together with their proofs will carry over without any changes to the new situation.

CHAPTER III:
The intuitionistic system of number theory

This is the last of the introductory chapters. In it we study the behaviour of intuitionistic proofs under the application of fork elimination. In addition, we prove a lemma which is crucial for the further development. The material presented here is essentially contained in [8] . As basis of our discussion, we take the theories $ZT(\subset_D)$ and $ZTi(\subset_D)$, respectively.

3.1. Elimination of forks in intuitionistic proofs

A. Let P be a proof in $ZTi(\subset_D)$, that is, a proof which contains only sequents having at most one formula on the right of the sequential arrow. One easily verifies the following fact: if we apply to P a preliminary reduction step, an induction reduction, a TI- or a TI_a-reduction step, then we obtain again an intuitionistic proof P* . If on, the other hand, we eliminate a fork in P, then it is clear by inspection that the resulting proof P' is no longer intuitionistic. However, as has been shown in [8] , it is sufficient to apply to P' a number of preliminary reduction steps in order to obtain again an intuitionistic proof P" , having the same endsequent as P' and, hence, as P . Below we will briefly describe how to get from P' to P" ; for a detailed treatment we refer to [8] .

To start with, let us call a proof P in $ZT(\subset_D)$ almost intuitionistic if there is a path $S_o,\dots,S_m$ (with S_m the endsequent) in the final part of P , which has the following properties: 1) S_o has the form $\Gamma_o \longrightarrow A$; 2) for $i \geq 1$, S_i has the form $\Gamma_i \longrightarrow A, \phi_i$ where ϕ_i may be empty; 3) ϕ_1 is not empty, and S_1 follows by right thinning from $\Gamma_o \longrightarrow A$; 4) the A indicated in $S_1,\dots,S_m$ is not side formula of any inference; 5) if S in P is different from $S_1,\dots,S_m$, then it contains at most one formula in the succedent. This definition of almost intuitionistic proof is a slightly more specialized version of that one given in [8] . For almost intuitionistic proofs, one can prove the following lemma:

Lemma 9: Let P be an almost intuitionistic proof of $\Gamma \longrightarrow A, \phi$ (where ϕ may be empty). P can be transformed into an intuitionistic proof P* of $\Gamma \longrightarrow A$ by means of a series of applications of preliminary reduction steps.

Proof: Let $S_o, \ldots, S_m$ be the path in P which satisfies the properties 1) - 5) mentioned above. As before, S_o is $\Gamma_o \longrightarrow A$, while S_i is $\Gamma_i \longrightarrow A, \phi_i$ for $i \geq 1$. Let k be the number of formulas among the ϕ_i's which are cut formulas; we call k the characteristic number of P. We prove the statement of the lemma by induction with respect to k. If $k = 0$, then ϕ_m is an image of ϕ_1. By cancelling all ϕ_i's and omitting the thinning S_o/S_1, one gets the desired proof P*.

If $k>0$, then there is a smallest i such that ϕ_i is the cut formula of a cut, which necessarily must look as follows:
$\Gamma_i \longrightarrow A, \phi_i \; ; \; \phi_i, \Sigma \longrightarrow \phi_{i+1} \; / \; \Gamma_i, \Sigma \longrightarrow a, \phi_{i+1}$.
We omit the thinning S_o/S_1 and cancel ϕ_1 together with all its images up to ϕ_i and derive $\Gamma_{i+1} \longrightarrow A, \phi_{i+1}$ (that is, $\Gamma_i, \Sigma \longrightarrow A, \phi_{i+1}$) by thinnings and interchanges from $\Gamma_i \longrightarrow A$. This transforms P into an almost intuitionistic proof P' whose characteristic number is k-1. The statement then follows from the induction hypothesis.

B. An immediate consequence of lemma 9 is

Theorem 12: Let P be an intuitionistic proof in $ZT(\subset_D)$ and let $\hat{P}$ be obtained from P by means of a logical reduction step (that is by means of an elimination of a fork). By a series of preliminary reduction steps, one can transform $\hat{P}$ into an intuitionistic proof P*, which has the same endsequent as $\hat{P}$ and hence as P.

Proof: We content ourself with the case where the fork in question is an $\supset$-fork. To this end we use the diagram introduced in chapter II, section 2.2., part C. in connection with the definition of $\supset$-reduction step. According to this definition, the altered proof $\hat{P}$ can be presented symbolically in the following way:

$$\dfrac{\dfrac{\dfrac{\begin{array}{c}\overline{P}_2\\ \vdots\\ \Gamma_3 \longrightarrow \Delta_3, A\end{array} \qquad \begin{array}{c}\overline{P}_1\\ \vdots\\ A, \Gamma_3 \longrightarrow \Delta_3, B\end{array}}{\Gamma_3, \Gamma_3 \longrightarrow \Delta_3, \Delta_3, B}\,\text{cut} \qquad \begin{array}{c}\overline{P}_3\\ \vdots\\ B, \Gamma_3 \longrightarrow \Delta_3\end{array}}{\Gamma_3, \Gamma_3, \Gamma_3 \longrightarrow \Delta_3, \Delta_3, \Delta_3}\,\text{cut}}{\Gamma_3 \longrightarrow \Delta_3}\,\text{interchanges, contractions}$$

$$\vdots$$

Since P is intuitionistic, it is evident from the definition of $\supset$-reduction step that Δ_3 is a single formula, say ϕ . Even more than this: an easy inspection shows that both $\overline{P}_1$ and $\overline{P}_2$ are almost intuitionistic proofs with ϕ playing the role of ϕ_m . $\overline{P}_3$ on the other hand is intuitionistic, as is evident from inspection. Now we apply lemma 9 to $\overline{P}_1$ and $\overline{P}_2$. It results that we can transform $\overline{P}_2$ and $\overline{P}_1$ by means of preliminary reduction steps only into proofs P^*_2 and P^*_1 of $\Gamma_3 \longrightarrow A$ and $A, \Gamma_3 \longrightarrow B$, respectively. This gives rise to a new proof P* which can symbolically be represented as follows:

$$\dfrac{\dfrac{\dfrac{\begin{array}{c}P^*_2\\ \vdots\\ \Gamma_3 \longrightarrow A\end{array} \qquad \begin{array}{c}P^*_1\\ \vdots\\ A, \Gamma_3 \longrightarrow B\end{array}}{\Gamma_3, \Gamma_3 \longrightarrow B}\,\text{cut} \qquad \begin{array}{c}\overline{P}_3\\ \vdots\\ B, \Gamma_3 \longrightarrow \phi_1\end{array}}{\Gamma_3, \Gamma_3, \Gamma_3 \longrightarrow \phi}\,\text{cut}}{\Gamma_3 \longrightarrow \phi}\,\text{interchanges, contractions}$$

$$\vdots$$

It is not difficult to verify that P* in turn can be obtained from $\hat{P}$ by means of a series of preliminary reduction steps. This concludes the proof of the theorem.

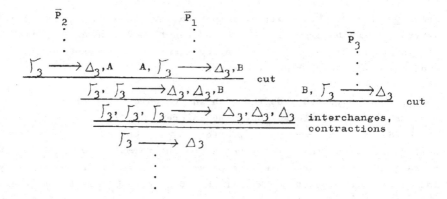

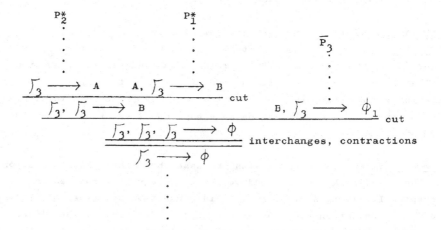

<u>Corollary 1:</u> Let P, P_1, P_2 be three proofs in ZT($\subset_D$) which satisfy the following conditions: a) P is intuitionistic; b) P_1 is obtained from P by means of a logical reduction step; c) P_2 is obtained from P_1 by a series of preliminary reduction steps; d) P_2 does not admit any preliminary reduction step. Then P_2 is intuitionistic.

<u>Proof:</u> The statement is an immediate consequence of lemma 9 and theorem 10.

The last corollary gives rise to
<u>Definition 16:</u> Let P, P' be two intuitionistic proofs (in ZT($\subset_D$)) . We say that P' is obtained from P by means of an intuitionistic logical reduction step if the following holds:
1) there is a proof P* which is obtained from P by means of a logical reduction step (in the sense of chapter II, section 2.2., part C.; 2) P' is obtained from P* by means of a series of preliminary reduction steps; 3) P' does not admit any preliminary reduction step.

The following statement is a trivial consequence of corollary 1, definition 10, and the results of chapter II:

<u>Corollary II:</u> a) Let P be a strictly normal intuitionistic proof containing a fork. Then we can apply an intuitionistic logical reduction step to P . b) If P' is the result of the application of this reduction step to P , then $O(P') \prec O(P)$.

A last result in this connection is
<u>Theorem 13:</u> Let P be a strongly normal intuitionistic proof in ZT($\subset_D$) which does not coincide with its final part. Assume that no preliminary reduction step, no intuitionistic logical reduction step, no induction reduction, no TI- and no TI_a-reduction step are applicable to P . Then P contains a critical logical inference whose principal formula has an image in the endsequent.

<u>Proof:</u> Since no intuitionistic logical reduction step is applicable to P, it follows from corollary II that no logical reduction step at all is applicable to P . The statement then follows from theorem 6, which, as noted earlier, holds also for ZT($\subset_D$).

In the chapters to follow we are mostly concerned with intuitionistic systems. Therefore, we will often simply speak of "logical reduction steps" instead of intuitionistic logical reduction steps" and speak of "classical logical reduction steps" if, for one reason or the other, we have to consider classical proofs in some classical system and logical reduction steps as introduced in chapter II, section 2.2., part C.

3.2. A basic lemma

A. In this section we prove a lemma of elementary combinatorial character which will play a crucial role throughout this work. It is responsible for the fact that the methods introduced by Gentzen in his second consistency proof can be extended to theories such as ZTi/I, ZTi/II etc.. There are two versions of this lemma. The first is very general and holds for almost every intuitionistic theory T, provided only that the notion of final part is defined in the same way as before. The second version improves the first one but applies only to those intuitionistic theories T, for which there exists an ordinal assignement to proofs which behaves more or less in the same way as the ordinal assignement introduced for proofs in $ZTi(\subset_D)$. We will prove both versions of this lemma; for simplicity we prove them for the case where T is $ZTi(\subset_D)$.

Basic lemma I: Let P be a proof in $ZTi(\subset_D)$ whose endsequent E has the form $\longrightarrow A$ and which does not contain any thinning in its final part. Let $S_1,\dots,S_m$ be the uppermost sequents of the final part, listed from left to right; let S_i be $\Gamma_i \longrightarrow A_i$. Then the following is true for every $i \leq m$: 1) there is a proof P_i of $\longrightarrow A_i$; 2) if B occurs in Γ_i, then there is a proof P' of $\longrightarrow B$.

Proof: We begin with two remarks concerning the concepts left-right. i) If S*, S** are two uppermost sequents in the final part of P, then S* is by definition on the left of S** if there is a cut S',S"/S in the final part of P, having the following properties: 1) S' is equal to S* or below S*; 2) S" is equal to S** or below S**. ii) Let S be any sequent in the final part and assume that S is $\Gamma \longrightarrow B$. Then there is an uppermost sequent S' in the final part having the following properties: 1) S' is equal to

S or situated above S ; 2) S' has the form $\Gamma' \longrightarrow B'$, and B is an image of B' . This statement is easily proved by "bar induction" over the final part. Now we prove the lemma by induction with respect to i.

Case 1: $i=1$. Since S is the leftmost one among the uppermost sequents of the final part, it must necessarily have the form $\longrightarrow A_1$. The statement of the lemma is therefore trivially satisfied.

Case 2: $i=k+1$. We assume that the statement of the lemma is true for $i \leq k$. We first prove part II of the lemma for S_{k+1} . Let B occur in Γ_{k+1}. Since the endsequent contains no formula on the left of the sequential arrow there must necessarily be a cut $S',S''/S$ in the final part of P having the following properties: a) S'' is equal to S_{k+1} or below S_{k+1} ; b) the cutformula F in S'' is an image of B and hence isomorphic with B . In view of remark ii) above, there is an uppermost sequent S_i , equal to S' or situated above S' , such that the cut formula F in S' is an image of A_i , and therefore isomorphic with A_i . In view of remark i) above, S_i is on the left of S_{k+1} , hence $i \leq k$. According to the induction hypothesis, there is a proof P_i of $\longrightarrow A_i$. Since A_i, B and F are all isomorphic with each other, we obtain a proof P' of $\longrightarrow B$ by adding, if necessary a conversion to P.

Now we prove part 1) of the lemma for S_{k+1} . Let Γ_{k+1} be $B_1, \ldots, B_N$. According to what has just been proved, there are proofs $P'_1, \ldots, P'_N$ of $\longrightarrow B_1, \ldots, \longrightarrow B_N$, respectively. On the other hand, there is a proof P^* of S_{k+1} , namely the subproof of S_{k+1} in P . By combining $P'_1, \ldots, P'_N$ and P^* in a suitable way by means of cuts, we obtain a proof P_{k+1} of $\longrightarrow A_{k+1}$ what concludes the proof.

It is clear from the proof of basic lemma I that no use has been made of the particular structure of $ZTi(\subset_D)$. We could replace $ZTi(\subset_D)$ by any intuitionistic theory T ; the proof of the basic lemma I would remain exactly the same. In particular, T can be any of the intuitionistic theories introduced so far (ZTi/I, ZTi/II, etc.) and any of the theories which will be introduced later (particular conservative extensions of ZTi/I, ZTi/II, etc.). This entitles us to make free use of the basic lemma I throughout the rest of this work. The second version of the basic lemma (called basic lemma II) is more special and has to do with ordinals. We first present the lemma and its proof and give a commentary afterwards.

<u>Basic lemma II:</u> Let P be a proof in $ZTi(\subset_D)$ whose endsequent has the form $\longrightarrow A$ and which does not contain any thinning in its final part. Let $S_1,\ldots,S_m$ be the uppermost sequents of the final part, listed from left to right; let S_i be $\Gamma_i \longrightarrow A_i$. Then the following is true: 1) for every $i < m$ there is a proof P_i of $\longrightarrow A_i$ for which $O(P_i) < O(P)$ holds; 2) for every $i \leq m$, if B occurs in Γ_i, then there is a proof P' of $\longrightarrow B$ for which $O(P') < O(P)$ holds.

<u>Proof:</u> i) We first prove 1) by constructing directly a proof P_i of $\longrightarrow A_i$. Since $i<m$, one must necessarily find a cut $S', S''/S$ in the final part having the following properties:
1) S' is equal to S_i or below S_i; 2) the cut formula F in S' is an image A_i. Let this cut be more explicitely $\Sigma \longrightarrow F$; $F, \Pi \longrightarrow G / \Sigma, \Pi \longrightarrow G$. Let in addition $P_{S'}$, $P_{S''}$ and P_S be the subproofs of S',S" and S in P respectively. Let us alter P as follows:

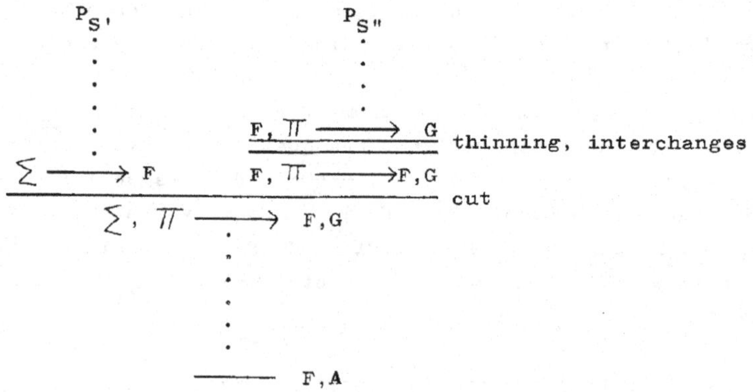

This proof, call it P* , has clearly the property that we can derive $\Sigma, \Pi \longrightarrow F, G$ from the left premiss of the cut indicated by thinning and interchanges. That is, we can apply to P* the operation called omission of a cut in order to obtain a new proof P** . We can arrange the thinnings and interchanges in a particular way so that P** has the following form:

$$P_{S'}$$
$$\vdots$$
$$\Sigma \longrightarrow F$$
$$\overline{\overline{\hspace{6em}}} \quad \text{thinnings, interchanges}$$
$$\Pi , \Sigma \longrightarrow F$$
$$\overline{\hspace{6em}} \quad \text{thinning}$$
$$\Pi , \Sigma \longrightarrow F,G$$
$$\vdots$$
$$\longrightarrow F,A$$

It is evident that P** is an almost intuitionistic proof. The path $\hat{S}_o,\dots,\hat{S}_n$ which is responsibel for P**, being an almost intuitionistic proof, is obviously that one beginning with $\Pi , \Sigma \longrightarrow F$ and ending with $\longrightarrow F,A$. According to lemma 9, we can transform P** into an intuitionistic proof $\breve{P}$ of $\longrightarrow F$. By adding a conversion if necessary to P, we finally obtain an intuitionistic proof P' of $\longrightarrow A_i$. The following equalities and inequalities are obviously satisfied in view of theorems 10 and 11:
a) $O(P)=O(P^*)$; b) $O(P^{**})<O(P^*)$; c) $O(\breve{P})\leq O(P^{**})$;
d) $O(P')=O(\breve{P})$. Hence, P' is the desired proof.

ii) In order to prove part 2) it is sufficient to show the following: if B occurs in Γ_i , then there is a $j<i$ such that A_j is isomorphic with B . The rest then follows from part 1), which has already been proved. But in order to prove the last statement, we proceed in exactly the same way as in the proof of the basic lemma I (the proof of part 2) under case 2)).

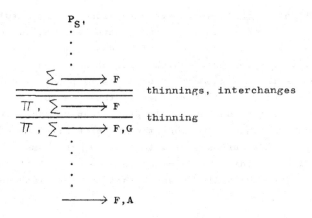

The construction of P' presented in the last proof could, of course, be used to prove basic lemma I . In the proof of basic lemma II no explicit use is made of the particular structure of $ZTi(\subset_D)$. We merely used the fact that lemma 9, theorems 10, 11, hold for $ZTi(\subset_D)$. Lemma 9 is rather a property of the final part and has nothing to do with the particular structure of $ZTi(\subset_D)$. Theorems 10 and 11, on the other hand, depend only on the definition of the final part and on the particular way to assign ordinals with proofs; the proofs of these theorems, too, do not depend on the particular structure of $ZTi(\subset_D)$. From the next chapter on, we will be concerned almost entirely with conservative extensions of the intuitionistic theories ZTi/I , ZTi/II , ... which have been introduced in chapter I, section 1.5. There will be ordinal assignements to proofs in these conservative extensions, which, from an abstract point of view, are the same as the assignement of ordinals to proofs in ZTi($\subset_D$) . It will be evident that lemma 9, theorems 10 and 11 will be true in all these cases and that their proofs can be taken over without any changes. In such situations, therefore, we will not give proofs for the statements corresponding to lemma 9, theorems 10, 11, and basic lemma II since this would amount to a mere repetition of arguments already given; we will content ourself instead with some relevant remarks.

CHAPTER IV:

A formally intuitionistic system as strong as classical analysis

In this chapter we present a proof theoretic of the theories ZTi/II_N and ZTi/II. Our aim will be to prove, eg. for ZTi/II, statements like the following: if A, B are closed formulas which do not contain special function constants, if, moreover, $ZTi/II \vdash \longrightarrow A \vee B$, then $ZTi/II \vdash \longrightarrow A$ or $ZTi/II \vdash \longrightarrow B$. We start with a treatment of ZTi/II_N, which is somewhat simpler than full ZTi/II, and extend the method afterwards to ZTi/II. The reasoning used in this chapter is essentially classical; some remarks on intuitionistic reasoning are presented in the last two sections. In particular, we consider ZTi/II as a subsystem of classical analysis having the property: if $\longrightarrow A$ is provable in ZTi/II, then A is true in the usual classical sense. For technical purposes it is very convenient, although not absolutely necessary, to include the corresponding classical systems ZT/II and ZT/II_N in our considerations.

4.1. A conservative extension of ZT/II_N

A. We start by reminding that ZT/II is the theory which is obtained from ZT by adding to it the new rule

$$\text{II.}\quad \frac{D(y),\ (x) \subset_D y\, A(x),\ \Gamma \longrightarrow \Delta, A(y)}{\mathfrak{W}(\subset_D),\ D(q),\ \Gamma \longrightarrow \Delta, A(q)}$$

where q is free for y in $A(y)$, and where y does not occur free in the conclusion, and where $u \subset_D v$ is an abbreviation for $u \subset_K v \wedge D(u) \wedge D(v)$. Here, $D(y)$ is a standard formula, that is, a formula of the form $R(y) \wedge seq(y)$ where $R(y)$ may be any formula; in particular, $R(y)$ may contain special function constants and additional free variables of any kind. If we restrict the above rule to the case where $D(y)$ (or what amounts to the same, $R(y)$) does not contain function parameters (in the sense of section 1.5., part A), we obtain a weaker rule, denoted by II_N. The theory which we obtain by adding II_N to ZT has been denoted by ZT/II_N. The corresponding intuitionistic theories have been denoted by ZTi/II and ZTi/II_N, respectively. They are characterised by the following requirement: a proof P with respect to ZT/II (with respect to ZT/II_N) is a proof with respect to ZTi/II if and only if every sequent which occurs in P contains at most one formula on the right of the sequential arrow. So much for repetition.

Now we extend the system ZT/II and ZT/II_N, respectively, by adding a set of new rules to each of them. The resulting new theories, which we will denote by ZTE/II and ZTE/II_N, respectively, will not be stronger than the old ones, because each of the new rules is derivable in the corresponding system ZT/II and ZT/II_N. In other words, the new theories are merely conservative extensions of the old ones; no more sequents are provable than before. It will also be evident from our definitions below, that if we restrict our attention to intuitionistic proofs in ZTE/II and ZTE/II_N, that we obtain intuitionistic theories ZTEi/II and $ZTEi/II_N$ which in turn are conservative extensions of ZTi/II and ZTi/II_N respectively. Actually, the theories ZTEi/II and $ZTEi/II_N$ are those which deserve our main attention since they are best suited for a proof theoretic treatment in Gentzen's spirit, as will be seen in the course of this chapter.

B. We begin by considering ZT/II_N and its conservative extension ZTE/II_N whose definition we are going to give. To this end, we are going to define a set of new rules. The first of these rules can be stated as follows: if P is a strictly normal proof in ZTi/II_N of $\longrightarrow \mathfrak{A}(\subset_D)$ where $\mathfrak{A}(\subset_D)$ does not contain special function constants nor free function variables, then we can infer from the premiss $D(y),\ (x) \subset_{D^y} A(x),\ \Gamma \longrightarrow \Delta, A(y)$ the conclusion $D(q),\ \Gamma \longrightarrow \Delta, A(q)$. A particular application of this rule is called Ti(P)-inference and is written as follows:

$$\mathrm{Ti}(P)\qquad \frac{D(y),\ (x) \subset_{D^y} A(x),\ \Gamma \longrightarrow \Delta, A(y)}{D(q),\ \Gamma \longrightarrow \Delta, A(q)}$$

Another rule can be described as follows: if P and $\mathfrak{A}(\subset_D)$ are as before, if P_1 is a strictly normal proof in ZTi/II_N of $\longrightarrow D(t)$, where t is a saturated term with $|t| = m$, then we can infer from the premiss $y \subset_D t,\ (x) \subset_{D^y} A(x),\ \Gamma \longrightarrow \Delta, A(y)$ the conclusion $q \subset_D t,\ \Gamma \longrightarrow \Delta, A(q)$.
A particular application of this rule is called $\mathrm{Ti}(P,P_1,m)$-inference

and is written as follows:

$$\mathrm{Ti}(P,P_1,m) \qquad \frac{y \subset_D t,\ (x) \subset_{D^y} A(x),\ \Gamma \longrightarrow \Delta, A(y)}{q \subset_D t,\ \Gamma \longrightarrow \Delta, A(q)}$$

The proof P in ZTi/II_N which appears in the definition of an inference

$$\mathrm{Ti}(P) \qquad \frac{S_1}{S_2}$$

is called side proof of this inference. The proof P which appears in the definition of an inference

$$\mathrm{Ti}(P,P_1,m) \qquad \frac{S_1}{S_2}$$

is called the first side proof of this inference, P_1 is called the second side proof of this inference, and $m = |t|$ is called the norm of the inference. Such inferences will also more conveniently be written by expressions such as $\mathrm{Ti}(P) : S_1/S_2$ and $\mathrm{Ti}(P,P_1,m) : S_1/S_2$ respectively. The variable y in both rules is not allowed to occur in the conclusion, and the term q has to be free for x in $A(x)$. Note that the proofs P and P_1 are required to be proofs in ZTi/II_N, that is intuitionistic proofs in ZT/II_N ! By adding the rules $\mathrm{Ti}(P)$ and $\mathrm{Ti}(P,P_1,m)$ to ZT/II_N, we obtain the extension ZTE/II_N of ZT/II_N. A proof tree in ZTE/II_N is again a finite tree whose nodes are sequents and which has the following properties: a) uppermost sequents are axioms; b) if S is not an uppermost node of the tree, then S has either one or two predecessors; c) if S is a node and S' its only predecessor, then S/S' is a one-premiss inference (with respect to the rules of ZTE/II_N); d) if S is a node and S_1, S_2 its predecessors from left to right, then $S_1, S_2/S$ is a two-premiss inference (with respect to the rules of ZTE/II_N). By an analysis of a proof P_o, we mean a specification which tells us for each node S of P_o: a) by which inference S follows from its predecessors (if S is not an uppermost node); b) if S follows from its predecessor S' by means of a

Ti(P)-inference, which is the side proof of this inference; c) if S follows from its predecessor by means of a Ti(P,P_1,m)-inference, which is its first side proof, which is its second side proof and which is its norm. In the following we always tacitly assume that, for each proof P_o in ZTE/II_N, such an analysis of P_o is effectively given. Such an analysis can, of course, be codified by means of Gödelnumbers: we can eg. associate with every inference in P_o a Gödelnumber which codifies the relevant information about this inference in a suitable way. A formula A is said to occur in P_o if it occurs in some node of P_o. A proof P' in ZTi/II_N is said to be a side proof of P if P contains a Ti(P)-inference or a Ti(P,P_1,m)-inference having P' as side proof (hence P=P' in the first case and P=P' or P_1=P' in the second case).

If we restrict our attention to those proofs P in ZTE/II_N which contain only sequents having at most one formula in the succedent, then we get the intuitionistic version of ZTE/II_N, to be denoted by ZTEi/II_N.

For proofs in ZTE/II we can introduce the notions of final part, successor, image, in the same way as in chapter II, sect. 2.1. In order to introduce the notion of normal proof for ZTE/II, one has to change clauses 3) and 5) in definition 8 slightly. In order to do this, let us call transfinite induction inference any particular application of one of the rules II, Ti(P), Ti(P,P_1,m). We call the variable y the critical variable of a transfinite induction inference if it is the y in the premiss, say,

$D(y),\ (x)\subset_{D^y} A(x),\ \Gamma \longrightarrow \Delta, A(y)$ or

$y \subset_D t,\ (x)\subset_{D^y} A(x),\ \Gamma \longrightarrow \Delta, A(y)$. Then clauses 3) and 5) in definition 8 have to be replaced by clauses 3*) and 5*) respectively: 3*) if S/S' is a quantifier inference, an induction or a transfinite induction inference, if y is the quantified variable, the induction variable or the critical variable of S/S', respectively, then y does not occur (free) below S; 5*) if y occurs free in a sequent S in P, then there is an inference S_1/S_2 with S_1 below or equal to S such that S_1/S_2 is a quantifier inference, an induction or a transfinite induction inference, and such that y is the quantified variable, the induction variable or the critical variable, respectively, of S_1/S_2. If we replace in def. 8 the clauses 3), 5) by 3*) and 5*), respectively, we obtain a new definition which will be referred to as definition 8*) . A matter of routine is the proof of the following statement: if P is a proof in ZTE/II_N (in $ZTEi/II_N$) and if no variable occurs both free and bound in the endsequent S of P, then there is a normal proof P* in ZTE/II_N (in $ZTEi/II_N$) of S . The proof is as usual by induction with respect to the longest path in P , by renaming eventually some free and bound variables in an appropriate way.

C. Our next task is to show that ZTE/II_N and $ZTEi/II_N$ are indeed conservative extensions of ZT/II_N and ZTi/II_N, respectively. Actually, we will obtain a slightly more sharp result. In order to prove it, we need

Definition 16: a) A proof P in ZT/II_N is said to have order n if every formula, which occurs in P contains at most n logical connectives. b) A proof P in ZTE/II_N is said to have degree n if every formula which occurs in P contains at most n/2 logical connectives and if every side proof P' of P has order n .
The result mentioned is given by

Theorem 14: a) If P is a proof in ZTE/II_N of degree n, then there exists a proof P' in ZT/II_N of order n, having the same endsequent as P . If P is intuitionistic then P' is intuitionistic.

Proof: The proof proceeds by induction with respect to the length of the longest path in P . If P consists of a single sequent S, then S is an axiom and we may choose for P' the proof P itself. Let P contain more than one sequent and let S be the endsequent of P. Let I be the lowest inference in P : the conclusion of I is necessarily S . Now we distinguish cases according to the type of I .

Case 1: I is a structural inference, a conversion, a logical inference, an induction, or a II_N-inference. Let, as an example, I be a cut $S_1,S_2/S_o$. Let furthermore P_1 and P_2 be the subproofs of S_1 and S_2 in P respectively. P_1 and P_2 both have degree n . By induction there are proofs P_1', P_2' in ZT/II_N of order n , having S_1 and S_2 as endsequents, respectively. Combining P_1' and P_2' by means of the same cut I : $S_1,S_2/S_o$, we obtain a proof P' in ZT/II_N of S_o which has degree n . If P is intuitionistic, then so are P_1 and P_2, and by induction P_1', P_2', and therefore P' is also intuitionistic.

Case 2: I is a $Ti(P_1)$-inference

$$Ti(P_1) \quad \frac{D(y),\ (x) \subset_{D^y} A(x),\ \Gamma \longrightarrow \Delta, A(y)}{D(q),\ \Gamma \longrightarrow \Delta, A(q)}$$

with P_1, as indicated, the side proof of this inference. Let P^* be the subproof of the premiss. P^* has degree n and therefore there exists a proof P^{**} in ZT/II_N of order n whose endsequent is the premiss of the above inference. Now we obtain the following proof P' in ZT/II_N of $D(q),\ \Gamma \longrightarrow \Delta, A(q)$:

$$\dfrac{\begin{matrix} P_1 \\ \vdots \\ \longrightarrow \mathfrak{W}(\subset_D) \end{matrix} \qquad II_N\ \dfrac{\begin{matrix} P^{**} \\ \vdots \\ D(y),\ (x) \subset_{D^y} A(x),\ \Gamma \longrightarrow \Delta, A(y) \end{matrix}}{\mathfrak{W}(\subset_D),\ D(q),\ \Gamma \longrightarrow \Delta, A(q)}}{D(q),\ \Gamma \longrightarrow \Delta, A(q)}\ \text{cut}$$

Since $\mathfrak{W}(\subset_D)$ contains no more logical connectives than $(x) \subset_{D^y} A(x)$, it follows that P' has order n ; moreover, if P is intuitionistic, then P^* is intuitionistic, P^{**} is intuitionistic in view of the induction hypothesis, and P is intuitionistic by assumption. Hence P' is intuitionistic.

Case III: I is a $Ti(P_1, P_2, m)$-inference

$$Ti(P_1, P_2, m) \quad \frac{y \subset_D t\ ,\ (x) \subset_{D^y} A(x),\ \Gamma \longrightarrow \Delta, A(y)}{q \subset_D t\ ,\ \Gamma \longrightarrow \Delta, A(q)}$$

with P_1 and P_2 first and second side proofs and $m = |t|$. Let us write $\subset$ for $\subset_D$. We start with the axiom

$(x)(x \subset y \supset .x \subset t \supset A(x)) \longrightarrow (x)(x \subset y \supset .x \subset t \supset A(x))$

and derive from it in a cut-free way, using only rules from intuitionistic predicate calculus the sequent S_1:

$s \subset t,\ s \subset y,\ (x)(x \subset y \supset .x \subset t \supset A(x)) \longrightarrow A(s)$.

In virtue of lemma 7 there is a cutfree derivation, using only rules of intuitionistic predicate calculus of S_2:
$s \subset y, y \subset t \longrightarrow s \subset t$. With the aid of a cut with leftpremiss S_1 and right premiss S_2, we derive first the sequent S_3 :
$s \subset y, y \subset t, (x)(x \subset y \supset .x \subset t \supset A(x)) \longrightarrow A(s)$
and then by two propositional operations the sequent S_4:
$y \subset t, (x)(x \subset y \supset .x \subset t \supset A(x)) \longrightarrow (x)(x \subset y \supset A(x))$.
The proof P_0 of S_4 so obtained is intuitionistic and of order n: the formula $(x)(x \subset y \supset .x \subset t \supset A(x))$ contains at most twice as many logical connectives as $(x)(x \subset y. \supset A(x))$, which in its turn contains at most n/2 logical symbols. On the other hand, it follows from our inductive assumption that there is a proof P* in ZT/II_N of order n of $(x) \subset_y A(x), y \subset t, \Gamma \longrightarrow \Delta, A(y)$. Combining P_0 and P* by means of a cut, whose left premiss is S_4, followed by an interchange, we obtain a proof P** of S_5 :
$y \subset t, (x)(x \subset y \supset .x \subset t \supset A(x)), \Gamma \longrightarrow \Delta, A(y)$. From S_5 we derive by means of an implicational inference $(\longrightarrow \supset)$ and left thinning the sequent S_6 :
$D(y), (x)(x \subset y \supset .x \subset t \supset A(x)), \Gamma \longrightarrow \Delta, y \subset t \supset A(y)$ and to S_6 we apply the rule II_N (with $x \subset t \supset A(x)$ in place of $A(x)$), obtaining thus S_7 : $\mathfrak{A}(\subset), D(q), \Gamma \longrightarrow \Delta, q \subset t \supset A(q)$.
The proof $\hat{P}$ of S_7 so obtained is still a proof in ZT/II_N of order n . At our disposal is in addition the proof P_1 of $\longrightarrow \mathfrak{A}(\subset)$ which by assumption is a proof in ZTi/II_N of order n . Combining P_1 and $\hat{P}$ by means of a cut, we obtain the sequent S_8 : $D(q), \Gamma \longrightarrow \Delta, q \subset t \supset A(q)$. Using lemma 7 (applied to $q \subset t \longrightarrow D(q)$), we finally obtain by a bit of intuitionistic predicate calculus a proof P' of S_9 : $q \subset t, \Gamma \longrightarrow \Delta, A(q)$.
P' is clearly a proof in ZT/II_N of order n . If the original proof is intuitionistic, then P* is intuitionistic in virtue of the induction hypothesis; then P' is also intuitionistic, as is evident from its construction. The theorem is thus proved.

4.2. Reduction steps

A. As already noted, we can carry over with almost no changes all definitions and notions introduced in sections 2.1 and 2.5 to the present situation. If eg. P is a proof in ZTE/II_N and S a sequent in P, then we say (again) that S belongs to the final part of P if the path leading from S to the endsequent of P does not encounter inferences other than conversions or structural inferences. With cuts, inductions, II_N-inferences, $Ti(P_1)$-inferences and $Ti(P_1,P_2,m)$-inferences we associate again natural numbers, called complexities. This assignement is defined in exactly the same way as in part B of section 2.5, treating thereby II_N-, $Ti(P_1)$- and $Ti(P_1,P_2,m)$-inferences in the same manner as TI- and TI_a-inferences: with a II_N-inference, for instance, we associate as complexity the number of logical connectives occuring in $(x)(x \subset_D y \supset A(x))$ and likewise with $Ti(P_1)$- and $Ti(P_1,P_2,m)$-inferences. Definition 12, as presented in section 2.5, serves again as definition of height; we merely have to replace the TI- and TI_a-inferences in clause 5) by the II_N-, $Ti(P_1)$- and $Ti(P_1,P_2,m)$-inferences. The definition of fork I_1, I_2, I_3 and of its associated cut are again given by definitions 11 and 12* in section 2.1. So, whenever we have to make allusion to the definitions of fork, height, etc., we will refer to sections 2.1 and 2.5 (and eventually to section 4.1 in case of definition 8*). Moreover, we will use all these notions freely and without further comments in connection with ZTE/II_N and $ZTEi/II_N$.

B. Our next task consists in defining reduction steps for ZTE/II_N and $ZTEi/II_N$. Actually, the syntactical transformations needed have already been introduced in chapter II (section 2.2 and 2.5); no new ones will appear. What we will do below is to fix the conditions under which this syntactical transformations are applicable to a proof in ZTE/II_N and $ZTEi/II_N$ respectively. To this end let P be a strictly normal proof in ZTE/II_N, that is, a normal proof (in the sense of definition 8*) whose endsequent does not contain free variables. For such a proof we are going to define a series of reduction steps.

a. Preliminary reduction steps: By preliminary reduction steps we understand again the step-by-step elimination of thinnings and logical axioms from the final part of P, as described in part A of section 2.2. Theorem 4 holds invariably in the present case.

b. Induction reduction: Let $A(x)$, $\Gamma \longrightarrow \Delta, A(x')/A(0)$, $\Gamma \longrightarrow \Delta, A(t)$ be a critical induction inference in P (that is with conclusion in the final part) such that t is saturated with value $|t| = n$. Then we apply to P the same syntactical transformation as described in part B of section 2.2, distinguishing thereby again between the cases $n=0$, $n=1$ and $1<n$. As before, we call such a transformation

an induction reduction.

<u>c. Logical reduction steps:</u> To begin with, let I_1, I_2, I_3 be a functional $\forall$-fork in P . Then we can apply to P the same syntactical transformation which has been described in part C of section 2.2 and which has been called functional $\forall$-reduction step. We thereby tacitly use the fact that lemma 4 and its corollary both hold invariably in the present case (but with def. 8* in place of def. 8); their proofs remain the same, hence we omit them.

If I_1, I_2, I_3 is an implicational fork in P, then we can perform on P that syntactical transformation which has been described in part C of section 2.2 and which we have called implicational reduction.

If, finally, I_1, I_2, I_3 is any other kind of fork ($\neg$-fork, numerical $\forall$-fork, etc.), then we proceed as before in the same way as in [1]. In each case we say accordingly that a functional $\forall$-reduction step, an implicational reduction step, etc. has been applied to P .

<u>d. II_N-reduction steps:</u> Let there be a critical II_N-inference in P, say

$$II_N \quad \frac{D(y), (x) \subset_D y A(x), \Gamma \longrightarrow \Delta, A(y)}{\mathfrak{A}(\subset_D), D(q), \Gamma \longrightarrow \Delta, A(q)}$$

Let the following two assumptions hold: 1) every constant term which occurs in $\mathfrak{A}(\subset_D)$ is saturated; 2) there is a strictly normal proof P* in ZTi/II_N of $\longrightarrow \mathfrak{A}(\subset_D)$. Since D contains no function parameters, it follows from assumption 1) that there is a formula D' which contains no special function constants and no free function variables at all, which is isomorphic with D . Therefore, by adding to P* a conversion, we obtain a strictly normal proof P_1 in ZTi/II_N of $\longrightarrow \mathfrak{A}(\subset_{D'})$. Now we can replace the above II_N-inference by the following series of inferences:

$$\mathrm{Ti}(P_1)\ \cfrac{\cfrac{\cfrac{D(y),(x)\subset_{D^y} A(x),\Gamma \longrightarrow \Delta, A(y)}{D'(y),(x)\subset_{D',y} A(x),\Gamma \longrightarrow \Delta, A(y)}\ \text{conversion}}{D'(y),\Gamma \longrightarrow \Delta, A(q)}}{\mathfrak{W}(\subset_D), D(q),\Gamma \longrightarrow \Delta, A(q)}\ \text{conversion, thinning}$$

This replacement transforms P into another proof P' in ZTE/II_N, whose endsequent is the same as that of P. We say that P' is obtained from P by means of a II_N-reduction step and that the reduction step has been applied to the particular II_N-inference above.

<u>e. Ti_1-reduction steps:</u> Let there be a critical $Ti(P_1)$-inference in P, say

$$\mathrm{Ti}(P_1)\ \frac{D(y),(x)\subset_{D^y} A(x),\Gamma \longrightarrow \Delta, A(y)}{D(q),\Gamma \longrightarrow \Delta, A(q)}$$

with P_1 a proof in ZTi/II_N of $\longrightarrow \mathfrak{W}(\subset_D)$; by assumption, $\mathfrak{W}(\subset_D)$ and therefore D do not contain free function variables or special function constants. Let the following two assumptions be satisfied: 1) q is saturated with value, say m, 2) there is a strictly normal proof P_2 in ZTi/II_N of $\longrightarrow D(q)$. The above $Ti(P_1)$-inference will be denoted briefly by $Ti(P_1)$: S/S'. As usual, P_S and $P_{S'}$ denote the subproofs of S and S' in P, respectively. By P_S^q we denote the proof which we obtain if we replace every occurence of y in P_S by q; by S^q we denote the endsequent of P_S^q. According to lemma 7 there is a proof P_o in ZTi of $y\subset_D q \longrightarrow D(y)$, which uses neither cuts nor inductions. Now we apply to P a syntactical transformation, which is an exact copy of the TI-reduction step, defined in part B of section 2.5 (chapter II). More precisely we replace the subproof $P_{S'}$ of S' in P by the following proof P^* of S':

$$
\begin{array}{lll}
P_o & P_S & \\
\vdots & \vdots & \\
y \subset_D q \longrightarrow D(y) & S & \\
\hline
y \subset_D q, (x) \subset_{D^y} A(x), \Gamma \longrightarrow \Delta, A(y) & & \text{cut} \\
\hline
s \subset_D q, \Gamma \longrightarrow \Delta, A(s) & & Ti(P_1,P_2,m) \\
\hline
\Gamma \longrightarrow \Delta, s \subset_D q \supset A(s) & P^q_S & \\
\hline
\Gamma \longrightarrow \Delta, (x) \subset_{D^q} A(x) & \vdots\; S^q & \\
\hline\hline
D(y), \Gamma \longrightarrow \Delta, A(q) & &
\end{array}
$$

A comparison shows that this diagram is merely a condensed version of the corresponding diagram in part B of section 2.5, which was used in order to explain the TI-reduction step; the only difference which shows up is that the index TI in the previous diagram is now replaced by the index $Ti(P_1,P_2,m)$. The proof P' which results from P by means of the above transformation is said to follow from P by means of a Ti_1-reduction step; we say that the Ti_1-reduction step has been applied to the $Ti(P_1)$-inference.

<u>f. Ti_2-reduction steps:</u> Let there be a critical $Ti(P_1,P_2,m)$ inference in P , say

$$
Ti(P_1,P_2,m) \quad \frac{y \subset_D t,\ (x) \subset_{D^y} A(x),\ \Gamma \longrightarrow \Delta, A(y)}{q \subset_D t,\ \Gamma \longrightarrow \Delta, A(q)}
$$

According to the definition of such inferences, $\mathfrak{A}(\subset_D)$ is a formula without function parameters, which does not contain free function variables nor special function constants, P_1 is a strictly normal proof in ZTi/II_N of $\longrightarrow \mathfrak{A}(\subset_D)$, t is saturated with value m and P_2 is a strictly normal proof in ZTi/II_n of $\longrightarrow D(t)$ (where $D(x)$ evidently does not contain free function variables nor special function constants). Let the following two assumptions be satisfied: 1) q is a saturated term with value, say, n ; 2) P_2' is a strictly normal proof in ZTi/II_N of $\longrightarrow q \subset_D t$. We denote the above inference more briefly by $Ti(P_1,P_2,m)$: S/S' . By P_S and P_S' we denote the subproofs of S and S' in P, respectively; P^q_S denotes the result of replacing every occurence of y in P_S by q and S^q denotes the endsequent of P_S^q . According to lemma 7 there are proofs P_o' and P_o in ZTi of $q \subset_D t \longrightarrow D(q)$ and $y \subset_D q,\ q \subset_D t \longrightarrow y \subset_D t$, respectively, which use neither cuts nor inductions. With the aid of P_2' and P_o' and a cut, we obtain a strictly normal proof of $\longrightarrow D(q)$ which we denote by P_3 . Now we apply to P syntactical transformation which in its turn is an exact copy of the TI_a-reduction step defined in part B of section 2.5. That is, we replace $P_{S'}$ in P by the following proof P^* of S' :

$$
\begin{array}{ll}
\begin{array}{cc}
P_o & P_S \\
\vdots & \vdots \\
y \subset_D q,\ q \subset_D t \longrightarrow y \subset_D t & S
\end{array} & \\
\hline\hline
y \subset_D q,\ (x)\subset_{D}{}^{y} A(x),\ q \subset_D t,\ \Gamma \longrightarrow \Delta, A(y) & \text{cut, interchanges} \\
\hline
s \subset_D t,\ q \subset_D t,\ \Gamma \longrightarrow \Delta, A(s) & Ti(P_1,P_3,n) \\
\hline
q \subset_D t,\ \Gamma \longrightarrow \Delta,\ s \subset_D q \supset A(s) & P_S^q \\
\hline
q \subset_D t,\ \Gamma \longrightarrow \Delta,\ (x)(x \subset_D q \supset A(x)) & S^q \\
\hline\hline
q \subset_D t,\ \Gamma \longrightarrow \Delta, A(q) &
\end{array}
$$

This diagram is just a condensed version of that one introduced in part B, section 2.5, in order to explain the TI_a-reduction step; again, the index $Ti(P_1,P_3,n)$ takes over the role of the index TI_a in the diagram in section 2.5. The proof P' , which is obtained from P by this transformation is said to follow from P by means of a Ti_2-reduction step; we also say that the Ti_2-reduction step has been applied to the $Ti(P_1,P_2,m)$ inference above.

This concludes our list of reduction steps. We note that, by an appropriate choice of the free variable s in the case of Ti_1- and Ti_2-reduction steps, we can always achieve that the altered proof P' is strictly normal, too; we always tacitly assume that s has been chosen in this way. All other reduction steps, applied to strictly normal proofs, yield automatically strictly normal proofs as results; this follows easily from inspection of their definitions.

Formally, the reduction steps are the same as those introduced in chapter II. Furthermore, given two strictly normal proofs P,P' in

ZTE/II_N , we can always decide in a recursive way whether P' follows from P by means of one of our reduction steps and, if so, by which one. However, the basic theorem 6 fails to hold in the present case. The reason for this failure is that in general we are not able to find proofs which satisfy the conditions 2) which appear in the definitions of II_N-, Ti_1- and Ti_2-reduction steps.

4.3. Ordinals

Now we are going to associate ordinals with proofs in ZTE/II_N in very much the same way as we have done with proofs in $ZT(\subset_D)$. Prior to this we need some preparations.

A. For formulas A which do not contain special function constants there is available a classical notion of truth which can roughly be described as follows: a) logical connectives are interpreted in the usual classical way, b) individual variables range over the set of natural numbers, c) function variables range over the full classical universe of number theoretic functions. We assume that the reader is familiar with this notion; we refer to it as "classical truth". All systems which have been introduced in chapter I are either particular formulations of what is known as classical analysis or (proper or improper) subsystems of this classical analysis (theorem 3). Let P be a proof in any of these systems of a sequent $\longrightarrow$ F , where F is supposed to be a closed formula not containing special function constants. If P contains special function constants then we can always replace them by appropriately chosen constant functors in order to obtain a proof P* of the same sequent, not containing special function constants. It is then clear that the formula F thus proved is classically true. In the particular case where F is $\mathfrak{W}(\subset_D)$, it follows that the partial ordering $R_D = \{ \langle p,q \rangle \, / \, p \subset_K q$ holds and both D(p), D(q) are classically true $\}$ is indeed wellfounded. This means that we can associate with every number a such that D(a) is classically true, an ordinal number, to be denoted by $\|a\|_D$. In addition we can associate with $\|R_D\|$ the smallest ordinal number which is greater than all ordinal numbers representable in the form $\|a\|_D$; we denote this ordinal number by $\|R_D\|$. If, in addition, there is another proof P_1 (in any of the systems introduced in chapter I) of $\longrightarrow b \subset_D a$, then we conclude that both a, b belong to the range of definition of R_D and that $R_D(b,a)$ holds; this clearly implies $\|b\|_D < \|a\|_D$. Now let Ω be the smallest among the ordinals, α having the following property: if P is a proof in ZTi/II_N with endsequent $\longrightarrow \mathfrak{W}(\subset_D)$ and with $\mathfrak{W}(\subset_D)$ not containing special function constants nor free function variables, then $\|R_D\| < \alpha$.

After this preliminaries we are ready to associate ordinals with proofs in ZTE/II_N .

B. Let P be any proof in ZTE/II_N ; we are going to associate with every sequent S occuring in P an ordinal, denoted by $o(S)$. The inductive definition of $o(S)$ goes as follows: 1) if S is an axiom, then $o(S)=1$; 2) if S is the conclusion of a structural inference, a conversion, a logical inference or an induction, then we proceed as in part A of section 2.4; 3) if S_1/S is a II_N-inference, then we put $o(S)=\omega_d((o(S_1)\#\omega^{\Omega+1})\omega^{\Omega+1})$ where $d=h(S_1)-h(S)$; 4) if S_1/S is a $Ti(P_1)$-inference, say

$$Ti(P_1) \quad \frac{D(y),(x) \subset_D^y A(x), \Gamma \longrightarrow \Delta, A(y)}{D(q), \Gamma \longrightarrow \Delta, A(q)}$$

then we put $o(S)=\omega_d((o(S_1)\#\omega^{\alpha+1})\omega^{\alpha+1})$ where $d=h(S_1)-h(S)$ and $\alpha = \|R_D\|$; 5) if S_1/S is a $Ti(P_1,P_2,m)$-inference, say

$$Ti(P_1,P_2,m) \quad \frac{y \subset_D t,\ (x) \subset_D^y A(x), \Gamma \longrightarrow \Delta, A(y)}{q \subset_D t, \Gamma \longrightarrow \Delta, A(q)}$$

(where $m=|t|$), then we put $o(S)=\omega_d((o(S_1)\#\omega^{\alpha+1})\omega^{\alpha+1})$ where $d=h(S_1)-h(S)$ and $\alpha = \|m\|_D$.

The ordinal of the endsequent is called the ordinal of the proof P. In order to summarize the properties of reduction steps and ordinal assignements, we call every reduction step which is not a preliminary one an essential reduction step. Furthermore, we remark that the operation "omission of a cut" defined in section 2.6, retains its meaning in the present context; its definition remains unaltered. Then we have

<u>Theorem 15:</u> a) Preliminary reduction steps do not increase the ordinal of a proof P . b) Omission of a cut lowers the ordinal of a proof P . c) Essential reduction steps lower the ordinal of P .

The proofs of part a) and b) are word by word the same as the proofs of theorems 11 and 10. Case c) splits up into two subcases: 1) the reduction step in question is a logical one or an induction reduction; 2) the reduction step in question is a II_N-, a Ti_1- or a Ti_2-reduction step. In the first case we proceed in exactly the same way as in the proof of theorem 7. In the second case we are in turn led to the calculations performed in part C of section 2.5. More explicitely, in order to verify that a II_N-reduction step lowers the ordinal of the proof to which it is applied, we are again led to the verification of an inequality
$\omega_d((\alpha \# m \# \omega^{\lambda+1})\omega^{\lambda+1} \# \alpha \# 2) < \omega_d((\alpha \# \omega^{\nu+1})\omega^{\nu+1})$ where ν is the ordinal Ω defined above, and where $\lambda = \|R_D\|$ for a D for which we have a proof P_1 in ZTi/II_N of $\longrightarrow \mathfrak{W}(\subset_D)$. By definition of Ω and $\|R_D\|$, we have $\lambda < \Omega$, and hence the inequality is true in virtue of the same reasoning as presented in part C of section 2.5.

The proof that a Ti_1-reduction step lowers the ordinal of the proof to which it is applied reduces again to the verification of the above inequality, but now with λ and ν given as follows: 1) ν is $\|R_D\|$ for a D for which we have a strictly normal proof P_1 in ZTi/II_N of $\longrightarrow \mathfrak{W}(\subset_D)$; 2) λ is $\|n\|_D$ for an n for which we have a strictly normal proof P_2 in ZTi/II_N of $\longrightarrow D(n)$. By definition of $\|n\|_D$ and $\|R_D\|$, we have $\lambda < \nu$, and the above inequality is again true in virtue of the arguments given in section 2.5.

The proof, finally, that a Ti_2-reduction step lowers the ordinal of the proof to which it is applied, leads again to a verification of the inequality $\omega_d((\alpha \# m \# \omega^{\lambda+1})\omega^{\lambda+1} \# \alpha \# 2) < \omega_d((\alpha \# \omega^{\nu+1})\omega^{\nu+1})$, but now with λ and ν given as follows: 1) λ is $\|m\|_D$ for a D for which proofs P_1 and P_2 (in ZTi/II_N) of $\longrightarrow \mathfrak{W}(\subset_D)$ and $\longrightarrow D(m)$ respectively are given; 2) ν is $\|n\|_D$ and a proof P_2' in ZTi/II_N of $\longrightarrow n \subset_D m$ is given. From our classical point of view, what is provable in ZT/II_N is true, hence $n \subset_D m$ is true, hence $\|n\|_D < \|m\|_D$, that is, $\lambda < \nu$ holds. As before, this implies the truth of the above inequality by the same arguments given in B, section 2.5.

For arbitrary proofs in ZTE/II_N, theorem 8 is of no use. For proofs P in $ZTEi/II_N$, however, whose endsequent contains nothing on the left side of the arrow, the situation is entirely different, as will be shown in the next section.

4.4. The system $ZTEi/II_N$

A. The passage from ZTE/II_N to $ZTEi/II_N$ is more or less the same as that from $ZT(\subset_D)$ to $ZTi(\subset_D)$, described in chapter III. One easily verifies that every reduction step which is not a logical reduction step transforms a strictly normal proof P in $ZTEi/II_N$ into another strictly normal proof P' in $ZTEi/II_N$. If, on the other hand, we apply to P a logical reduction step, then we obtain a proof P' which is still strictly normal, but no longer intuitionistic. However, it is trivial to verify that theorem 12 invariably holds in the present case, that is, we have

Theorem 16: Let P be an intuitionistic proof in ZTE/II_N and let $\hat{P}$ be obtained from P by means of a logical reduction step. By a series of preliminary reduction steps one can transform $\hat{P}$ into an intuitionistic proof P* , which has the same endsequent as P .

The proof remains exactly the same. Corollary 1 of theorem 12 remains of course, true in the present case and so we can use definition 16 as it stands as definition of intuitionistic logical reduction step. Finally, it is clear in virtue of theorem 15 that corollary II of theorem 12 remains true. For the sake of completeness, we formulate a variant of theorem 15 which summarizes the properties of reduction steps and ordinal assignements for intuitionistic proofs.

Theorem 15*: Let P be a strictly normal proof in $ZTEi/II_N$.
a) A preliminary reduction step, applied to P , transforms P into a strictly normal proof P' in $ZTEi/II_N$, whose ordinal o(P') is not larger than o(P) . b) Omission of a cut transforms P into a strictly normal proof P' in $ZTEi/II_N$ whose ordinal o(P') is smaller than o(P) . c) An essential reduction step other than fork elimination transforms P into a strictly normal proof P' in $ZTEi/II_N$, whose ordinal is smaller than that of P. d) An intui-

tionistic logical reduction step (in the sense of def. 16) applied to P transforms P into a strictly normal proof P' in $ZTEi/II_N$, whose ordinal is smaller than that of P.

If no danger of confusion arises, we omit the attribute "intuitionistic" and speak merely of logical reduction step.

B. In section 3.2 we have proved for the theory $ZTi(\subset_D)$ two lemmas, or rather two variants of one and the same lemma, which we have called there Basic lemma I and Basic lemma II. As we have already mentioned there, this lemmas hold for a large class of intuitionistic theories; the theory $ZTEi/II_N$ is no exception in this respect. The proof of Basic lemma I presented in section 3.2 applies to $ZTEi/II_N$ without any changes, as an easy inspection shows. The same is true of the proof of Basic lemma II in section 3.2: all we have to do is to refer to theorem 15* instead of theorems 11 and 12. Actually, if we inspect the proof of basic lemma II, then we see that it yields a slightly more sharp statement, which in the present case reads as follows:

Basic lemma II: Let P be a strictly normal proof in $ZTEi/II_N$ of degree n ; assume that it has no thinning in the final part and that its endsequent has the form $\longrightarrow A$. Let $S_1, S_2, \ldots, S_m$ be the uppermost sequents of the final part, listed from left to right; let S_i be $\Gamma_i \longrightarrow A_i$. Then the following is true: 1) for every $i < m$ there is a strictly normal proof P_i (in $ZTEi/II_N$) of degree n whose endsequent is $\longrightarrow A_i$ and for which $o(P_i) < o(P)$ holds; 2) for every $i \leq m$, if B occurs in Γ_i, then there is a strictly normal proof P' (in $ZTEi/II_N$) of degree n whose endsequent is $\longrightarrow B$ and for which $o(P') < o(P)$ holds.

Proof: Exactly the same as that of Basic lemma II in section 3.2.

If we drop in Basic lemma II_1 the reference to ordinals, then we obtain a sharpening of Basic lemma I, which could, of course, be obtained directly from the proof of Basic lemma I; we merely have to sharpen slightly the induction hypothesis used in the proof of Basic lemma I (part 2)). Actually, all we need in this chapter is this sharpened version of Basic lemma I; we do not use the fact that the ordinals of $o(P_i)$ and $o(P')$ are smaller than $o(P)$. Now let P be a strictly normal proof in $ZTEi/II_N$, whose endsequent has the particular form $\longrightarrow A$, whose degree is n, and which contains only saturated terms in its final part. Assume that no thinning occurs in the final part of P and let there be a critical II_N-inference in P, say

$$II_N \qquad \frac{D(y), (x) \subset_D y\, A(x), \Gamma \longrightarrow A(y)}{\mathfrak{A}(\subset_D), D(q), \Gamma \longrightarrow A(q)}$$

Without loss of generality, we can assume that the formula $D(x)$ does not contain special function constants and that x is its only free variable; otherwise we would replace the II_N-inference above by a conversion, followed by another II_N-inference and a second conversion. The formula $\mathfrak{A}(\subset_D)$ in particular does not contain free variables and no special function constants. From Basic lemma II_1 it follows that we can extract from P a proof P^* of $\longrightarrow \mathfrak{A}(\subset_D)$ which still has degree n. From theorem 14 it follows that P^* can be transformed into a proof P' in ZTi/II_N of $\longrightarrow \mathfrak{A}(\subset_D)$, whose order is n. Since there is no variable, which occurs free in $\mathfrak{A}(\subset_D)$, we can transform P' into a strictly normal proof P_1 in ZTi/II of $\longrightarrow \mathfrak{A}(\subset_D)$, whose order is still n: we merely have to rename eventually free and bound variables in a suitable way. An inspection shows that the conditions which appear in the definition of II_N-reduction step are satisfied: P_1 is the proof required by them. Therefore, we can apply to the II_N-inference above a II_N-reduction step: we can replace the original II_N-inference by a $Ti(P_1)$ inference in the way described in the definition of this reduction step.

The situation is similar if P contains a critical $Ti(P_1)$ inference, say

$$Ti(P_1) \qquad \frac{D(y), (x) \subset_D y\, A(x), \Gamma \longrightarrow A(y)}{D(q), \Gamma \longrightarrow A(q)}$$

By assumption, q is saturated and has a value $|q| = m$. As before, we apply Basic lemma II_1 and extract a subproof P^* of $\longrightarrow D(q)$ which still has degree n. Then we transform P^* with the aid of theorem 14 into a proof P' in ZTi/II_N of $\longrightarrow D(q)$, whose order is n. Finally, by renaming eventually free and bound variables in an appropriate way we transform P' into a strictly normal proof P in ZTi/II_N of $\longrightarrow D(q)$, whose order is still n . An inspection shows that all conditions, stated in the definition of Ti_1-reduction step, are satisfied: P_2 is the proof required by them. Hence we can apply to the above $Ti(P_1)$ inference a Ti_1-reduction step by replacing the $Ti(P_1)$ inference above by a $Ti(P_1,P_2,m)$ inference in the way described in the definition of Ti_1-reduction step.

Finally, let P contain a critical $Ti(P_1,P_2,m)$ inference, say

$$Ti(P_1,P_2,m) \quad \frac{y \subset_D q,\ (x) \subset_{Dy} A(x),\ \Gamma \longrightarrow A(y)}{p \subset_D q,\ \Gamma \longrightarrow A(q)}$$

(with $m = |q|$) . By assumption, p is saturated with value say r . Then, by proceeding as in the previous cases, we can find effectively a strictly normal proof P_2' in ZTi/II_N of $\longrightarrow p \subset_D q$, whose order is n . Using lemma 7, we obtain a proof P_3 in ZTi/II of $\longrightarrow D(p)$ which is still strictly normal and has order n . An inspection shows that the two conditions stated in the definition of Ti_2 reduction step are both satisfied: P_2' in particular is the proof whose existence is required by the second of these conditions. This means that we can apply a Ti_2-reduction step to the above $Ti(P_1,P_2,m)$ inference by replacing it by a $Ti(P_1,P_3,n)$ inference in the way described in the definition of Ti_2-reduction step. These facts are summarized by the following

<u>Theorem 17:</u> Let P be a strictly normal proof in $ZTEi/II_N$ whose degree is n , whose endsequent has the form $\longrightarrow A$ and which does not contain thinnings in the final part. Assume, that every constant term in the final part is saturated. Then the following holds: 1) if there is a critical II_N inference in P, then we can effectively apply a II_N-reduction step to this inference; 2) if there is a critical $Ti(P_1)$ inference in P, then we can effectively apply a Ti_1-reduction step to this inference; 3) if there is a critical $Ti(P_1,P_2,m)$ inference in P, then we can effectively apply a Ti_2-reduction step to this inference. In each of these three cases we obtain as result a strictly normal proof P* of degree n .

From the above it follows that we can reobtain suitably formulated variants of theorems 5 and 6 for $ZTEi/II_N$ if we restrict our attention to proofs P whose endsequent has the particular form $\longrightarrow$ A . In view of their importance, we introduce a name for such proofs:

Definition 17: A proof P is called standard if its endsequent has the particular form $\longrightarrow$ A . As abbreviation for "strictly normal standard proof" we use the expression "s.n.s. proof".

In order to obtain appropriate versions of theorems 5,6, we restrict the class of II_N- , Ti_1- and Ti_2-reduction steps.

Definition 18: Let P be a saturated s.n.s. proof in ZTEi/II_N which does not contain thinnings in its final part. If P contains a critical II_N-inference then we can apply to it that particular II_N-reduction step which is described in the proof of theorem 17: we call this particular reduction step the canonical reduction step associated with the critical II_N inference in question. Similarly, in case of a critical $Ti(P_1)$ inference or a critical $Ti(P_1,P_2,m)$ inference in P .

That is, among all possible reduction steps which can eventually be applied to the critical II_N inference in question, we select a particular one: that one described in the considerations preceeding theorem 17.

Theorem 5 can now be restated as follows:

Theorem 18: Let W be the twoplace relation which applies to proofs P,P' in ZTEi/II_N if and only if the following holds: 1) P,P' are saturated s.n.s. proofs which do not contain thinnings and logical axioms in the final part; 2) P' can be obtained from P by application of a logical reduction step, an induction reduction or a canonical II_N- , Ti_1- or Ti_2-reduction step. Then W is decidable. Moreover, if W(P,P') holds, then we can effectively determine the reduction step which, applied to P , yields P' . Finally, there is a recursive function Θ having the property: if W(P,P') holds, then there are at most Θ(P) symbols which occur either in P' or in one of its side proofs.

As mentioned earlier, theorem 4 remains true as it stands for all proofs and hence in particular for standard proofs; we will not restate it again. The basic theorem 6 on the other hand now reads as follows:

Theorem 19: Let P be a saturated s.n.s. proof in $ZTEi/II_N$, which does not contain thinnings and logical axioms in the final part and which is different from its final part. Assume that no logical reduction step, no induction reduction and no canonical II_N-, Ti_1- and Ti_2-reduction step is applicable to P. Then there is a critical logical inference in P whose principal formula has an image in the endsequent.

Proof: From theorem 17, it follows that P does not contain any critical II_N-, $Ti(P_1)$- or $Ti(P_1,P_2,m)$-inference. Then we obtain the statement of the theorem by proceeding in the same way as in the proof of theorem 6.

Definition 19: A reduction step will be called canonical if it is a canonical II_N-, Ti_1- or Ti_2-reduction step. A reduction step will be called strictly essential if it is a logical reduction step, an induction reduction or a canonical reduction step.

C. Before coming to applications, there is still a point to consider. Let P be an s.n.s. proof in $ZTEi/II_N$ which does not contain thinnings and logical axioms in its final part, and assume a) that no strictly essential reduction step is applicable to P; b) that there is no critical logical inference whose principal formula has an image in the endsequent; c) that P does not coincide with its final part. A comparison with theorem 19 shows that P necessarily must have the following properties: 1) there are constant terms in the final part shich are not saturated; 2) there is at least one critical induction inference, II_N inference, $Ti(P_1)$ inference or $Ti(P_1,P_2,m)$ inference in P. That 1) holds is a consequence of theorem 19: otherwise we would obtain a contradiction in view of assumption b). In order to prove 2), we prove the following lemma:

Lemma 9: We can effectively decide whether a proof P in ZTE/II_N is saturated or not. If it is not saturated and if $\alpha_{u_1}^{i_1}, \ldots\ldots, \alpha_{u_s}^{i_s}$ is a given listing of the distinct special function constants occuring in P , then we can find effectively a p.r. continuity function $\tau(x_1,\ldots,x_s)$ having the following property: if $\tau(v_1,\ldots,v_s)\neq 0$ and if P* results from P by replacing every $\alpha^{i_k}_{u_k}$ by $\alpha^{i_k}_{u_k * v_k}$, then P* is saturated. The proof of this lemma is an immediate consequence of the definitions of term and saturated term and is omitted. In order to show that P has property 2) stated above, let $\alpha^{i_1}_{u_1},\ldots\ldots,\alpha^{i_s}_{u_s}$ be the distinct special function constants occuring in P and let $\tau(x_1,\ldots,x_s)$ be the continuity function associated with P according to the lemma. Let $v_1,\ldots,v_s$ be such that $\tau(v_1,\ldots,v_s)\neq 0$ and denote by P* the result of replacing every $\alpha^{i_k}_{u_k}$ in P by $\alpha^{i_k}_{u_k * v_k}$. Now it is evident that the following statements are true: α) if there is a fork in P*, then there is a fork in P ; β) if there is a critical induction in P*, there is a critical induction in P ; γ) if there is a critical II_N-, $Ti(P_1)$- or $Ti(P_1,P_2,m)$-inference in P*, then there is such an inference in P ; δ) if there is a critical logical inference in P* whose principal formula has an image in the endsequent, then there is such an inference in P . Moreover, P* is clearly a saturated s.n.s. proof in $ZTEi/II_N$ which does not contain thinnings and logical axioms in its final part. In virtue of theorem 19, the assumptions about P and the list α)- δ), it follows that P* must contain either a critical induction, a critical II_N-inference, a critical $Ti(P_1)$-inference or a critical $Ti(P_1,P_2,m)$-inference. Therefore, in view of α)- δ), the same is true for P , what proves that P has property 2). Consider e.g. the case where the inference stated in 2) is an induction:

$A(x), \Gamma \longrightarrow A(x')/A(0), \Gamma \longrightarrow A(q)$. The reason why we cannot apply an induction reduction to P, and to this inference in particular, is that q is not saturated; hence it cannot be replaced by a numeral with the aid of a conversion. The situation is similar in case of a critical II_N-, $Ti(P_1)$- or $Ti(P_1,P_2,m)$-inference.

Remark: In virtue of lemma 9, we can associate with every s.n.s. proof P which is not saturated in an effective way a continuity function τ which is related to P in the way described by lemma 9; we denote this continuity function by τ_P and call it the continuity function associated with P . Finally, we need

Definition 20: a) Let P be a s.n.s. proof and $\alpha^{i_1}_{u_1}, \ldots, \alpha^{i_s}_{u_s}$ the critical special function constants which occur in P . Let $v_1, \ldots, v_s$ be sequence numbers all having the same length $\neq 0$. If the s.n.s. proof P* has been obtained from P by replacing every occurence of $\alpha^{i_k}_{u_k}$ in P by $\alpha^{i_k}_{u_k * v_k}$ $(k \leq s)$, then we call P* a substitution instance of P . b) If, in particular, $v_i = \overline{\alpha}_i(n)$ $(i \leq s)$ are such that $\tau_p(v_1, \ldots, v_s) \neq 0$, while $\tau_p(\overline{\alpha}_1(m), \ldots, \overline{\alpha}_s(m)) = 0$ for $m < n$, then we say that P* has been obtained from P by means of an inessential reduction step.

The above considerations may be summed up with the aid of this definitions as follows:

Theorem 20: Let P be a s.n.s. proof in $ZTEi/II_N$ having the following properties: a) no strictly essential reduction step is applicable to P ; b) there is no critical logical inference whose principal formula has an image in the final part; c) P does not coincide with its final part. Then P is not saturated and contains either a critical induction inference, a critical II_N-inference, a critical $Ti(P_1)$-inference or a critical $Ti(P_1, P_2, m)$-inference.

D. In connection with theorem 19, there is a last syntactical operation to be considered. To this end let P be a saturated s.n.s. proof in $ZTEi/II_N$, which satisfies the conditions of theorem 19. We distinguish a number of cases according to the form of the endsequent of P .

Case 1: The endsequent of P is $\longrightarrow A \wedge B$. Since P is an intuitionistic proof whose endsequent has empty antecedent, it follows that the critical inference given by theorem 19 must necessarily have the form

$$\frac{\Gamma \longrightarrow A' \qquad \Gamma \longrightarrow B'}{\Gamma \longrightarrow A' \wedge B'}$$

with A' and B' isomorphic with A and B, respectively. It furthermore follows from the intuitionistic structure of P that this inference is the rightmost one among all critical inferences in P, and that the path leading from $\Gamma \longrightarrow A' \wedge B'$ to the endsequent is the rightmost one among all the paths in the final part of P . Therefore we have two possibilities: we can omit the inference in question and cancel its right premiss, obtaining thus a proof P_1 of $\longrightarrow A$, or we can omit the inference and cancel its left premiss, obtaining thus a proof P_2 of $\longrightarrow B$. It goes without saying that both proofs P_1 and P_2 are s.n.s. proofs in ZTEi/II whose ordinals $o(P_1)$, $o(P_2)$ are smaller than $o(P)$.

Case 2: The endsequent of P is $\longrightarrow A \vee B$. The critical inference given by theorem 19 must be of the form $\Gamma \longrightarrow A' / \Gamma \longrightarrow A' \vee B'$ or $\Gamma \longrightarrow B' / A' \vee B'$ with A',B' isomorphic with A,B, respectively. Again the inference in question is the right-most one among all critical inferences. In either case we can omit the inference, obtaining a proof P_1 of $\longrightarrow A'$ or of $\longrightarrow B'$. As before, P_1 is an intuitionistic s.n.s. proof and $o(P_1) < o(P)$ holds.

Case 3: The endsequent of P is $\longrightarrow (x)A(x)$. Then the critical inference given by theorem 19 has the form

$$\frac{\Gamma \longrightarrow A'(z)}{\Gamma \longrightarrow (x)A'(x)}$$

where $A'(z)$ is isomorphic with $A(z)$. Let $\hat{P}$ be the subproof of $\Gamma \longrightarrow A'(z)$ in P . Now we replace every occurence of z in $\hat{P}$

by n and obtain a proof $\hat{P}_n$ of $\longrightarrow A'(n)$. Next we replace $\hat{P}$ in P by $\hat{P}_n$ and omit the quantifier inference in question: this yields a proof P_1 of $\longrightarrow A(n)$ which is still a s.n.s. proof in $ZTEi/II_N$; its ordinal $o(P_1)$ is clearly smaller than $o(P)$.

Case 4: The endsequent of P is $\longrightarrow (\forall \xi)A(\xi)$. In this case the critical inference given by theorem 19 must have the form

$$\frac{\Gamma \longrightarrow A'(\xi)}{\Gamma \longrightarrow (\forall \xi)A'(\xi)}$$

where $A'(\alpha)$ is isomorphic with $A(\alpha)$. Let $\alpha^i_{<\,>}$ be any special function constant associated with the empty sequent which does not occur in P. We replace every occurence of α in $\Gamma \longrightarrow A'(\alpha)$ or above by $\alpha^i_{<\,>}$ and omit the quantifier inference $\Gamma \longrightarrow A'(\alpha)/\ \Gamma \longrightarrow (\forall \xi)A'(\xi)$. The result is a proof P_1 of $\longrightarrow A(\alpha^i_{<\,>})$; P is clearly an s.n.s. proof in $ZTEi/II_N$ whose ordinal is smaller than that of P.

Case 5: The endsequent of P is $\longrightarrow (E\xi)A(\xi)$. The critical inference given by theorem 19 must have the form

$$\frac{\Gamma \longrightarrow A'(F)}{\Gamma \longrightarrow (E\xi)A'(\xi)}$$

where $A'(\xi)$ is isomorphic with $A(\xi)$. Since P is a s.n.s. proof, it follows that F is a constant functor. By omitting the inference $\Gamma \longrightarrow A'(F)/\ \Gamma \longrightarrow (E\xi)A'(\xi)$, we obtain a proof P_1 of $\longrightarrow A(F)$. As before, P_1 is a s.n.s. proof in $ZTEi/II_N$ and its ordinal is smaller than that of P.

Case 6: The endsequent of P is $\longrightarrow (Ex)A(x)$. The critical inference given by theorem 19 has the form

$$\frac{\Gamma \longrightarrow A'(t)}{\Gamma \longrightarrow (Ex)A'(x)}$$

where $A'(x)$ is isomorphic with $A(x)$. Since P is normal, it follows that t is a constant term. By omitting the above critical inference, we obtain a proof P_1 of $\longrightarrow A(t)$. P is, of course, a s.n.s. proof in $ZTEi/II_N$ whose ordinal $o(P_1)$ is smaller than $o(P)$.

Case 7: The endsequent of P is $\longrightarrow A \supset B$. The critical inference given by theorem 19 must have the form

$$\frac{A', \Gamma \longrightarrow B'}{\Gamma \longrightarrow A' \supset B'}$$

where A' and B' are isomorphic with A and B, respectively. By omitting this inference, we obtain a proof P_1 of $A \longrightarrow B$. P_1 is still a strictly normal proof in $ZTEi/II_N$ and its ordinal is still smaller than that of P . However, P_1 is no longer a standard proof since its endsequent has an antecedent which is not empty.

Case 8: The endsequent of P is $\longrightarrow \neg A$. The critical inference given by theorem 19 must be

$$\frac{A', \Gamma \longrightarrow}{\Gamma \longrightarrow \neg A'}$$

where A' is isomorphic with A . By omitting this inference, we obtain a proof P of $A \longrightarrow$. P is still a strictly normal proof in $ZTEi/II_N$ but it is no longer standard since its endsequent has a nonempty antecedent. The above considerations give rise to the definition below.

Definition 21: Let 1) - 8) denote the cases 1) - 8) which have just been discussed above. Let P be a saturated s.n.s. proof in $ZTEi/II_N$ which does not admit preliminary nor strictly essential reduction steps and which does coincide with its final part. Let S be the endsequent of P . A proof P^* is said to follow from P by application of a subformula reduction step if one of the following alternatives holds: a) S is $\longrightarrow A \wedge B$ and P^* is one of the proofs P_1 or P_2 in 1) ; b) S is $\longrightarrow A \vee B$ and P^* is the proof P_1 in 2) ; c) S is $\longrightarrow (x)A(x)$ and P^* is one of the proofs P_1 defined in 3) ; d) S is $\longrightarrow (\forall \xi)A(\xi)$ and P^* is the proof P_1 of 4), while $\alpha \overset{i}{<} >$

in 4) is the first in the list $\alpha^1_{< >}$, $\alpha^2 < >$, ... which does not occur in P ; e) S is $\longrightarrow (E\mathfrak{F})A(\mathfrak{F})$ and P* is the proof P_1 in 5) ; f) S is $\longrightarrow (Ex)A(x)$ and P* is the proof P_1 defined in 6) ; g) S is $\longrightarrow A \supset B$ and P* is the proof P_1 in 7) ; h) S is $\longrightarrow \neg A$ and P* is the proof P_1 in 8) .

With the aid of definition 21, we can sum up the above considerations as follows:

Theorem 21: Let P be a saturated s.n.s. proof in $ZTEi/II_N$ which does not coincide with its final part and which does not admit preliminary nor strictly essential reduction steps. Then we can effectively apply to P a subformula reduction step; the resulting proof P* is a strictly normal proof in $ZTEi/II_N$ whose ordinal o(P*) is smaller than o(P) .

Corollary: Let P,P* be as in theorem 21 and let S,S* be their endsequents respectively. If S is $\longrightarrow A \vee B$, then S* is $\longrightarrow A$ or $\longrightarrow B$, if S is $\longrightarrow (E\mathfrak{F})A(\mathfrak{F})$ then S* is $\longrightarrow A(F)$ for some constant functor F, if S $\longrightarrow (Ex)A(x)$ then S* is $\longrightarrow A(t)$ for some constant term t .

Remark: The functor and the term t may of course contain special function constants.

4.5. Applications

A. Applications of our analysis of the system $ZTEi/II_N$ are most immediately obtained by introducing two wellfounded relations R,L which are both intimately connected with our reduction steps.

Definition 22: Let the two-place relation R hold for s.n.s. proofs P,P' in $ZTEi/II_N$ (in symbols R(P,P')) if and only if one of the following two conditions A,B below are satisfied. A. P is not saturated and P' follows from P by means of an inessential reduction step. B. P is saturated and there is a list $P_1,\dots,P_s,P_{s+1}$ (s=1 admitted) of proofs having the following properties: 1) $P=P_1$, $P'=P_{s+1}$, 2) for $i \leq s$ P_i follows from P_{i-1} by means of a preliminary reduction step, 3) no preliminary reduction step is applicable to P_s, 4) P_{s+1} follows from P_s by means of a strictly essential reduction step.

The second relation, denoted by L, is introduced by the following <u>Definition 23:</u> The two-place relation L holds between s.n.s. proofs P,P' in $ZTEi/II_N$ if and only if one of the three conditions A,B,C below are satisfied.

<u>A.</u> P is not saturated and R(P,P') holds. <u>B.</u> P is saturated and R(P,P') holds. <u>C.</u> P is saturated and there is a list $P_1,\ldots,P_s,P_{s+1}$ $(1 \leqq s)$ of proofs having the following properties: 1) $P=P_1$, $P'=P_{s+1}$, 2) for $i \leqq s$ P_i follows from P_{i-1} by means of a preliminary reduction step, 3) no preliminary reduction step is applicable to P_s, 4) P_{s+1} follows from P_s by means of a subformula reduction step.

The main properties of R,L are described by the following <u>Theorem 22:</u> a) R,L both are decidable, b) given P, the predicates (EX)R(P,X), (EX)L(P,X) are decidable, c) R and L are wellfounded, that is, no infinite sequence $P_1,\ldots$ such that $R(P_i,P_{i+1})$ for all i or $L(P_i,P_{i+1})$ for all i exists.

<u>Proof:</u> The proof of a) is rather routine and hence omitted. We sketch the proof of b). Given a s.n.s. proof P in $ZTEi/II_N$, we first decide whether P is saturated or not. If not, then we can apply to P an inessential reduction step in order to obtain a proof P' with R(P,P'). Hence (EX)R(P,X) holds. If P is saturated, then there are finitely many chains $P_1,\ldots,P_s$ with the property: 1) $P_1=P$, 2) P_{i+1} follows from P_i by means of a preliminary reduction step, 3) no preliminary reduction step is applicable to P_s. For each such chain we take the corresponding P and check whether an essential reduction step is applicable to P or not. If there is such a chain, then (EX)R(P,X) holds, if not, then (EX)R(P,X) is false. The argument for L is quite similar. In order to prove c), assume that $P_1,P_2,\ldots$ is such an infinite chain with respect to R; that is, $R(P_i,P_{i+1})$ is assumed to hold for all i. Obviously, $o(P_{i+1}) \leq o(P_i)$. However, it is easy to see that there must be an infinite subsequence $i_1 < i_2 < i_3 \ldots$ such that P_{i_k+1} follows from P_{i_k} by means of a strictly essential reduction step. Hence $o(P_{i_k}) > o(P_{i_{k+1}})$ in virtue of theorem 15, what leads to a contradiction. The argument is quite the same in the case of the

relation L .

The applications of the previous theorem are now immediate:

Theorem 23: Let $A \vee B$, $(Ex)A(x)$, $(E\mathfrak{f})A(\mathfrak{f})$ be formulas which do not contain free variables nor special function constants. a) Given a proof P in $ZTEi/II_N$ of $\longrightarrow A \vee B$, we find effectively a proof P' in $ZTEi/II_N$ of $\longrightarrow A$ or $\longrightarrow B$. b) Given a proof P in $ZTEi/II_N$ of $\longrightarrow (Ex)A(x)$, we effectively find an n and a proof P' in $ZTEi/II_N$ of $\longrightarrow A(n)$. c) Given a proof P in $ZTEi/II_N$ of $\longrightarrow (E\mathfrak{f})A(\mathfrak{f})$, we effectively find a constant functor F not containing special function constants and a proof P' in $ZTEi/II_N$ of $\longrightarrow A(F)$.

Proof: We content ourself with the proof of c) . The other cases are treated in exactly the same way. Since $(E\mathfrak{f})A(\mathfrak{f})$ does not contain free variables at all, there is no variable which occurs both free and bound in P . Hence there is a normal proof P* of $\longrightarrow (E\mathfrak{f})A(\mathfrak{f})$ (see part B of this section) and by replacing those special function constants which eventually may occur in P* by suitably chosen constants for p.r. functions, we get a s.n.s. proof P in $ZTEi/II_N$ of $\longrightarrow (E\mathfrak{f})A(\mathfrak{f})$ which does not contain special function constants at all. In virtue of theorem 23, we effectively find a chain $P_0, P_1, \ldots, P_N$ such that $\neg(EX)R(P_N, X)$ holds. The endsequent of P_N is, of course, still $\longrightarrow (E\mathfrak{f})A(\mathfrak{f})$ and one easily verifies that P_N is saturated and does not contain special function constants. Now we apply as many preliminary reduction steps as possible to P_N ; we obtain in this way a proof P_N^* of $\longrightarrow (E\mathfrak{f})A(\mathfrak{f})$ which is saturated and does not admit preliminary reduction steps. No strictly essential reduction step is applicable to P_N^*, since otherwise $\neg(EX)R(P_N, X)$ would be false. On the other hand P_N^* cannot coincide with its final part, since in this case only prime formulas would occur in P_N^* . Hence, in virtue of theorem 21 it follows that a subformula reduction step is applicable to P_N^* . The result of this reduction step is a proof $\hat{P}$ of $\longrightarrow A(F)$, as is clear from the corollary of theorem 21. F is a constant functor and, since $\hat{P}$ does not contain special function constants, it follows that also F does not contain special function constants. Since, moreover, $\hat{P}$ is a proof in $ZTEi/II_N$ the statement c) of the theorem is proved.

Remark: We note that in the above proof we have heavily used the fact that $ZTEi/II_N$ is consistent: a successive application of preliminary reduction steps to a standard proof does not affect its endsequent.

The result above can be generalized. In order to obtain this generalization, we note a lemma which has been used implicitely several times, in particular also in the proof of theorem 23, namely

Lemma 10: Let P be a s.n.s. proof in ZTEi/II of $\longrightarrow A$ and let $\alpha_{u_1}^{i_1}, \ldots, \alpha_{u_s}^{i_s}$ be those special function constants which occur in P but not in A. Then we can replace the constants $\alpha_{u_1}^{i_1}, \ldots, \alpha_{u_s}^{i_s}$ by suitably chosen constants for primitive recursive functions in order to obtain a s.n.s. proof P' of $\longrightarrow A$ which contains only those special function constants which occur in $\longrightarrow A$. We have $o(P)=o(P')$.

We omit the trivial proof of this lemma. Another evident lemma whose routine proof is omitted is the following

Lemma 11: Let P be a s.n.s. proof in $ZTEi/II_N$ of $\longrightarrow A$ which has the following property: every special function constant which occurs in P occurs in A. If $R(P,P')$ holds then P' still has this property.

In order to have a word at hand let us call a s.n.s. proof P stratified if every special function constant which occurs somewhere in P already occurs in its endsequent.

Definition 24: Let P be a stratified s.n.s. proof in $ZTEi/II_N$ and $\alpha_{u_1}^{i_1}, \ldots\ldots, \alpha_{u_s}^{i_s}$ the special function constants occuring in P, listed in some fixed way. Let $w_1, \ldots, w_s$ be sequence numbers all having length > 0. A substitution of

$\alpha^{i_1}_{u_1*v_1},\ldots\ldots,\ \alpha^{i_s}_{u_s*v_s}$ for $\alpha^{i_1}_{u_1},\ldots\ldots,\ \alpha^{i_s}_{u_s}$ is said to be compatible with $w_1,\ldots\ldots,w_s$ if $w_i \subseteq {}_K u_i*v_i$ for $1 \leq i \leq s$. A pair P,P' is said to be compatible with $w_1,\ldots\ldots,w_s$ if P' is a substitution instance of P and if the substitution which transforms P into P' is compatible with $w_1,\ldots\ldots,w_s$. A chain $P_o,\ldots\ldots,P_N$ with $P_o=P$ is said to be compatible with $w_1,\ldots\ldots,w_s$ if a) $R(P_i,P_{i+1})$ for all $i<N$, b) P_i,P_{i+1} is compatible with $w_1,\ldots\ldots,w_s$ whenever P_i is not saturated. A chain $P_o,\ldots\ldots,P_N$ with $P_o=P$ is said to be compatible with functions $\xi',\ldots\ldots,\ \xi^s$ if a) $R(P_i,P_{i+1})$ for all $i<N$, b) there is a sufficiently large K such that for all i P_i,P_{i+1} is compatible with $\bar{\xi}'(K),\ldots\ldots,\ \bar{\xi}^s(K)$ whenever P_i is not saturated.

Remark: For use below, we mention the following easily provable fact: if P, $\alpha^{i_1}_{u_1},\ldots\ldots,\ \alpha^{i_s}_{u_s}$ and $w_1,\ldots\ldots,w_s$ are as in definition 24, then there is at most one P' such that $R(P,P')$ holds and such that the pair P,P' is compatible with $w_1,\ldots\ldots,w_s$; moreover, we can effectively decide if there is such a P' and if so we can find this P' effectively. Now we are able to state the generalization of theorem 23, namely

Theorem 24: a) Let P be a s.n.s. proof in $ZTEi/II_N$ of $\longrightarrow (E\xi)A(\alpha^i_u, \xi)$ where α^i_u is the only special function constant occuring in the endsequent of P. Then there exists a recursive continuity function $\delta(x)$ with the property: if $\delta(v)\neq 0$, then one effectively finds a functor F, containing at most α^i_{u*v} as special function constant, and a proof P' (in $ZTEi/II_N$) of $\longrightarrow A(\alpha^i_{u*v},F)$. b) Similarly, if P is a proof of $\longrightarrow (Ex)A(\alpha^i_u,x)$ but with a term t in place of the functor F. c) If P is a s.n.s. proof in $ZTEi/II_N$ of

$\longrightarrow A(\alpha^i_u) \vee B(\alpha^i_u)$ where α^i_u is the only special function constant in A,B, then there is a continuity function $\delta(x)$ having the property: if $\delta(v) \neq 0$, then one effectively finds a proof P' of either $\longrightarrow A(\alpha^i_{u*v})$ or of $\longrightarrow B(\alpha^i_{u*v})$. d) An analogous statement holds if the special function constants which appear in the endsequent of P are $\alpha^{i_1}_{u_1}, \ldots\ldots, \alpha^{i_s}_{u_s}$; the continuity function $\delta(x)$ has then to be replaced accordingly by $\delta(x_1, \ldots, x_s)$.

<u>Proof:</u> We prove only the first case; the three other cases are treated in exactly the same way. In view of lemma 10, we can assume without loss of generality that P is stratified. Let us call a sequence number v secured if the following is true: there is a chain $P_o, \ldots, P_N$ with $P_o = P$ which is compatible with u*v and such that $\neg(EX)R(P_N, X)$ holds. We want to show that the property of a sequence number to be secured is decidable. First, we note that, given any chain of proofs $P_o, \ldots, P_N$, it is decidable whether this chain is compatible with u*v or not. Next, we look at the set B of chains which are compatible with u*v . We claim that this set is finite. To this end, given any chain $P_o, \ldots, P_N$ with $R(P_i, P_{i+1})$ $(i<N)$, let us call $P_o, \ldots, P_N, P_{n+1}$ a successor of this chain if also $R(P_N, P_{N+1})$ holds. Now we apply the fan theorem and show: 1) there is no infinite chain $P_o, P_1, \ldots$ such that for every N $P_o, \ldots, P_N$ is a chain in the set B ; 2) a chain $P_o, \ldots, P_N$ in B has at most finitely many successors in B . Now 1) is a consequence of the fact (already noted earlier) that no infinite sequence $P_o, P_1, P_2 \ldots$ with $R(P_i, P_{i+1})$ exists. On the other hand, given a chain $P_o, \ldots, P_N$ of the set B there are two possibilities: either P_N is saturated and there are at most finitely many P*'s with $R(P_N, P^*)$, as noted earlier, or P_N is not saturated and there is at most one P* such that $R(P_N, P^*)$ holds and

such that the pair P_N, P^* is compatible with $u*v$ (see remark following definition 24). In both cases $P_o, \ldots, P_N$ has at most finitely many successors in B . Now we call a set M admissible if its elements are chains $P_o, \ldots, P_N$ which are compatible with $u*v$. Clearly, B is admissible and every other admissible set M is a subset of B ; in other words, B is the largest admissible set. Our proof is essentially finished if we can show that given an admissible set M we can decide whether M is maximal or not. To this end, let $C_o, \ldots, C_A$ be the chains in M . As in the application of the fan theorem above, we conclude that each C has at most finitely many successors which are compatible with $u*v$. In virtue of theorem 4, theorem 18, theorem 22 and the remark following definition 24, it follows that for each i we can decide whether C_i has successors in B and, if so, we can find them all in an effective way. Let $M(C_i)$ be the set of successors of C_i which are in B (empty if there are none). All we have to do is to check whether $M \bigcup_i M(C_i)$ is a proper extension of M or not. But this is obviously a decidable problem. To sum up: 1) given v, we can effectively decide whether a finite set M of chains is admissible (with respect to v); 2) given an admissible set M , we can decide whether it is maximal or not; 3) there is precisely one maximal admissible set (the B above). From this it follows that, given v , the maximal admissible set B can effectively be found. In order to decide whether v is secured or not, we only have to check whether B contains a chain $P_o, \ldots, P_N$ such that $\neg(EX)R(P_N, X)$ holds. Hence, we can effectively decide whether v is secured or not. Now we define a recursive function as follows: 1) if $\delta(v) \neq 0$, then v is a sequence number of length > 0; 2) $\delta(v) \neq 0$ iff v is secured; 3) if v is secured, then $\delta(v)=1$. It remains to verify that $\delta(x)$ is the continuity function we are looking for. To this end we note that, given a function ξ, we can effectively find a chain $P_o, \ldots, P_N$

which is compatible with $u*\xi$ and for which $\neg(EX)R(P_N,X)$ holds; this is an easy consequence of theorems 4, 18 and the remark following definition 24 . By definition, this means that there is a K such that $u*\bar{\xi}(K)$ is secured. Hence σ is continuous.

Finally, let v be secured, that is, $\sigma(v)=1$. Then we effectively find a chain $P_o,\ldots\ldots,P_N$ compatible with $u*v$ for which $\neg(EX)R(P_N,X)$ holds. P_N is, of course, saturated, does not admit any preliminary nor strictly essential reduction step. The endsequent of P_N has the form $\longrightarrow (E\xi)A(\alpha^{i}_{u*w},\xi)$ where w is a certain sequence number for which $v \subseteq w$ holds. Clearly, P_N does not coincide with its final part. By theorem 21, we can apply a subformula reduction step to P_N , obtaining thus a proof $\hat{P}$ of $\longrightarrow A(\alpha^{i}_{u*w},F_o)$ where F_o is a certain constant functor, effectively determined by P_N, which does not contain special function constants other than eventually α^{i}_{u*w} . If $w=v$, we are finished. Otherwise we replace α^{i}_{u*w} by α^{i}_{u*v} in P , obtaining thus a proof of $\longrightarrow A(\alpha^{i}_{u*v},F)$, where F is a constant functor which does not contain other special function constants than eventually α^{i}_{u*v}. This concludes the proof.

<u>B.</u> Another kind of application is connected with the notion of constructive, infinite ω-proof, introduced by Schütte in [10] . We content ourself with a rather superficial treatment of this matter. A rigorous treatment would involve a precise definition of constructive cut-free ω-proof and several applications of the fixed point theorem for partial recursive functions. As an intuitive substitute for partial recursive functions and the fixed point theorem, we use the notion "effective" in about the same way as Schütte in [10] . To this end, we introduce a certain infinitary rule, which we call constructive ω-rule, and a semiformal system S_ω containing this

rule. In this connection we use the following <u>notation</u>: if S is a sequent whose special function constants are among $\alpha_{u_1}^{i_1},\ldots,\alpha_{u_s}^{i_s}$, whose free function variables are among $\xi_1,\ldots,\xi_t$ and whose free number variables are among $x_1,\ldots,x_r$, then we express this by writing $S(\alpha_{u_1}^{i_1},\ldots,\alpha_{u_s}^{i_s},\xi_1,\ldots,\xi_t,x_1,\ldots,x_r)$, or in a more condensed form $S(\alpha_{u_1}^{i_1},\ldots,\alpha_{u_s}^{i_s},\vec{\xi},x_1,\ldots,x_r)$, or $S(\alpha_{u_1}^{i_1},\ldots,\alpha_{u_s}^{i_s},\vec{\xi},\vec{x})$, respectively. We remind that if $\tau(x_1,\ldots,x_s)$ is a continuity function of type $[s,0]$, then $\bar{\alpha}_1(n),\ldots,\bar{\alpha}_s(n)$ is called immediately secured with respect to τ if $\tau(\bar{\alpha}_1(n),\ldots,\bar{\alpha}_s(n))\neq 0$, and if $\tau(\bar{\alpha}_1(i),\ldots,\bar{\alpha}_s(i))=0$ for all $i<n$. The fact that $v_1,\ldots,v_s$ is immediately secured with respect to τ will be expressed by writing $\tau(v_1,\ldots,v_s)\stackrel{*}{\neq}0$.

<u>Definition 25:</u> The constructive ω-rule is determined by the clauses a), b) below. a) Let $S(\alpha_{u_1}^{i_1},\ldots,\alpha_{u_s}^{i_s},\vec{\xi},\vec{x})$ be a sequent and assume that we are effectively given a continuity function τ of type $[s,0]$, having the following property: if $\tau(v_1,\ldots,v_s)\stackrel{*}{\neq}0$, then we are effectively given a proof $P_{v_1\ldots v_s}$ (in some suitable system) of $S(\alpha_{u_1*v_1}^{i_1},\ldots,\alpha_{u_s*v_s}^{i_s},\vec{\xi},\vec{x})$. b) Let $S(\alpha_{u_1}^{i_1},\ldots,\alpha_{u_s}^{i_s},\vec{\xi},x_1,\ldots,x_r)$ be a sequent and assume that for each r-tuple $n_1,\ldots,n_r$ we are effectively given a proof $P_{n_1\ldots n_r}$ of $S(\alpha_{u_1}^{i_1},\ldots,\alpha_{u_s}^{i_s},\vec{\xi},n_1,\ldots,n_r)$. In each of these cases we are permitted to infer $S(\alpha_{u_1}^{i_1},\ldots,\alpha_{u_s}^{i_s},\vec{\xi},x_1,\ldots,x_r)$ from the premisses.

<u>Notation:</u> An application of the constructive ω-rule will be written as follows: $\tau(v_1,\ldots,v_s)\stackrel{*}{\neq}0$: $S(\alpha_{u_1*v_1}^{i_1},\ldots,\alpha_{u_s*v_s}^{i_s},\vec{\xi},\vec{x})/S(\alpha_{u_1}^{i_1},\ldots,\alpha_{u_s}^{i_s},\vec{\xi},\vec{x})$ in the case a) of definition 25 and

$n_1,\dots,n_r<\omega$: $S(\alpha^{i_1}_{u_1},\dots,{}^{i_s}_{u_s},\vec{\xi},n_1,\dots,n_r)/S(\alpha^{i_1}_{u_1},\dots,\alpha^{i_s}_{u_s},\vec{\xi},\vec{x})$ in case b) of definition 25.

The system S mentioned above is introduced by the following

Definition 26: The language and the axioms of S_ω are the same as those of ZTi (and hence as those of ZT, ZTEi/II_N etc.). The rules of S are: 1) the structural rules except cut; 2) the conversion rule; 3) the logical rules of sequential calculus; 4) the constructive ω-rule; 5) an additional rule, denoted by C , whose definition is as follows: if $\alpha^i_{<\,>}$ is a special function constant, S a sequent and α a function variable free for $\alpha^i_{<\,>}$ in S , then we can infer S' from S where S' is obtained from S by replacing every occurence of $\alpha^i_{<\,>}$ in S by α .

The notion of infinitary proof tree (with respect to S_ω) can be introduced in the usual way (see [10]), and with every such infinitary proof we can associate in a natural way an ordinal, called its tree ordinal. For details we refer to [10] . Our ω-rule is only seemingly more general than ω-rule introduced in [10]. It would, in fact, be easy to show that our ω-rule is derivable by means of the usual ω-rule; by adopting definition 26, however, we can save a few lemmas. Notation: the fact that S is provable in S_ω will be expressed by the notation $S_\omega \vdash S$.

Theorem 25: Let A be a formula with the properties: 1) neither $\supset$ nor $\neg$ occur in A ; b) no variable occurs both free and bound in A . Let P be a proof in ZTEi/II_N of $\longrightarrow A$. Then one effectively finds a proof P_ω in S_ω of $\longrightarrow A$.

<u>Proof:</u> <u>A.</u> First we observe that it is sufficient to prove the statement for the case where P is an s.n.s. proof. In order to see this, let P be an arbitrary proof in $ZTEi/II_N$ and assume for simplicity that A contains precisely two free variables, namely, α and x ; we indicate this by writing $A(\alpha, x)$. Since by assumption neither α nor x occurs bound in A , there is a normal proof P* of $\longrightarrow A(\alpha, x)$. Let $\alpha^{i}_{< >}$ be a special function constant, associated with the empty sequence, which does not occur in P* ; let n be an arbitrary, but fixed numeral. By replacing every occurence of α and x by $\alpha^{i}_{< >}$ and n, respectively, we get a proof P'_n of $\longrightarrow A(\alpha^{i}_{< >}, n)$. According to earlier remarks, there exists a s.n.s. proof P_n of $\longrightarrow A(\alpha^{i}_{< >}, n)$. Since, by assumption, the theorem holds for s.n.s. proofs it follows that we effectively find proofs P^{ω}_n in S_ω of $\longrightarrow A(\alpha^{i}_{< >}, n)$. By means of the constructive ω-rule (clause b) of definition 25, we can piece the P^{ω}_n's together in order to get a proof $\hat{P}$ in S_ω of $\longrightarrow A(\alpha^{i}_{< >}, x)$. Now we apply to $\longrightarrow A(\alpha^{i}_{< >}, x)$ an inference of type C (see clause 5) of definition 26 and obtain a proof P_ω of $\longrightarrow A(\alpha, x)$.

<u>B.</u> In order to prove the theorem for s.n.s. proofs, we proceed by bar induction over the relation L , introduced by definition 23. To this end, let P be an s.n.s. proof of $\longrightarrow A$ where A has the properties stated in the theorem; according to the definition of P, there are no free variables in A . The proof by transfinite induction over L is essentially accomplished if we can show that the theorem holds for P in each of the following two cases:
a) $\neg(EX)L(P,X)$ holds; b) if $L(P,P')$ holds, then the theorem is true for P' . <u>Case 1:</u> $\neg(EX)L(P,X)$ holds. Then P is a saturated s.n.s. proof which does not admit any kind of reduction step. In virtue of theorem 19, it follows that P coincides with its final part.

Since no logical axioms and no thinnings occur in P, it follows that A must be a saturated prime formula and since cuts, contractions, interchanges and conversions are the only inferences in P, it follows that A is true. Hence, $\longrightarrow A$ is an axiom of S what proves the theorem in this case. <u>Case 2:</u> Assume (EX)L(P,X), and assume furthermore that the theorem is true for all proofs P' for which L(P,P') holds. We have to consider subcases. For simplicity, we assume that A contains exactly one special function constant, say α^i_u ; we express this by writing $A(\alpha^i_u)$. The case where A contains more than one special function constant is treated in exactly the same way. <u>Subcase 1:</u> P is not saturated. Let $\tau(x)$ be the continuity function associated with P according to lemma 9 and the remark preceeding definition 20, and let P_v be the proof which we obtain from P by replacing every occurence of α^i_u in P by α^i_{u*v}. According to the definition of τ and of the inessential reduction steps, we have $L(P,P_v)$ for all v for which $\tau(v) \neq 0$ holds and, conversely, if L(P,P') holds, then P' is P_v for some v, according to the definition of L. By induction, we are effectively given proofs P^ω_v in S_ω of $\longrightarrow A(\alpha^i_{u*v})$. The proofs P^ω_v can be pieced together by means of the following application of the constructive ω-rule: $\tau(v) \neq 0 : \longrightarrow A(\alpha^i_{u*v}) / \longrightarrow A(\alpha^i_u)$. The result is a proof P_ω in S_ω of $\longrightarrow A(\alpha^i_u)$.

<u>Subcase 2:</u> P is saturated and L(P,P') holds in virtue of clause B of definition 23, that is, R(P,P') holds. Then P' has the same endsequent as P. According to the induction hypothesis, we effectively find a proof P_ω in S_ω of $\longrightarrow A$, that is, the theorem applies to P. <u>Subcase 3:</u> P is saturated and L(P,P') holds in virtue of clause C in definition 23. Then P' is obtained from P by means of a subformula reduction step, and we have to distinguish subsubcases according to the outermost logical symbol in A. We content ourself with the treatment of two cases where the outermost

logical symbol is a universal quantifier applied to a function variable and a universal quantifier applied to a number variable, respectively. a) Let A have the form $(\forall \xi)B(\alpha^i_u, \xi)$. According to the definition of subformula reduction step, it follows that P' is a s.n.s. proof whose endsequent has the form $\longrightarrow B(\alpha^i_u, \alpha^k_{<\,>})$, where $\alpha^k_{<\,>}$ is a special function constant, associated with the empty sequent, which does not occur in P and hence not in A . According to the induction hypothesis, there is a proof P'_ω in S_ω of $\longrightarrow B(\alpha^i_u, \alpha^k_{<\,>})$. From P'_ω and an application of rule C, we get a proof P_ω in S_ω of $\longrightarrow A$ as follows:

$$
\begin{array}{ccc}
 & & P'_\omega \\
 & & \vdots \\
 & C & \dfrac{\longrightarrow B(\alpha^i_u, \alpha^k_{<\,>})}{\longrightarrow B(\alpha^i_u, \alpha)} \\
\longrightarrow & \forall & \dfrac{}{\longrightarrow (\forall \xi)B(\alpha^i_u, \xi)}
\end{array}
$$

b) Let A have the form $(\forall z)B(\alpha^i_u, z)$. According to the definition of subformula reduction step and clause C of definition 23, it follows that there is a denumerable list of proofs $P_o, P_1, P_2, \ldots$ having the following properties: o) if $L(P,P')$ holds, then P' occurs in the list; 1) $L(P,P_n)$ holds for $n < \omega$; 2) the endsequent of P_n has the form $\longrightarrow B(\alpha^i_u, n)$. By the induction hypothesis we are effectively given proofs P^n_ω in S_ω of $\longrightarrow B(\alpha^i_u, n)$. Combining these proofs with the aid of the constructive ω-rule followed by a universal quantification, we get a proof P_ω in S_ω of $\longrightarrow A$ as follows:

$$\dfrac{\dfrac{\begin{array}{c}P^{n}_{\omega}\\ \vdots\\ \longrightarrow B(\alpha^{i}_{u},n)\end{array}}{\longrightarrow B(\alpha^{i}_{u},x)}\;\; n<\omega}{\longrightarrow (\forall z)B(\alpha^{i}_{u},z)}\;\; \longrightarrow \forall$$

what proves the statement for this case.

The last theorem and its proof are nothing else than appropriate generalizations of theorem 6 and its proof presented in [8].

<u>C.</u> There is another application, intimately connected with the last theorem and which we will discuss only superficially. To this end, let A be a closed formula, not containing special function constants and having prenex normal form. In order to fix the ideas, we assume that A is, say, $(\alpha)(E\beta)(\gamma)(Ex)B(\alpha,\beta,\gamma,x)$, B quantifierfree. We say that A has a constructive model if we find recursive functionals $F[\alpha]$, $G[\alpha,\gamma]$ and a recursive function $\Delta(\alpha,\gamma)$ such that $B(\alpha, F[\alpha], G[\alpha,\gamma], \Delta(\alpha,\gamma))$ is an identically true formula (thereby using the notion "formula" in a slightly more general sense than in chapter I. This concept can be generalized in a natural and rather obvious way to arbitrary closed formulas not containing the signs $\supset$ and $\neg$ and not containing special function constants. Finally, let A be a formula which does not contain $\supset$ nor $\neg$, whose special function constants are among $\alpha^{i_1}_{u_1},\dots,\alpha^{i_s}_{u_s}$ and whose free variables are among $\xi_1,\dots,\xi_t,x_1,\dots,x_q$. As usual, we write $A(\alpha^{i_1}_{u_1},\dots,\alpha^{i_s}_{u_s},\xi_1,\dots,\xi_t,x_1,\dots,x_q)$ in place of A. We say that A admits a constructive model if the formula

$(\forall \eta_1, \ldots, \eta_s, \xi_1, \ldots, \xi_t, x_1, \ldots, x_q) A(u_1 * \eta_1, \ldots, u_s * \eta_s, \xi_1, \ldots, \xi_t, x_1, \ldots, x_q)$ admits a constructive model. The main result then says: if A is a formula which does not contain $\supset$ nor $\neg$, and if P is a proof in $ZTEi/II_N$ of $\longrightarrow A$, then we effectively find a constructive model of A. Here "effective" means that the Goedel numbers of the recursive functions and functionals whose existence is claimed can be found effectively from the Goedelnumber of the proof P. There are two possibilities to prove this statement:
a) by transfinite induction over the wellfounded relation L, using thereby the fact that the statement follows for formulas containing free variables if it has been proved for closed formulas;
b) by transfinite induction over the proof P_ω in S_ω of $\longrightarrow A$ which is provided by the last theorem. In both cases the fixpoint theorems for partial recursive functions have to be used in an essential way.

It is interesting in this connection to consider the simplest case, namely, that one where the formula A in question has the form $(x)(Ey)B(x,y)$, where B is prime, without special function constants and without free variables other than x,y. Let P be a proof in $ZTEi/II_N$ of $\longrightarrow A$. From P we obtain for each numeral n in an effective way a s.n.s. proof P in $ZTEi/II_N$ of $\longrightarrow (Ey)B(n,y)$. In order to find an m such that $B(n,m)$ is true, we construct a chain $P^n_o, \ldots, P^n_N$ such that a) $P^n_o = P$, b) $R(P^n_i, P^n_{i+1})$ for all $i < N$, c) $\neg(EX)R(P^n_N, X)$. In virtue of the properties of R, such a chain can always effectively be found. The endsequent of P^n_N is still $\longrightarrow (Ey)B(n,y)$. Since P^n_N is saturated and admits neither preliminary nor strictly essential reduction steps, it follows in virtue of theorem 19 that a subformula reduction step is applicable to P^n_N. The result is a s.n.s. proof P^*_n in $ZTEi/II_N$ whose endsequent has the form $\longrightarrow B(n,t)$, where t is a constant term.

By applying eventually an inessential reductionstep to P_n^*, we get a proof $\tilde{P}_n$ of $\longrightarrow B(n,t^*)$, where t^* is saturated with value, say, m. By means of a conversion, we finally get a proof $\hat{P}_n$ of $\longrightarrow B(n,m)$. The procedure described is effective, that is, given P, we can find for each n effectively a proof $\hat{P}_n$ of $\longrightarrow B(n,m)$ for some m. The m depends, of course, on n, hence it may be written as $\varphi(n)$. That is, from P we have extracted a recursive function $\varphi(x)$ such that $B(n, \varphi(n))$ is true for each n, that is, such that $B(x, \varphi(x))$ is identically true. In this connection we may ask the following question: if $\longrightarrow (x)(Ey)B(x,y)$ (with B prime) has been proved in $ZTEi/II_N$, can we then prove $\longrightarrow (E\xi)(x)B(x, \xi(x))$? In virtue of theorem 23 the answer is clearly negative. The reason is that from a proof of $\longrightarrow (E\xi)(x)B(x, \xi(x))$ we can find, according to this theorem, a functor F and a proof of $\longrightarrow B(x,F(x))$; this implies that there is a p.r. function φ such that $B(n, \varphi(n))$ is true for all n. On the other hand, it is not difficult to find a prime formula $B(x,y)$ having the following properties: a) for each primitive recursive function φ there is an n with $B(n, \varphi(n)) \neq 0$; b) $ZTEi/II_N \vdash \longrightarrow (x)(Ey)B(x,y)$ holds. A consequence of this argument is

<u>Theorem 26:</u> There is a prime formula $B(x,y)$ for which the following sequent is unprovable in $ZTEi/II_N$:

$$(x)(Ey)B(x,y) \longrightarrow (E\xi)(x)B(x, \xi(x)).$$

As corollary we immediately obtain the

<u>Corollary:</u> The axiom of choice for primitive recursive formulas is not provable in $ZTEi/II_N$.

C. Up to now we have formulated all results for the theory $ZTEi/II_N$. But, since $ZTEi/II_N$ is merely a conservative extension of ZTi/II_N, it follows immediately that these results hold invariably for ZTi/II_N. On the other hand, if A is a formula without special function constants, if P is a proof in $ZTEi/II_N$ of $\longrightarrow A$, then there is a proof P* in ZTi^*/II_N of $\longrightarrow A$ that is a proof not containing special function constants at all. This implies that the theorems 23 and 25 remain true for ZTi^*/II_N. There is also a suitable transformation of theorem 24 into the language L* which is true for ZTi^*/II_N : all we have to do is to replace the special function constants $\alpha_{u_1}^{i_1}, \ldots\ldots, \alpha_{u_s}^{i_s}$ by functors $u_1{}^* \alpha_1, \ldots\ldots, u_s{}^* \alpha_s$ where the α_i's are suitably chosen free function variables. Finally, it presents no difficulties to pass from ZTi/II_N and ZTi^*/II_N to corresponding Hilbert-type systems ZHi/II_N and ZHi^*/II_N with the aid of theorem 0. It is clear, that theorems 23 - 26, suitably reformulated, remain true for these Hilbert-type systems. We do not pursue the details of these passages from one system to the other, since they involve only routine techniques of a rather trivial nature.

4.6. The system ZTi/II and its conservative extension ZTEi/II

In this section we consider a conservative extension ZTE/II of ZT/II which is related to the latter in the same way as ZTE/II_N to ZT/II_N. The intuitionistic version of ZTE/II, to be denoted by ZTEi/II, is in its turn a conservative extension of ZTi/II. To ZTEi/II we apply a treatment which parallels that one of $ZTEi/II_N$. In order to avoid a repetition of the arguments presented in the last section, we content ourself in pointing out the changes which have to be made in passing from $ZTEi/II_N$ to ZTEi/II.

A. According to the definition of ZT/II , we obtain this system by adding to ZT the new rule

$$\text{II.}\quad \frac{D(y),\ (x) \subset_{D^y} A(x),\ \Gamma \longrightarrow \Delta, A(y)}{\mathfrak{W}(\subset_D),\ \Gamma \longrightarrow \Delta, A(q)}$$

where, as before, $\mathfrak{W}(\subset_D)$ and $(x)\subset_{D^y}A(x)$ are abbreviations for the formulas $(\alpha)\neg(x)(\alpha(x+1)\ {}_{K}\ \alpha(x) \wedge D(\alpha(x+1)) \wedge D(\alpha(x)))$ and $(x)(x \subset_D y . \supset . A(x))$, respectively, while q and y are subject to the stipulations stated in part B of section 1.5. Here, in contrast to ZTi/II_N , the formula $\mathfrak{W}(\subset_D)$ is not required to be a formula "without function parameters"; that is, free function variables and special function constants may occur in $\mathfrak{W}(\subset_D)$ in a quite essential way. In order to obtain a conservative extension ZTE/II of ZT/II which corresponds to ZTE/II_N, we need new rules which correspond to the rules Ti(P) and $Ti(P,P_1,m)$ introduced in section 4.1. To this end, let $v_1,\ldots,v_s$ and $w_1,\ldots,w_s$ be two lists of sequence numbers such that $l(v_1)=\ldots=l(v_s)$ and $l(w_1)=\ldots=l(w_s)$ holds. Let $D(\alpha^{i_1}_{u_1},\ldots,\alpha^{i_s}_{u_s},x)$ be a formula whose only free variable is x and whose distinct special function constants are $\alpha^{i_1}_{u_1},\ldots,\alpha^{i_s}_{u_s}$; we denote this formula briefly by $D(x)$. Let $G(x)$ and $H(x)$ be $D(\alpha^{i_1}_{u_1*v_1},\ldots,\alpha^{i_s}_{u_s*v_s},x)$ and $D(\alpha^{i_1}_{u_1*v_1*w_1},\ldots,\alpha^{i_s}_{u_s*v_s*w_s},x)$, respectively. The first of the above-mentioned rules is defined as follows: if P is a strictly normal proof in ZTi/II of $\longrightarrow \mathfrak{W}(\subset_D)$, then we are allowed to infer from the premiss $G(y),(x)\subset_{G^y}A(x),\ \Gamma \longrightarrow \Delta, A(y)$ the conclusion $G(q),\ \Gamma \longrightarrow \Delta, A(q)$. This rule is denoted by TI(P) and written as follows:

$$\mathrm{TI(P)} \quad \frac{G(y),\ (x)(x \subset_G y. \supset A(x)),\ \Gamma \longrightarrow \Delta, A(y)}{G(q),\ \Gamma \longrightarrow \Delta, A(q)}$$

Here y is not allowed to occur free in the conclusion and q is assumed to be free for y in $A(y)$.

The second rule is given as follows: if P_1 is a s.n.s. proof in ZTi/II of $\longrightarrow \mathfrak{W}(\subset_D)$, if P_2 is a s.n.s. proof in ZTi/II of $\longrightarrow G(t)$, where t is a saturated term, then we are allowed to infer from the premiss $y \subset_H t,\ (x)(x \subset_H y. \supset A(x)), \Gamma \longrightarrow \Delta, A(y)$ the conclusion $q \subset_H t,\ \Gamma \longrightarrow \Delta, A(q)$ where q, y are subject to the same stipulations as before. We write this rule as follows:

$$\mathrm{TI}(P_1,P_2,m) \quad \frac{y \subset_H t,\ (x)(x \subset_H y. \supset A(x)),\ \Gamma \longrightarrow \Delta, A(y)}{q \subset_H t,\ \Gamma \longrightarrow \Delta, A(q)}$$

where $m = |t|$.

By adding the just defined rules TI(P) and $\mathrm{TI}(P_1,P_2,m)$ to ZT/II, we obtain the system ZTE/II.

<u>Remark:</u> In the case of the TI(P) rule above it is evident that by replacing every occurence of $\alpha^{i_1}_{u_1}, \ldots\ldots, \alpha^{i_s}_{u_s}$ in P by $\alpha^{i_1}_{u_1 * v_1}, \ldots\ldots, \alpha^{i_s}_{u_s * v_s}$, respectively, we obtain a s.n.s. proof P' in ZTi/II of $\longrightarrow \mathfrak{W}(\subset_G)$. Similarly, by replacing $\alpha^{i_1}_{u_1}, \ldots\ldots, \alpha^{i_s}_{u_s}$ in P_1 and $\alpha^{i_1}_{u_1 * v_1}, \ldots\ldots, \alpha^{i_s}_{u_s * v_s}$ in P_2 by $\alpha^{i_1}_{u_1 * v_1 * w_1}, \ldots\ldots, \alpha^{i_s}_{u_s * v_s * w_s}$, we get proofs P_1' and P_2' of $\longrightarrow \mathfrak{W}(\subset_H)$ and $\longrightarrow H(t)$, respectively, in case of a $\mathrm{TI}(P_1,P_2,m)$ inference. Thus, TI(P) and $\mathrm{TI}(P_1,P_2,m)$ are generalisations of the rules Ti(P) and $\mathrm{Ti}(P_1,P_2,m)$ defined in section 4.1.

The systems ZTE/II and ZTE/II_N look clearly very much the same and it is to be expected that what we have done for ZTE/II_N can be done in more or less the same way for ZTE/II . This is indeed rather evident for the content of sections 4.1.: all statements, definitions and results carry over to ZTE/II with almost no changes. Thus we can e.g. introduce the notion of side proof, degree and order in exactly the same way as in section 4.1. Theorem 14 remains true for ZTE/II ; its proof remains essentially the same except that the last remark has to be used at a few places. Of course, we can pass from ZTE/II to its intuitionistic version ZTEi/II which in virtue of theorem 14 is a conservative extension of ZTi/II . To sum up: we will apply all notions and results given in section 4.1. without further comments to ZTE/II and ZTEi/II . To the notions defined in section 4.1. we add a new one, namely, that of the <u>index</u> of a $TI(P_1,P_2,m)$ inference. To this end, let $\alpha_{u_1}^{i_1},\ldots,\alpha_{u_s}^{i_s}, v_1,\ldots,v_s,w_1,\ldots,w_s$ and D,G,H have the same meaning as above in the definition of TI(P)- and $TI(P_1,P_2,m)$-inferences. The list $v_1,\ldots,v_s$ of sequence numbers, which is determined by P_1 and P_2, will be called the <u>index</u> of the $TI(P_1,P_2,m)$ inference in question. The index will play an important role in connection with the ordinal assignement which will be discussed below.

<u>B.</u> With a cut, an induction, a TI(P) inference or a $TI(P_1,P_2,m)$ inference, we can, of course, associate a natural number, called its complexity, in exactly the same way as in part B of section 2.5. Based on the notion of "complexity" we can associate with each sequent S in a proof P in ZTE/II another natural number, called its height, and denoted again by h(S) ; the definition of height, too, is, of course, the same as the definition of height in part B of section 2.5. With the notion of height at hand, we can now define reduc-

tion steps for proofs P in ZTE/II in almost the same way as we have done it for proofs P in ZTE/II_N . In particular, we can introduce preliminary reduction steps, induction reduction steps and logical reduction steps in precisely the same way as before. In order to introduce the notions "substitution instance" and "inessential reduction step", we can, of course, use definition 20 without any change. Minor differences appear in the definition of II-, TI_1- and TI_2-reduction step which correspond to the II_N-, Ti_1- and Ti_2-reduction steps, respectively, defined in section 4.2.

a) <u>II-reduction steps.</u> Let

$$\text{II} \qquad \frac{D(y),\ (x) \subset_D y\, A(x),\ \Gamma \longrightarrow \Delta, A(y)}{\mathfrak{W}(\subset_D),\ D(q),\ \Gamma \longrightarrow \Delta, A(q)}$$

be a critical II-inference in a strictly normal proof P in ZTE/II . Let P_1 be a strictly normal proof in ZTi/II of $\longrightarrow \mathfrak{W}(\subset_D)$. Finally, let q be saturated. According to the definition of "strictly normal", it follows automatically that y is the only free variable in D(y) . A II-reduction step consists in replacing the above inference by the following inferences:

$$\text{TI}(P_1) \qquad \cfrac{\cfrac{D(y),(x) \subset_D y\, A(x),\ \Gamma \longrightarrow \Delta, A(y)}{D(q),\ \Gamma \longrightarrow \Delta, A(q)}}{\mathfrak{W}(\subset_D),\ D(q),\ \Gamma \longrightarrow \Delta, A(q)} \ \text{thinning}$$

The proof P' so obtained is said to follow from P by means of a II-reduction step; we say that the reduction step has been applied to the II-inference above.

<u>b) TI-reduction steps.</u> Let $D(\alpha_{u_1}^{i_1},\ldots,\alpha_{u_s}^{i_s},x)$ be a formula containing only x free and whose special function constants are precisely $\alpha_{u_1}^{i_1},\ldots,\alpha_{u_s}^{i_s}$. Let $v_1,\ldots,v_s$ be a list of sequence numbers all of the same length, and let $G(x)$ be $D(\alpha_{u_1*v_1}^{i_1},\ldots,\alpha_{u_s*v_s}^{i_s},x)$. Let P_1 be a strictly normal proof in ZTi/II of $\longrightarrow W(\subset_D)$. Let there be a critical $TI(P_1)$ inference in the strictly normal proof P in ZTEi/II, namely

$$TI(P_1)\quad \frac{G(y),\ (x)\subset_G y\, A(x),\ \Gamma\longrightarrow\Delta,A(y)}{G(q),\ \Gamma\longrightarrow\Delta,A(q)}$$

and let q be a saturated term with value $|q|$, say m. Finally, assume that we have at disposal a strictly normal proof P_2 in ZTi/II of $\longrightarrow G(q)$. Then we apply to P the same syntactical transformation as in the case of Ti_1-reduction step, that is, we alter the $TI(P_1)$ inference as follows:

$$\begin{array}{ll}
 & \qquad\qquad\qquad P_o \qquad\qquad\qquad\qquad P_S \\
 & \qquad\qquad\qquad\ \vdots \qquad\qquad\qquad\qquad\ \vdots \\
\text{cut} & y\subset_G q\longrightarrow G(y) \qquad\qquad S \\ \cline{2-2}
 & y\subset_G q,\ (x)\subset_G y\, A(x),\ \Gamma\longrightarrow\Delta,A(y) \\ \cline{2-2}
TI(P_1,P_2,m) & s\subset_G q,\ \Gamma\longrightarrow\Delta,A(q) \qquad\qquad P_S^q \\ \cline{2-2}
 & \Gamma\longrightarrow\Delta,\ s\subset_G q\supset A(s) \qquad\qquad \vdots \\ \cline{2-2}
 & \Gamma\longrightarrow\Delta,(x)\subset_G q\, A(x) \qquad\qquad S^q \\ \cline{2-2} \cline{2-2}
 & D(q),\ \Gamma\longrightarrow\Delta,A(q)
\end{array}$$

Here, S, P_o, P_S^q and S^q have the same meaning as in the definition of Ti_1-reduction step in section 4.2. The resulting proof P' is said to be obtained from P by means of a TI_1-reduction step; we also say that the TI_1-reduction step has been applied to the above

$TI(P_1)$ inference.

<u>c) TI_2-reduction steps.</u> Let D , $\alpha^{i_1}_{u_1},\ldots,\alpha^{i_s}_{u_s}$, $v_1,\ldots,v_s$ and G be as before and let H be $D(\alpha^{i_1}_{u_1 * v_1 * w_1},\ldots,\alpha^{i_s}_{u_s * v_s * w_s},x)$ where $w_1,\ldots,w_s$ is a second list of sequence numbers all having the same length. Let there be a critical $TI(P_1,P_2,m)$ inference in P , say

$$TI(P_1,P_2,m) \quad \frac{y \subset_H t,\ (x) \subset_H y\, A(x),\ \Gamma \longrightarrow \Delta, A(y)}{q \subset_H t,\ \Gamma \longrightarrow \Delta, A(q)}$$

where $|t| = m$, P_1 is a strictly normal proof in ZTi/II of $\longrightarrow \hat{W}(\subset_D)$ and P_2 is a strictly normal proof in ZTi/II of $\longrightarrow G(t)$. Assume that q is saturated with $|q| = n$, and that we have at disposal a strictly normal proof P_3 in ZTi/II of $\longrightarrow q \subset_H t$. Finally, let P_3^* be a cut-free proof in intuitionistic predicate calculus of $q \subset_H t \longrightarrow H(q)$ and P_4 the following proof:

$$\frac{\begin{array}{c} P_3 \\ \vdots \\ \longrightarrow q \subset_H t \end{array} \qquad \begin{array}{c} P_3^* \\ \vdots \\ q \subset_H t \longrightarrow H(q) \end{array}}{\longrightarrow H(q)}$$

Then we apply to P a syntactical transformation which is just a copy of the TI_a-reduction step defined in section 2.5, namely

$$\begin{array}{lcc}
 & P_o & P_S \\
 & \vdots & \vdots \\
 & y \subset_H q,\ q \subset_H t \longrightarrow y \subset_H t & S \\
\text{cut} & \overline{\qquad\qquad\qquad\qquad\qquad\qquad\qquad\qquad} & \\
 & y \subset_H q,\ q \subset_H t, (x) \subset_{H^y} A(x), \Gamma \longrightarrow \Delta, A(y) & \\
TI(P_1,P_4,n) & \overline{\qquad\qquad\qquad\qquad\qquad\qquad\qquad\qquad} & \\
 & s \subset_H q, q \subset_H t, \Gamma \longrightarrow \Delta, A(s) & \\
 & \overline{\qquad\qquad\qquad\qquad\qquad\qquad\qquad\qquad} & P^q_S \\
 & q \subset_H t, \Gamma \longrightarrow \Delta, s \subset_H q . \supset A(s) & \vdots \\
 & \overline{\qquad\qquad\qquad\qquad\qquad\qquad\qquad\qquad} & \\
 & q \subset_H t, \Gamma \longrightarrow \Delta, (x) \subset_{H^q} A(x) & S^q \\
 & \overline{\overline{\qquad\qquad\qquad\qquad\qquad\qquad\qquad\qquad\qquad\qquad}} & \\
 & q \subset_H t, \Gamma \longrightarrow \Delta, A(q) &
\end{array}$$

Here P_o, P_S, P^q_S, S and S^q have the same meaning as in the definition of Ti_2-reduction step in section 4.2. We say that the resulting proof P' has been obtained from P by means of a TI-reduction step, and we also say that this reduction step has been applied to the given $TI(P_1,P_2,m)$ inference.

C. Next, we want to associate ordinals with proofs P in ZTE/II. To start with, let $D(\alpha^{i_1}_{u_1},\ldots\ldots,\alpha^{i_s}_{u_s},x)$ be a standard formula (that is of the form $R(\alpha^{i_1}_{u_1},\ldots\ldots,\alpha^{i_s}_{u_s},x) \wedge seq(x))$ whose special function constants are precisely those indicated and whose only free variable is x. Let $\alpha_1,\ldots,\alpha_s$ be pairwise distinct function variables free for $\alpha^{i_1}_{u_1},\ldots\ldots,\alpha^{i_s}_{u_s}$ and let $G(\alpha_1,\ldots,\alpha_s,x)$ be the formula $D(u_1 * \alpha_1,\ldots,u_s * \alpha_s,x)$. Let $(\forall \alpha_1,\ldots\ldots,\alpha_s)\hat{W}(\subset_G)$ be true in the usual classical sense; this is classically, of course, the same as to say that $(\forall \alpha_1,\ldots\ldots,\alpha_s)W(\subset_G)$ is true. This means that, for every s-tuple of number theoretic functions $f_1,\ldots,f_s$, the set $\{n/G(f_1,\ldots\ldots,f_s,n) \text{ true}\}$ is wellordered by the relation $\{\langle n,m\rangle / n \subset_K m \text{ and } G(f_1,\ldots,f_s,n), G(f_1,\ldots\ldots,f_s,m) \text{ both true}\}$. The set $\{n/G(f_1,\ldots,f_s,n)\}$ will be denoted by $\hat{D}(f1,\ldots,fs)$, the relation $\{\langle n,m\rangle / n \subset_K m$ and $G(f_1,\dots,f_s,n), G(f_1,\dots,f_s,m)$ true$\}$ by $R_D(f_1,\dots,f_s/x,y)$. Since D is a standard formula, it follows that every $n \in \hat{D}(f_1,\dots,f_s)$ is a sequence number. Now let Q be the set of ordered pairs $\langle\langle v_1,\dots,v_s\rangle ,n\rangle$ (written more briefly as $\langle v_1,\dots,v_s/n\rangle$) whose first component is an s-tuple $v_1,\dots,v_s$ of sequence numbers v_i all having the same length (length zero thereby admitted), while the second component is an arbitrary natural number. We remind at this place that $\langle n,m\rangle = (n+m)^2+3n+m$ and $\langle n_1,\dots,n_s\rangle = \langle\langle n_1,\dots,n_{s-1}\rangle ,n_s\rangle$; the elements of Q in particular are themselves natural numbers. By Q_D we denote the subset of Q which is defined as follows: $\langle v_1,\dots,v_s/n\rangle$ Q_D iff $n \in \hat{D}(v_1*f_1,\dots,v_s*f_s)$ for every choice $f_1,\dots,f_s$ of numbertheoretic functions. Now we are going to define a partial ordering $\tilde{L}_D$ of the elements of Q_D . We put $\langle v_1,\dots,v_s/n\rangle \tilde{L}_D \langle w_1,\dots,w_s/m\rangle$ if and only if the following holds: 1) $\langle v_1,\dots,v_s/n\rangle$ and $\langle w_1,\dots,w_s/m\rangle$ are both in Q_D ; 2) $v_i \subseteqq_K w_i$ for all $i \leq s$; 3) $R_D(w_1*f_1,\dots,w_s*f_s/n,m)$ holds for all s-tuples $f_1,\dots,f_s$ of numbertheoretic functions. The so defined relation $\tilde{L}_D$ is a wellfounded partial ordering; we omit the easy verification of this statement. From the partial ordering $\tilde{L}_D$ we now pass to a total ordering L_D of Q_D . To this end, we note that in view of the wellfoundedness of L_D there is a mapping φ which associates with every element $e \in Q_D$ an ordinal $\varphi(e)$ in such a way that the following holds: if eL_De' holds, then $\varphi(e)$ is smaller than $\varphi(e')$. Now we define a relation L_D as follows: 1) if eL_De' then $e,e' \in Q_D$; 2) if e and e' are in Q_D and if $\varphi(e)$ is smaller than $\varphi(e')$, then eL_De' ; 3) if e and e' are in Q_D , if $\varphi(e) = \varphi(e')$ and if $e < e'$, then eL_De' . The relation L_D is a wellfounded, total ordering of Q_D , as is easy to verify. Therefore we can associate with every $e \in Q_D$ inductively an ordinal $\|e\|$ in the following way: $\|e\|$ is the smallest ordinal greater than all ordinals $\|e'\|$ for which $e'L_De$ holds. Finally we can also associate with the relation L_D itself an ordinal, to be denoted by $\|L_D\|$: it is the smallest ordinal greater than all ordinals $\|e\|$, $e \in Q_D$.

So, whenever we are given a formula $\mathfrak{A}(\subset_D)$, with D as above and such that $(\forall \alpha_1, \dots, \alpha_s)\mathfrak{A}(\subset_D)$ is true, then we can associate with this formula the wellordering L_D of Q_D as described above.

Now let, conversely, $D(\alpha^{i_1}_{u_1}, \dots, \alpha^{i_s}_{u_s}, x)$ be a standard formula (denoted more briefly by D) whose special function constants are precisely those indicated and whose only free variable is x . Assume that we have a proof P in ZTi/II of $\longrightarrow \mathfrak{A}(\subset_D)$. Now let $\alpha_1, \dots, \alpha_s$ be suitably chosen pairwise distinct function variables. Then by replacing every occurence of $\alpha^{i_k}_{u_k}$ in P by $u_k * \alpha_k$ we get a proof P' of $\longrightarrow \mathfrak{A}(\subset_G)$ where $G(\alpha_1, \dots, \alpha_s, x)$ is $D(u_1 * \alpha_1, \dots, u_s * \alpha_s, x)$. If there are other special function constants which occur in P', we replace them by suitably chosen constants for primitive recursive functions, obtaining thus a proof P" in ZTi/II of $\longrightarrow \mathfrak{A}(\subset_G)$ which does not contain special function constants at all. This means that we can associate with $\mathfrak{A}(\subset_G)$ the set Q_D and the wellordering L_D of Q_D which we have described above.

<u>Definition 27:</u> Let $D(\alpha^{i_1}_{u_1}, \dots, \alpha^{i_s}_{u_s}, x)$ be a standard formula, containing precisely $\alpha^{i_1}_{u_1}, \dots, \alpha^{i_s}_{u_s}$ as distinct special function constants, and whose only free variable is x . Let P be a strictly normal proof in ZTi/II of $\longrightarrow \mathfrak{A}(\subset_D)$. Then we call the wellordering L_D described above the wellordering induced by P ; Q_D is called the domain of L_D and $\|e\|$ (for $e \in Q_D$) and $\|L_D\|$ have the meaning described above.

After these preliminaries we are ready to associate ordinals with proofs in ZTE/II .

Definition 28: By Ω we denote the smallest ordinal λ having the following property: for any proof P in ZTi/II of $\longrightarrow \mathfrak{W}(\subset_D)$ (with D as in definition 27) the relation $\|L_D\| < \lambda$ holds.

Now let P be a fixed proof in ZTE/II . With each sequent S in P we associate a certain ordinal, to be denoted by o(S) . If S is an axiom of P, then o(S)=1 . If S is the conclusion of a conversion or a one-premiss structural rule S'/S, then o(S)=o(S') . If S is the conclusion of a one-premiss logical inference S'/S, or a two-premiss logical inference S',S"/S, then we put $o(S') \# 1 = o(S)$ in the first case and $o(S)=o(S') \# o(S") \# 1$ in the second case. If S is the conclusion of an induction S'/S, then we put $o(S)= \omega_d(\omega . o(S'))$ where d=h(S')-h(S) . If S is the conclusion of a cut S',S"/S then we put $o(S)= \omega_d(o(S') \# o(S"))$ where d=h(S')-h(S) . It remains to describe the ordinal assignement in the case where S is the conclusion of a II-, $TI(P_1)$- or $TI(P_1,P_2,m)$-inference S'/S respectively.

Case a): S'/S is a II-inference. Then we put $o(S)= \omega_d((o(S') \# \omega^{\Omega+1})\, \omega^{\Omega+1})$. Case b): S'/S is a $TI(P_1)$-inference, say

$$TI(P_1) \quad \frac{G(y),(x) \subset_G y A(x), \Gamma \longrightarrow \Delta, A(y)}{G(q), \Gamma \longrightarrow \Delta, A(q)}$$

Let P_1 be a proof of $\longrightarrow \mathfrak{W}(\subset_D)$, where D is the formula $D(\alpha^{i_1}_{u_1}, \ldots, \alpha^{i_s}_{u_s}, x)$ and G the formula

$D(\alpha^{i_1}_{u_1*v_1},\ldots,\alpha^{i_s}_{u_s*v_x},x)$ for some list $v_1,\ldots,v_s$ of sequence numbers all having the same length. Then we put $o(S)=\omega_d((o(S')\#\omega^{\alpha+1})\,\omega^{\alpha+1})$ where $\alpha=\|L_D\|$ and $d=h(S')-h(S)$. <u>Case c):</u> S'/S is a $TI(P_1,P_2,m)$-inference, say

$$TI(P_1,P_2,m)\qquad \frac{y\subset_H t,(x)\subset_H y A(x),\Gamma\longrightarrow\Delta,A(y)}{q\subset_H t,\Gamma\longrightarrow\Delta,A(q)}$$

Here P_1 is a proof (in ZTi/II) of $\longrightarrow \mathfrak{W}(\subset_D)$, where D is a standard formula $D(\alpha^{i_1}_{u_1},\ldots,\alpha^{i_s}_{u_s},x)$, containing precisely $\alpha^{i_1}_{u_1},\ldots,\alpha^{i_s}_{u_s}$ as distinct special function constants, and whose only free variable is x. P_2 in its turn is a proof of $\longrightarrow G(t)$ where G is the formula $D(\alpha^{i_1}_{u_1*v_1},\ldots,\alpha^{i_s}_{u_s*v_s},x)$, while $v_1,\ldots,v_s$ is a list of sequence numbers all having the same length. t is by definition saturated and has value m. Clearly, $(\forall\alpha_1,\ldots,\alpha_s)D(u_1*v_1*\alpha_1,\ldots,u_s*v_s*\alpha_s,m)$ is a true formula, hence $\langle v_1,\ldots,v_s/m\rangle$ an element of Q_D. We put $o(S)=\omega_d((o(S')\#\omega^{\beta+1})\quad{}^{+1})$ where $\beta=\|\langle v_1,\ldots,v_s/m\rangle\|$ and $d=h(S')-h(S)$. This concludes our definition of ordinal assignement. As ordinal of a proof P we take as usual the ordinal of its endsequent.

<u>D.</u> From now on we can apply to ZTE/II, and in particular to ZTEi/II, essentially the same treatment as to ZTE/II_N and $ZTEi/II_N$, respectively. We do not consider the details of this treatment, since this would amount to a mere repetition of the considerations contained in the sections 4.3. up to 4.5. In particular, theorems 23-26 remain true for ZTEi/II and hence for ZTi/II without any changes. The same can be said about the proofs of these theorems which depend essentially on the wellfoundedness of two relations $\check{R}$

and $\check{L}$, whose definitions are, of course, copies of the definitions of R and L given in section 4.5. and which behave in any respect like R and L .

4.7. Some remarks on the proof theoretic treatment of $ZTEi/II_N$ and $ZTEi/II$

Most of the results mentioned in this section will not be proved; none of the proofs omitted requires a new technique or a new mathematical idea but all of them are rather lengthy if done in detail. For these reasons we prefer to call the results mentioned in this section (apart from some exceptions) statements rather than theorems.

A. To start with, let us look at $ZTEi/II_N$ and its proof theoretic treatment presented in sections 4.1. - 4.5. An easy inspection of the arguments presented in these sections shows that they can be formalized in full Zermelo-Fränkel set theory (to be denoted by ZF). Immediately the question comes up whether the content of 4.1. - 4.5. can already be formalized in ZF^- , that is, the theory obtained from ZF by omitting the powerset axiom. Now a second inspection shows that we used at some central places the assumption that, if $\longrightarrow \mathfrak{W}(\subset_D)$ has been proved in ZTi/II_N, then $\mathfrak{W}(\subset_D)$ is true; below we will refer to this assumption as assumption (A). On the other hand, we know that ZTi/II has proof theoretically the same strength as ZT/II, and that ZT/II in turn is as strong as classical analysis, that is as ZF^- . This makes it very plausible that already ZT/II_N and hence ZTi/II_N has proof theoretically the same strength as ZF^- . Now the author has learned from H. Friedmann that this is indeed the case. So assumption (A) is evidently not provable in ZF^- , as some routine Goedel arguments show. However, by refining the reasoning presented in sections 4.1. - 4.5. slightly, it is possible to reduce $ZTEi/II_N$ to ZF^- . To this end, let us denote by $ZTEi_n/II_N$ the subtheory of $ZTEi/II_N$ which we obtain by restricting our attention only to proofs of degree n , that is, $ZTEi_n/II_N \vdash S$, if and only if there is a proof P in $ZTEi/II_N$ whose degree is n and whose endsequent is S . Similarly, ZTi_n/II_N is related to ZTi/II_N as $ZTEi_n/II_N$ to $ZTEi/II_N$. Let us denote by (A_n) the following assumption: if $ZTi/II_N \vdash \longrightarrow \mathfrak{W}(\subset_D)$ then $\mathfrak{W}(\subset_D)$ is true. The relation between the theories ZTi_n/II_N and ZF^- is described by the following

Statement I: For each n we can prove a (suitably formalized version of) the hypothesis (A_n) in ZF^-.

Although the proof of this statement is routine and does not involve difficulties of particular interest, it is quite long and hence we omit it. The next step consists in relativizing the content of sections 4.1. - 4.5. to the theories ZTi_n/II_N and $ZTEi_n/II_N$. In particular, we replace the ordinal Ω by the ordinals Ω_n whose definition is as follows: Ω_n is the smallest ordinal ξ for which $\|R_D\| < \xi$ holds whenever $\longrightarrow \mathfrak{W}(\subset_D)$ has been proved in ZTi_n/II_N. Furthermore, we replace the relations R and L introduced by definitions 22, 23 by corresponding relations R^n and L^n, respectively, whose definition is as follows: R^n and L^n are the restrictions of R and L, respectively, to proofs P in $ZTEi/II_N$ having degree n. Then, making use of statement I, one can show that for each fixed n we can translate the relativizations of sections 4.1. - 4.5. to $ZTEi/II_N$ into ZF^-. As a result one obtains the following

Statement II: For each n we can prove in ZF^- the wellfoundedness of R^n and L^n respectively.

If we refine the proofs of the above two statements somewhat, then we get a still sharper result, namely

Statement III: a) For each n we can prove hypothesis (A_n) in ZT/II_N, b) for each n we can prove the wellfoundedness of R^n and L^n, respectively, in ZT/II_N.

What has been done for $ZTEi/II_N$ and ZTi/II_N can, of course, be done in the same way for $ZTEi/II$ and ZTi/II, respectively. That is, if we work out for $ZTEi/II$ and ZTi/II the program outlined above, then we obtain a statement IV which corresponds to the conjunction of statements I and II. In order to formulate it, let ZTi_n/II and $ZTEi_n/II$ be the subsystems obtained from ZTi/II and $ZTEi/II$, respectively, by restricting attention to proofs of degree n; let $\check{R}^n$ and $\check{L}^n$ be the restrictions of $\check{R}$ and $\check{L}$, respectively, to proofs of degree n; and let finally $(\check{A}_n)$ be the following hypothesis: if $\longrightarrow \mathfrak{W}(\subset_D)$ has been proved in ZTi_n/II, then $\mathfrak{W}(\subset_D)$ is true. Then we have

Statement IV: a) For each n we can prove in ZF^- a suitably formalized version of the hypothesis $(\check{A}_n)$; b) for each n one can prove in ZF^- the wellfoundedness of $\check{R}^n$ and $\check{L}^n$ respectively.

By using a similar refinement as that one which leads from statements I and II to statement III, one obtains a corresponding

Statement V: a) For each n one can prove in ZT/II a suitably formalized version of the hypothesis $(\check{A}_n)$; b) for each n one can prove in ZT/II the wellfoundedness of $\check{R}_n$ and $\check{L}_n$, respectively. The most important of these results is part b) of statement V. Another, more elegant way of obtaining part b) of statement V is to use a result which has been communicated to the author by G. Kreisel and which seems to be contained implicitely in several papers. In order to state this result, let ZT/CA be that version of second-order analysis which we obtain by adding to ZT all instances of the following form of the comprehension axiom:

$$\longrightarrow (\forall \vec{\alpha})(E\beta)(x)(\beta(x)=0 \longleftrightarrow A(\vec{\alpha},x))$$

(where $\vec{\alpha}$ is a list $\alpha_1, \ldots\ldots, \alpha_s$ of function variables which may occur as parameters in A and where β does not occur free in A). This result, which will be referred to as

Statement VI, says: if a Σ^1_3-formula G without free variables is provable in ZF^- then $\longrightarrow$ G is provable in ZT/CA . As we have already mentioned in the proof of theorem 3 (section 1.5.), it follows from work of W. Howard that ZT/II is as strong as classical analysis. More precisely, he shows among others that if ZT/CA $\vdash$ S holds then ZT/II $\vdash$ S holds. By combining this with statement VI, one immediately gets

Statement VII: If G is a Σ^1_3-formula without free variables such that $ZF^- \vdash G$ holds then ZT/II $\vdash \longrightarrow$ G holds.

Now, the formalized versions of the sentences „$\check{R}^n$ is wellfounded" and „$\check{L}^n$ is wellfounded" are clearly Σ^1_3-formulas, say, P_n and Q_n, respectively, which do not contain free variables nor special function constants. By combining statement VII with part b) of statement IV we obtain

Statement VIII: For each n we have $ZT/II \vdash \longrightarrow P_n$ and $ZT/II \vdash \longrightarrow Q_n$. Since P_n and Q_n do not contain special function constants, it is clear that we obtain as an immediate conclusion of statement VIII the

Statement IX: $ZT^*/II \vdash \longrightarrow P_n$ and $ZT^*/II \vdash \longrightarrow Q_n$ hold for all n . Finally, using theorem 1) and its corollary (section 1.5. in chapter I) we obtain immediately

Statement X: $ZTi^*/II \vdash \longrightarrow \mathfrak{P}_n$ and $ZT^*i/II \vdash \longrightarrow \mathfrak{Q}_n$ hold for all n . However, this is not yet all. As we will show below, the following theorem is true.

Theorem 27: Let $R(x)$ be a prime formula, which contains x among its free variables and which does not contain special function constants. In ZT^*i/II we can prove the following sequent:

$$(\forall\beta)\neg(\forall\ y)\neg R(\bar{\beta}(y)) \longrightarrow (\forall\beta)(Ey)R(\bar{\beta}(y))$$

Before coming to the proof of this theorem, we will quickly draw some conclusions which interest us. Since these conclusions depend on the statements I - X for which we did not give proofs, we prefer to call these conclusions again "statements" instead of "theorem" or "corollary".

Statement XI: If $\varphi(x)$ is a primitive recursive function of one argument and, if $(\alpha)(Ey)\varphi(\bar{\alpha}(y))=0$ is provable in ZF^-, then $\longrightarrow (\alpha)(Ey)\varphi(\bar{\alpha}(y))=0$ is provable in ZTi^*/II.

Proof: This statement is an immediate consequence of theorem 27, theorem 1 and its corollary.

Statement XII: For all n , $ZTi^*/II \longrightarrow P_n$ and $ZTi^*/II \vdash \longrightarrow Q_n$ hold.

Proof: This is an immediate consequence of statements IV, XII and a result of Kleene, according to which every Π^1_1-statement can be brought into the form $(\alpha)(Ey)\varphi(\bar{\alpha}(y))=0$ with φ primitive recursive.

<u>Statement XIII:</u> If $\varphi(x,y)$ is a twoplace primitive recursive function and if $(x)(Ey)\varphi(x,y)=0$ is provable in ZF^-, then $\longrightarrow (x)(Ey)\varphi(x,y)=0$ is provable in ZTi*/II .

<u>Proof:</u> First, we note that $(x)(Ey)\varphi(x,y)=0$ is a very special case of a Π^1_1-statement. According to statement VI, it follows that $\longrightarrow (x)(Ey)\varphi(x,y)=0$ is provable in ZT*/II ; from theorem 1 and its corollary, it follows that $\longrightarrow (x)\neg(y)\neg\varphi(x,y)=0$ is provable in ZTi*/II . Next, let $b(x)$ be the primitive recursive function defined as follows: 1) $b(n)=0$ if n is not a sequence number; 2) $b(n)=m$ if $n=\bar{\alpha}(m)$ (in particular $b(1)=0$) . The function b is, of course, available in ZTi*/II in form of a suitable constant which we also denote by b . The defining axioms of b , which are at hand in ZTi*/II , permit us to prove $\longrightarrow b(\bar{\alpha}(y))=y$ and hence $\varphi(x,y)=0 \longrightarrow \varphi(x,b(\bar{\alpha}(y)))=0$ and $\varphi(x,b(\bar{\alpha}(y)))=0 \longrightarrow \varphi(x,y)=0$ in ZTi*/II . From the last two sequents we can derive in ZTi*/II by means of a little bit of intuitionistic predicate calculus the following sequents:

a) $(x)\neg(y)\neg\varphi(x,y)=0 \longrightarrow (x)(\beta)\neg(y)\neg\varphi(x,b(\bar{\beta}(y)))=0,$

b) $(x)(\beta)(Ey)\varphi(x,b(\bar{\beta}(y)))=0 \longrightarrow (x)(Ey)\varphi(x,y)=0.$

Since $\longrightarrow (x)\neg(y)\neg\varphi(x,y)=0$ is provable in ZTi*/II , it follows that $\longrightarrow (x)(\beta)\neg(y)\neg\varphi(x,b(\bar{\beta}(y)))=0$ is provable in ZTi*/II . From theorem 27 and another bit of intuitionistic predicate calculus, it follows that $\longrightarrow (x)(\beta)(Ey)\varphi(x,b(\bar{\beta}(y)))=0$ is provable and from b), finally, we conclude that $\longrightarrow (x)(Ey)\varphi(x,y)=0$ is provable in ZTi*/II .

From the last statement it follows that if a recursive function can be proved in ZF^- to exist, then one can "compute" its value for any given argument in the sense described in part C of section 4.5.

Before coming to the proof of theorem 27, we would like to make a last remark. As noted above, the wellfoundedness of the recursive relation is not provable in ZF^- ; however, we can prove in ZF^- the wellfoundedness of L for each fixed number n . This makes it very plausible that the ordinal associated with L is the least upper bound of the provable recursive wellorderings of ZF^- , or, what amounts to the same, that if λ is the ordinal associated with some provable recursive wellordering, then $\lambda < \|\check{L}_n\|$ for some n, where $\|\check{L}_n\|$ is the ordinal associated with the wellfounded relation $\check{L}_n$. Now this can indeed be proved. One possible way to prove this runs as follows: a) one adds to number theory ZT the rule of transfinite induction with respect to $\check{L}$, obtaining thus an extension of ZT, to be denoted by $ZT(\check{L})$; b) one proves in $ZT(\check{L})$ by transfinite induction over $\check{L}$ the following reflection principle: "if $\prec$ is a recursive linear ordering for which $\longrightarrow W(\prec)$ is provable in ZTi/II, then $\prec$ is a wellordering"; c) by using b) one constructs in $ZT(\check{L})$ a linear wellordering $\prec_o$ which is essentially the sum of all recursive linear orderings which can be proved in ZTi/II to be wellordered; d) using e.g. cut elimination methods as in [10], one proves the inequality $\|\prec_o\| \in \epsilon_\xi$ where $\|\prec_o\|$ is the ordinal of $\prec_o$, where $\xi = \|\check{L}\|$ and where ϵ_ξ is the smallest fixpoint of $\omega^x = x$ which exceeds ξ; e) using the connection between ZF^- and ZTi/II given by statement XI, one shows that, if $\lambda < \|\prec_o\|$, then $\epsilon_\lambda < \|\prec_o\|$; f) combining d) and e), we obtain $\|\prec_o\| \leq \xi$ what is essentially what we are looking for. There are other, more direct ways to prove the above statement; we do not discuss them here.

Now let us conclude with the

Proof of theorem 27: We prove a variant of the theorem which, in virtue of the relationship between wellfounded recursive trees and their corresponding Brower-Kleene partial orderings, is easily seen to be equivalent to the theorem. That is, we want to prove the following: if $D(x)$ is a quantifierfree formula, then we can prove in ZTi/II the sequent $\mathfrak{P}(\subset_D) \longrightarrow W(\subset_D)$. Instead of giving a formal derivation of this sequent, we prefer to give an informal proof; but it will be clear that this informal proof can be formalized in ZTi/II almost as it stands. We start by noting that, since D is quantifierfree, the tertium non datur holds for D. Now we assume $\mathfrak{P}(\subset_D)$. Then transfinite induction over $\subset_D$ is available in ZTi/II in the following form:

$$(y)(D(y) \wedge (x)(x \subset_D y . \supset A(x)) . \supset A(y)) . \supset (z)(D(z) \supset A(z))$$

where A may be any formula. Let us, in particular, choose for $A(x)$ the formula $W(\subset_D^x)$, where $y \subset_D^x z$ is an abbreviation for $y \subset_D z \wedge y \subset_D x \wedge z \subset_D x$. Our first aim is to prove the left side of the transfinite induction statement, that is, $(y)(D(y) \wedge (x)(x \subset_D y. \supset W(\subset_D^x)) \supset W(\subset_D^y))$. To this end, let n be any number for which $D(n)$ holds and assume that for any m with $m \subset_D n$ the statement $W(\subset_D^m)$ is true. Now let α be any numbertheoretic function; we have to find an i such that $\neg\, \alpha(i+1) \subset_D^n \alpha(i)$ is true. Such an i can be found by distinguishing a number of cases. <u>Case 1:</u> $\neg\, \alpha(0) \subset_D n$ holds. Then clearly $\neg\, \alpha(1) \subset_D \alpha(0)$ holds, since $\alpha(1) \subset_D \alpha(0)$ implies among others $\alpha(0) \subset_D n$, contradicting the assumption. <u>Case 2:</u> $\alpha(0) \subset_D n$ holds. Then by assumption $W(\subset_D^{\alpha(0)})$ is true. Let β be defined as follows: $\beta(x) = \alpha(x+1)$. Since $W(\subset_D^{\alpha(0)})$ is true, it follows that there is a j with $\neg\, \beta(j+1) \subset_D^{\alpha(0)} (j)$, and so there is a smallest k such that $\neg\, \beta(k+1) \subset_D^{\alpha(0)} \beta(k)$ holds. Now we distinguish subcases. <u>Subcase 1:</u> $\beta(k) \subset_D \alpha(0)$ holds. Then $\neg\, \beta(k+1) \subset_D^n \beta(k)$ is true since otherwise $\beta(k+1) \subset_D \beta(k)$ and therefore $\beta(k+1) \subset_D \beta(0)$ would hold, what would imply $\beta(k+1) \subset_D^{\alpha(0)} \beta(k)$, contradicting the assumption. Hence, for $i=k+1$ we have $\neg\, \alpha(i+1) \subset_D^n \alpha(i)$. <u>Subcase 2:</u> $\neg\, \beta(k) \subset_D \alpha(0)$ holds. Then k is necessarily 0, since otherwise $\neg\, \beta(k) \subset_D^{\alpha(0)} \beta(k-1)$ would hold, contradicting the minimality of k . Hence $\alpha(1) \subset_D \alpha(0)$ holds, and therefore also $\neg\, \alpha(1) \subset_D \alpha(0)$. Hence we can take $i=1$. Since n was arbitrary, we have proved $(y)((D(y) \wedge (x)(x \subset_D y. \supset W(\subset_D^x)). \supset W(\subset_D^y))$, and so we can conclude $(z)(D(z) \supset W(\subset_D^z))$. It remains to see that the latter formula implies $W(\subset_D)$. That is, given any numbertheoretic function α, we have to find an i such that $\neg\, \alpha(i+1) \subset_D \alpha(i)$ holds. Let again β denote the function defined by $\beta(x) = \alpha(x+1)$. We make a distinction of cases very similar to that one above. <u>Case 1:</u> $D(\alpha(0))$ is false. Then $\neg\, \alpha(1) \subset_D \alpha(0)$ is true, and we can put $i=1$.

<u>Case 2:</u> $D(\alpha(0))$ is true. Then there is a j such that $\neg\beta(j+1) \subset_D^{\alpha(0)} \beta(j)$ holds. Let k be the smallest number such that $\neg\beta(k+1) \subset_D^{\alpha(0)} \beta(k)$ holds.

<u>Subcase 1:</u> $\beta(k) \subset_D \alpha(0)$ is true. Then $\beta(k+1) \subset_D \beta(k)$ is false, since otherwise $\beta(k+1) \subset_D \alpha(0)$, and hence $\beta(k+1) \subset_D^{\alpha(0)} \beta(k)$ would follow, contradicting the assumption. Hence we can put $i=k+1$. <u>Subcase 2:</u> $\beta(k) \subset_D \alpha(0)$ is false. Then necessarily $k=0$, since otherwise $\beta(k) \subset_D^{\alpha(0)} \beta(k-1)$ would be false, contradicting the minimality of k. So again we can take $i=0$.

This concludes the proof.

<u>Corollary:</u> In ZTi/II the following form of Markov's principle is provable: $\longrightarrow (x)(\neg(y)\neg D(x,y) \longrightarrow (Ey)D(x,y))$ where D is quantifierfree.

<u>Proof:</u> We use the same argument as in the proof of statement XIII, that is, we use the fact that the following two sequents are provable in ZTi/II : a) $D(x,b(\bar{\beta}(y))) \longrightarrow D(x,y)$, b) $D(x,y) \longrightarrow D(x,b(\bar{\beta}(y)))$ (with b again given by $b(\bar{\beta}(y))=y$). Then we continue in the same way as in the proof of statement XIII.

<u>Corollary:</u> Theorem 1 and the above corollary remain true if we replace ZTi/II by ZTi/V .

<u>Proof:</u> An inspection of the proof of theorem 27 and its first corollary shows that we have used the rule of transfinite induction only in the form available in ZT/V .

This concludes temporarily our investigations about the theories ZTi/II and ZTi/II_N . We will encounter them again in chapter VIII.

CHAPTER V:

Transfinite induction with respect to recursive wellorderings without function parameters

5.1. A conservative extension of ZTi/IV_N

A. We recall theorem 2 in chapter I which states that for every $Q \in PR$ there is a prime formula t_Q such that: a) t_Q has exactly the same free variables and the same special function constants as Q, b) the sequents $t_Q = 0 \longrightarrow Q$, $Q \longrightarrow t_Q = 0$ and $\longrightarrow t_Q = 0 \vee t_Q = 1$ are provable in ZTi . For quantifierfree Q there is a sharper statement, namely

Theorem 2*: For every quantifierfree formula Q one effectively finds a prime formula t_Q such that a),b),c) above and the following additional property d) are satisfied: d) if $z_1,\dots,z_s$ are distinct, free number variables in Q , if $r_1,\dots,r_s$ are any terms free for $z_1,\dots,z_s$ in Q , if V is $S^{r_1\dots r_s}_{z_1\dots z_s}Q$, then $S^{r_1\dots r_s}_{z_1\dots z_s}t_Q$ is t_V .

Proof: Instead of giving the proof for the general case, we treat a particular case which makes it fully clear how to proceed in the general case. Let δ , τ and μ be fixed p.r. functions such that: 1) $\delta(x)=0$ if $x=0$ and 1 otherwise, 2) $\tau(x)=1-\delta(x)$, 3) $\mu(x,y)=\delta(x-y)$. We can assume that Q has conjunctive normal form. Let Q eg. be:
$(a_1=a_1' \vee a_2=a_2' \vee a_3 \neq a_3') \wedge (b_1=b_1' \vee b_2 \neq b_2' \vee b_3 \neq b_3')$. As term t_Q we take:
$$(\delta(a_1,a_1').\mu(a_2,a_2').\tau(\mu(a_3,a_3'))+\mu(b_1,b_1').\tau(\mu(b_2,b_2')).\tau(\mu(b_3,b_3')))$$
. The proof that t_Q has the properties a),b),c) above is an easy exercise in formalized primitive recursive function theory and the proof that d) holds is evident from the construction of t_Q .

If, in particular, R is a standard formula $R_o(x) \wedge seq(x)$, with $R_o(x)$ quantifierfree, then $x \subset_K y \wedge R(x) \wedge R(y)$ and $\neg(x \subset_K y \wedge R(x) \wedge R(y))$ both are quantifierfree, and we effectively find terms $p_R(x,y)=0$ and $q_R(x,y)=0$ such that the above sequents are provable in ZTi , once with $x \subset_R y$ and $p_R(x,y)$ in place of Q and t_Q, respectively, and once with $\neg x \subset_R y$ and $q_R(x,y)$ in place of t_Q and t , respectively. The two formulas $p_R(x,y)=0$ and

$q_R(x,y)=0$ which are welldetermined by R have been denoted by $x <_R y$ and $x \nless_R y$ respectively (chapter I) . By $W'(\prec_R)$ we have denoted the formula $(\alpha)(Ex)\ (x+1) \nless_R \alpha(x)$. In the sequel we also use $(x) <_{R^y} A(x)$ as abbreviation for $(x)(x <_R y \supset A(x))$. In order to state a corollary of theorem 2* we introduce the following

<u>Definition 29:</u> A quantifierfree formula R without free variables is called saturated if, for every prime formula p=q occuring in R, both p and q are saturated. A saturated prime formula p=q is by definition true or false according to whether $|p| = |q|$ or $|p| \neq |q|$. Based on truth and falsity of prime formulas, we associate in an obvious way a truth value ("true" or "false") with every saturated quantifierfree formula R by interpreting the propositional connectives in the usual way.

The proof of the following corollary of theorem 2* can easily be obtained either via theorem 2* or by using directly the construction of t_Q outlined in the proof of theorem 2*.

<u>Corollary:</u> a) Let R(x) be quantifierfree and not contain function parameters. Then $t_R(y)=0$, $x <_R y$, $x \nless_R y$ do not contain function parameters. b) Let R(x) be as before but with x as its only free variable. If $p <_R q$ is saturated, then R(p) and R(q) are saturated, and conversely. If $p <_R q$ is saturated and true, then $p \subset_R q$ is saturated and true, and conversely. Similarly with $t_R(p)=0$ and R(p) . c) If $W'(<_R)$ is saturated, then there is effectively a quantifierfree Q(x) not containing free function variables other than x nor special function constants such that: 1) R(x), $t_R(x)$ and $W'(<_R)$ are isomorphic with Q(x), $t_Q(x)=0$ and $W'(<_Q)$ respectively; 2) $W'(<_Q)$ does not contain free variables nor special function constants.

The system ZT/IV is obtained from ZTi by adding to it the rule

$$\text{IV} \qquad \frac{t_R(y)=0,\ (x) <_{R^y} A(x),\ \Gamma \longrightarrow \Delta, A(y)}{W'(<_R),\ t_R(q)=0\ ,\ \Gamma \longrightarrow \Delta\ , A(q)}$$

where q and y are subject to the usual stipulations. The system ZT/IV_N is obtained by restricting the above rule to the case where

R does not contain function parameters. In virtue of theorem 2* $\sphericalangle_R$ and t_R do not contain special function constants either. The intuitionistic versions of ZT/IV and ZT/IV_N are denoted by ZTi/IV and ZTi/IV_N respectively. In the sequel we are mainly concerned with ZTi/IV_N .

<u>B.</u> In what follows we introduce a certain conservative extension $ZTFi/IV_N$ of ZTi/IV_N . This extension is known if we know what its proofs are. This will be done by introducing certain proof trees, called <u>intuitionistic proofs of type (m,n)</u> . They are defined inductively by means of the clauses I, II below.

<u>I.</u> P is an intuitionistic proof of type (m.0) if and only if it is a proof (-tree) in ZTi/IV_N , whose formulas contain at most m logical symbols.

<u>II.</u> Assume that for all $s \leq i$ and all m we know what proofs of type (m,s) are. Intuitionistic proof trees of type (m,i+1) and their nodes are defined inductively by means of the clauses 1) - 5) below. 1) If S is an axiom of ZTi/IV_N containing only formulas with at most m logical symbols, then S is an intuitionistic proof P of type (m,i+1) . The only node of P is S . 2) Let P be an intuitionistic proof of type (m,i+1) and S' its endsequent; let S be a sequent whose formulas do not contain more than m logical symbols and which contains at most one formula in the succedent. The tree

$$\begin{array}{c} P \\ \vdots \\ \underline{S'} \\ S \end{array}$$

denoted by P' , is said to be an intuitionistic proof of type (m,i+1) in any of the following cases: a) S'/S is a conversion; b) S'/S is a one-premiss structural inference; c) S'/S is a one-premiss logical inference; d) S'/S is an induction; e) S'/S is a IV_N-inference. A sequent S* is a node of P' if it is a node of P or if it is S . 3) Let P_1, P_2 be intuitionistic proofs of type (m,i+1) and S_1,S_2 its respective endsequents. Let S be a sequent whose formulas do not contain more than m logical symbols and which contains at most one formula in the succedent. The tree

$$\begin{array}{ccc} P_1 & & P_2 \\ \vdots & & \vdots \\ S_1 & & S_2 \\ \hline & S & \end{array}$$

to be denoted by P' is said to be an intuitionistic proof of type $(m,i+1)$ in any of the following cases: a) $S_1, S_2/S$ is a cut, b) $S_1, S_2/S$ is a two-premiss logical inference. A sequent S^* is said to be a node of P' if it is a node of P_1 or P_2 or if it is S. 4) Let P be an intuitionistic proof of type $(m,i+1)$ of $t_R(y)=0,\ (x) <_R y A(x),\ \Gamma \longrightarrow A(y)$ and let P_1 be an intuitionistic proof of type (m,i) of $\longrightarrow W'(<_R)$ where $W'(<_R)$ does not contain free variables or special function constants. The tree

$$T(P_1) \qquad \frac{\begin{array}{c} P \\ \vdots \\ t_R(y)=0,\ (x) <_R y A(x),\ \Gamma \longrightarrow A(y) \end{array}}{t_R(q)=0,\ \Gamma \longrightarrow A(q)}$$

to be denoted by P' is an intuitionistic proof of type $(m,i+1)$. A sequent S^* is said to be a node of P' if it is a node of P or if it is the sequent $t_R(q)=0,\ \Gamma \longrightarrow A(q)$. This sequent is said to follow from the premiss $t_R(y)=0,\ (x) <_R y A(x),\ \Gamma \longrightarrow A(y)$ by means of a $T(P_1)$-inference. The term q and the variable y are subject to the usual stipulations. 5) Let P be an intuitionistic proof of type $(m,i+1)$ of $x <_R t,\ (x) <_R y A(x),\ \Gamma \longrightarrow A(y)$ where t is saturated with value a. Let P and $W'(<_R)$ be as in clause 4) and assume that $t_R(y)=0$ is true. The tree

$$T(P_1,a) \qquad \frac{\begin{array}{c} P \\ \vdots \\ y <_R t,\ (x) <_R y A(x),\ \Gamma \longrightarrow A(y) \end{array}}{q <_R t,\ \Gamma \longrightarrow A(q)}$$

to be denoted by P' is an intuitionistic proof of type $(m,i+1)$. A sequent S^* is said to be a node of P' if it is a node of P or if it is $q <_R t, \vdash\!\longrightarrow A(q)$. The latter sequent is said to follow from the premiss $y <_R t, (x) <_{R^y} A(x), \vdash\!\longrightarrow A(y)$ by means of a $T(P_1,a)$-inference. The term q and the variable y are subject to the usual stipulations.

Remarks and definitions. a) The formula $W'(<_R)$ in clauses 4),5) of II does by definition not contain special function constants nor free variables. Since $x \not<_R y$ contains in virtue of theorem 2* the same free variables and special function constants as $x \subset_K y \wedge R(x) \wedge R(y)$, it follows that the only free variable in $R(x)$ is x and that $R(x)$ does not contain special function constants. $R(x)$ is thus automatically a formula without function parameters. b) Since $R(x)$ contains no special function constants and has x as only free variable, the same is true for $t_R(x)$; hence $t_R(a)$ in clause 5) is automatically saturated and the value is 1 or 0. The assumption $t_R(a)=0$ true thus implies that a belongs to the domain of the partial ordering $\subset_R$. c) The proof P_1 which appears in the clauses 4),5) above is said to be the side proof of the $T(P_1)$- and the $T(P_1,a)$-inference respectively. We also call P_1 a side proof of the proof tree P' in which the $T(P_1)$- and $T(P_1,a)$-inference respectively occur.

Definition 30: A sequent S is said to be provable in $ZTFi/IV_N$ if there is an intuitionistic proof of type (m,i) (for some m,i) having S as endsequent. In this case we write $ZTFi/IV_N \vdash S$.

For technical purposes we also need the notion of classical proof of type (m,i). Its inductive definition is given by clauses I*, II* below.

I*. P is a classical proof of type $(m,0)$ if it is a proof in ZT/IV_N whose formulas contain at most m logical symbols.

II*. Assume that for all m and all $s \leq i$ we know what a classical proof of type (m,s) is. Classical proof trees of type $(m,i+1)$ and their nodes are defined inductively by means of clauses 1*)-5*) where 1*)-5*) follow from 1)-5) by means of the following modifications: a) the proof P in 1),2) is assumed to be a classical proof of type $(m,i+1)$ and S is allowed to contain more than one formula in the succedent; b) the proofs P_1, P_2 in 3) are assumed to be classical proofs of type $(m,i+1)$ and S is allowed to contain more than one formula in the conclusion; c) the proof P in 4) is assumed to be a classical proof of type $(m,i+1)$ with endsequent $t_R(y)=0,\ (x) <_R y A(x),\ \Gamma \longrightarrow \Delta, A(y)$; the proof P_1, however, is still an intuitionistic proof of type (m,i) of $\longrightarrow W'(<_R)$, and as conclusion of the (classical) $T(P_1)$-inference we take $t_R(q)=0,\ \Gamma \longrightarrow \Delta, A(q)$; d) to 5) we apply the same modifications as to 4), described in c) .

The remarks made above in connection with intuitionistic proofs of type (m,i) apply essentially also to classical proofs of type (m,i) .

There is a more compact, but slightly less precise way to define the system $ZTFi/IV_N$. That is, we can obtain $ZTFi/IV_N$ by adding to ZTi/IV_N two rules to be defined below. The first of these is given as follows: if P is a proof in $ZTFi/IV_N$ already at hand, whose endsequent is $\longrightarrow W'(<_R)$ with $W'(<_R)$ not containing free variables nor special function constants, then we can infer from the premiss $t_R(y)=0,\ (x) <_R y A(x),\ \Gamma \longrightarrow A(y)$ the conclusion $t_R(q)=0,\ \Gamma \longrightarrow A(q)^R$. Written more symbolically this rule looks as follows:

$$T(P_1) \quad \frac{t_R(y)=0,\ (x) <_R y A(x),\ \Gamma \longrightarrow A(y)}{t_R(q)=0,\ \Gamma \longrightarrow A(q)}$$

where y and q are subject to the usual stipulations. The rule is called $T(P_1)$-rule and a special application of it $T(P_1)$-inference. P_1 is called side proof of the inference. The second rule is defined similarly. Let P_1 be a proof in $ZTFi/IV_N$ already at hand of $\longrightarrow W'(<_R)$; let $W'(<_R)$ be as before. Let t be a saturated term with value a such that $t_R(a)=0$ is true. Then we are allowed to infer from the premiss $y <_R t,\ (x) <_R y A(x),\ \Gamma \longrightarrow A(y)$ the conclusion $q <_R t,\ \Gamma \longrightarrow A(q)$. More formally, the rule is written as follows:

$$T(P_1, a) \quad \frac{y <_R t,\ (x) <_R y A(x),\ \Gamma \longrightarrow A(y)}{q <_R t,\ \Gamma \longrightarrow A(q)}$$

The rule is called $T(P_1,a)$-rule, a particular application of it $T(P_1,a)$-inference. P_1 is called side proof of this inference. This new definition of $ZTFi/IV_N$ is equivalent to the old one, as is easily established, although we lose in this way the notion of type of a proof. Correspondingly, we get back to the notion of classical proof of type (m,i) for some m,i by generalizing the above rules as follows: in the first case we allow premiss and conclusion to be of the form $t_R(y)=0,\ (x)<_R y A(x),\ \Gamma \longrightarrow \Delta, A(y)$ and $t_R(q)=0,\ \Gamma \longrightarrow \Delta, A(q)$, in the second case we allow them to be of the form $y <_R t,\ (x) <_R y A(x),\ \Gamma \longrightarrow \Delta, A(y)$ and $q <_R t,\ \Gamma \longrightarrow \Delta, A(q)$ respectively. In both cases, however, P_1 must still be a proof in $ZTFi/IV_N$. The system so obtained is again ZTF/IV_N.

C. Simple properties of $ZTFi/IV_N$ and ZTF/IV_N are given by the following

Lemma 12: An intuitionistic proof of type (m,i) is also an intuitionistic proof of type (m',i') for $m \leq m'$, $i \leq i'$. Similarly with classical proofs.

The proof is by induction with respect to i and is omitted in view of its triviality. The fact that $ZTFi/IV_N$ is a conservative extension of ZTi/IV_N is given by

Theorem 28: a) An intuitionistic proof of type (m,i) can be transformed effectively into a proof P' in ZTi/IV_N of order $2m$, having the same endsequent as P. b) Similarly with classical proofs.

Proof: We merely sketch the proof. One starts with a) and proceeds by induction with respect to i. If $i=0$, the statement is trivially true. If P has type $(m,i+1)$, then all its side proofs have type (m,i) and the induction hypothesis applies to them. Then we proceed essentially in the same way as in the proof of theorem 14. In order to prove b) we use a), and proceed then essentially in the same way as in the proof of thm. 14.

5.2. Reduction steps

A. For proofs (intuitionistic or classical) of type (m,i) we can introduce all the syntactical notions introduced in earlier cases. So we have the notion of final part, normal proof, strictly normal proof and standard proof. Their definitions parallel the definitions of the corresponding notions for ZTE/II_N in chapter IV. Moreover, we can associate a number, called complexity, with every cut, induction, IV_N-inference, $T(P_1)$- and $T(P_1,a)$-inference. The definition is exactly the same as in the case of ZTE/II_N. With the aid of this complexity we can associate with every sequent S in P another natural number, called its height and denoted by $h(S)$. The definition of height is of course the same as in all previous cases. An inference other than a conversion or structural rule is again called critical if its conclusion belongs to the final part. The notion of fork and of cut associated with a given fork I_1, I_2, I_3 is introduced in the usual way. Moreover, basic lemmas I and II remain true and there proofs remain the same. There is a variant of basic lemma I, which reads as follows:

Basic lemma I: Let P be a strictly normal proof in $ZTFi/IV_N$ of type (m,i). Assume that no thinning occurs in the final part and that its endsequent has the form $\longrightarrow A$. Let $S_1, \ldots\ldots, S_n$ be the uppermost sequents of the final part, listed from left to right; let S_j be $\Gamma_j \longrightarrow A_j$. Then: 1) for $j < n$ there is a strictly normal intuitionistic proof P_j of type (m,i) whose endsequent is $\longrightarrow A_j$; 2) for $j \leq n$, if B occurs in Γ_j, then there is a strictly normal proof P' of type (m,i) of $\longrightarrow B$.

Proof: Take the subproofs P and P' provided by the construction described in the proof of basic lemma II.

Below, after having introduced ordinals, we will formulate a sharpening of basic lemma II, which corresponds to the variant of basic lemma II mentioned in section 4.4.

B. We start by introducing reduction steps for intuitionistic proofs of type (m,i). Their definition is up to one minor point the same as in all previous cases. That is, we have preliminary reduction steps, intuitionistic logical reduction steps (definition 16) and induction reductions. They are defined in the same way as before. Next we have, what we call T_1- and T_2-reduction steps. Their definition parallels that one of TI- and TI_a-reduction steps.

<u>T_1-reduction steps.</u> Let P be an intuitionistic proof of type (m,i) containing a critical $T(P_1)$-inference, say

$$T(P_1) \qquad \frac{t_R(y)=0\ ,\ (x) <_{Ry} A(x),\ \Gamma \longrightarrow A(y)}{t_R(q)=0\ ,\ \Gamma \longrightarrow A(q)}$$

Let q be saturated with value, say, a . Then $t_R(q)$ is saturated. <u>Case 1:</u> $t_R(q)$ has value 1 . Then $t_R(q)=0 \longrightarrow$ is an axiom and so we can derive the conclusion of the above $T(P_1)$-inference by thinning and interchange from this axiom. <u>Case 2:</u> $t_R(q)$ has value 0 . Let S and S' be premiss and conclusion of the above $T(P_1)$-inference. Let P_S and $P_{S'}$ be their respective subproofs. Let P^q_S be the result of replacing every occurence of y in P_S by q ; let S^q be the endsequent of P^q_S . In virtue of the assumption D in chapter I, the sequent $y <_R q \longrightarrow t_R(y)=0$ is an axiom of ZTi . We replace $P_{S'}$ in P by the following derivation:

$$\begin{array}{ll}
 & \qquad\qquad P_S \\
 & \qquad\qquad \vdots \\
y <_R q \longrightarrow t_R(y)=0 & \qquad\qquad S \\
\hline
y <_R q,\ (x) <_{Ry} A(x),\ \Gamma \longrightarrow A(y) & \text{cut} \\
\hline
s <_R q,\ \Gamma \longrightarrow A(q) & T(P_1) \\
\hline
\Gamma \longrightarrow s <_R q \supset A(q) & \qquad\qquad P^q_S \\
\hline
\Gamma \longrightarrow (x) <_{Rq} A(x) & \qquad\qquad S^q \\
\hline\hline
t_R(q)=0,\ \Gamma \longrightarrow A(q) & \text{cut, interchanges}
\end{array}$$

The resulting proof P' is said to follow from P by means of a T_1-reduction step. We also say that the T_1-reduction step has been applied to the particular $T(P_1)$-inference above.

<u>T_2-reduction steps.</u> Let P be an intuitionistic proof of type (m,i) which contains a critical $T(P_1,a)$-inference, say

$$T(P_1,a)\quad \frac{y <_R t\ ,\ (x)_{<_R y} A(x)\ ,\ \Gamma \longrightarrow A(y)}{q <_R t\ ,\ \Gamma \longrightarrow A(q)}$$

where t is saturated with value a. Let q be saturated with value b.

<u>Case 1:</u> $q <_R t$ is false. Then $q <_R t \longrightarrow$ is an axiom and we can derive the conclusion by thinning and interchange from this axiom. <u>Case 2:</u> $q <_R t$ is true, hence $t_R(q)=0$, that is $t_R(a)=0$ true. Let S, S' be premiss and conclusion of the $T(P_1,a)$-inference, let P_S, $P_{S'}$, P_S^q and S^q have the same meaning as before. By assumption D, the sequent $y <_R q\ ,\ q <_R t \longrightarrow y <_R t$ is an axiom. Now we replace P_S in P by the following derivation:

$$\begin{array}{ll}
 & \qquad\qquad\qquad P_S \\
 & \qquad\qquad\qquad \vdots \\
y <_R q\ ,\ q <_R t \longrightarrow y <_R t & \qquad\qquad\qquad S \\
\hline
y <_R q\ ,\ q <_R t\ ,\ (x)_{<_R y} A(x),\ \Gamma \longrightarrow A(y) & \text{cut} \\
\hline
s <_R q\ ,\ q <_R t\ ,\ \Gamma \longrightarrow A(s) & T(P_1,b) \\
\hline
q <_R t\ ,\ \Gamma \longrightarrow s <_R q \supset A(s) & \qquad\qquad\qquad\qquad \vdots\ P_S^q \\
\hline
q <_R t\ ,\ \Gamma \longrightarrow (x)_{<_R q} A(x) & \qquad\qquad\qquad\qquad S^q \\
\hline\hline
q <_R t\ ,\ \Gamma \longrightarrow A(q) & \text{cut, interchanges}
\end{array}$$

The result P' of this operation is said to follow from P by means of a T_2-reduction step. We also say that the T_2-reduction step is applied to the above $T(P_1,a)$-inference.

<u>IV_N-reduction step.</u> Let P be an intuitionistic proof of type (m,i) which contains a critical IV_N-inference, say

$$IV_N\quad \frac{t_R(y)=0,\ (x)_{<_R y} A(x)\ ,\ \Gamma \longrightarrow A(y)}{W'(<_R),\ t_R(q)=0,\ \Gamma \longrightarrow A(q)}$$

whose endsequent has the form $\longrightarrow B$ and assume that $W'(<_R)$ is saturated. From the corollary of theorem 2* it follows that there is a quantifierfree $Q(x)$ not containing special function constants, such that $R(x)$, $t_R(x)$, $W'(<_R)$ are isomorphic with $Q(x)$, $t_Q(x)$ and $W'(<_Q)$, respectively, and such that $W'(<_Q)$ does not contain free variables or special function constants. According to the variant of basic lemma I, cited in this section, one effectively can extract from P a certain proof P' of $\longrightarrow W'(<_R)$ which is again an intuitionistic proof of type (m,i) . By adding to P' a suitable conversion, we obtain in virtue of the above remarks an intuitionistic proof P of type (m,i) of $\longrightarrow W'(<_Q)$. Now we alter P as follows:

$$
\begin{array}{ll}
t_R(y)=0,\ (x) <_R y\, A(x),\ \Gamma \longrightarrow A(y) & \\
\hline
t_Q(y)=0,\ (x) <_Q y\, A(x),\ \Gamma \longrightarrow A(y) & \text{conversion} \\
\hline
t_Q(q)=0,\ \Gamma \qquad A(q) & T(P_1) \\
\hline
W'(<_R),\ t_R(q)=0,\ \Gamma \longrightarrow A(q) & \text{Thinning}
\end{array}
$$

The resulting proof so obtained is intuitionistic of type $(m,i+1)$. We say that P^* has been obtained from P by means of a IV_N-reduction step and that the IV_N-reduction step has been applied to the above IV_N-inference.

<u>Remark:</u> The side proof P_1 which appears in the definition of IV_N-reduction step is uniquely determined by the procedure described in the proof of basic lemma II and by the critical IV_N-inference, to which the reduction step is applied. We call P_1 the <u>side proof determined by the critical IV_N-inference</u>. Similarly the reduction step is entirely determined once the critical IV_N-inference is given. We call this reduction step <u>the IV_N-reduction step determined by the critical IV_N-inference</u>.

The logical reduction steps, the induction reductions, the IV_N- , T_1- and T_2-reduction steps are also called strictly essential reduction steps. The notions "substitution instance" and "inessential reduction step" are introduced in precisely the same way as in section 4.4., (def. 20) of the last chapter. The reduction steps so introduced have the same properties as the corresponding reduction steps in earlier cases. The main properties of preliminary reduction steps are again given by theorem 4. In order to describe the properties of strictly essential reduction steps, we introduce a relation W by means of the following variant of def. 14, stated in section 2.2:

Definition 31: The two place relation W applies to intuitionistic s.n.s. proofs (of some type (m,i)) iff the following holds: 1) there is a list $P_o, \dots, P_N$ of proofs such that $P_o = P$ and such that P_{i+1} follows from P_i by means of a preliminary reduction step $(i<N)$; 2) no preliminary reduction step is applicable to P_N ; 3) P' follows from P_N by means of a strictly essential reduction step.

Theorem 29: 1) W is recursive; 2) given P, there are at most finitely many P' with W(P,P') and, if so, they can be found effectively; 3) (EX)W(P,X) is decidable. The strictly essential reduction steps in turn have the properties described by theorem 6, that is, we have

Theorem 30: Let P be a saturated intuitionistic s.n.s. proof of some type different from its final part whose final part does not admit preliminary or essential reduction steps. Then there is a critical logical inference whose principal formula has an image in the endsequent.

The proof is practically the same as that of the corresponding theorem 19. Finally, we can introduce the notion of subformula reduction step in exactly the same way as in part D of section 4.4. of the preceeding chapter. Corresponding to theorem 21 we have

Theorem 31: Let P be a saturated intuitionistic s.n.s. proof of some type which does not coincide with its final part. Assume that no preliminary and no strictly essential reduction step is applicable to P . Then we can effectively apply to P a subformula reduction step. The resulting proof P* is again a strictly normal intuitionistic proof of the same type.

With respect to inessential reduction steps, the situation is the same as earlier. That is, given intuitionistic proofs P,P' of type (m,i), we can effectively decide whether P is saturated or not, and if not, we can effectively decide whether P' follows from P by means of an inessential reduction step or not.

C. Classical proofs of type (m,i) do not play an important role in our considerations. For technical reasons, we introduce two kinds of reduction steps for them: 1) preliminary reduction steps,

2) logical reduction steps (fork elimination). Their definitions are the same as usual. As described by definition 16, we can decompose an intuitionistic logical reduction step into a classical logical reduction step followed by some preliminary reduction steps. The classical logical reduction step transforms the intuitionistic proof P to which it is applied into a classical proof P' , the preliminary reduction steps transform P' back into an intuitionistic proof P". It is this fact which will be used below.

5.3. Ordinals

A. In order to associate ordinals with certain proofs in ZTF/IV_N and $ZTFi/IV_N$, we introduce two relations R* and L* whose definitions are given by definitions 22 and 23, respectively. More precisely we can use definition 22 in order to introduce a relation R*, using thereby the notion "strictly essential reduction step" in the sense defined in section 5.2. Similarly we can use definition 2.3. in order to introduce a relation L* , replacing thereby R by R* . The relations R* , L* are counterparts of R and L and have similar properties; in particular, theorem 22, part a) (with R* , L* in place of R,L) and its proof holds invariably in the present case. For simplicity, we omit the star and write R and L in place of R* and L* , without danger of confusion. Of basic importance are certain subtrees of L .

Definition 32: Let P be an intuitionistic s.n.s. proof of type (m,i) . A sequence $P_o,\ldots,P_s$ is called a P-chain in each of the following cases: 1) s=0 and $P_o=P$; 2) $s>0$, $P_o=P$ and $L(P_i,P_{i+1})$. The set D_P is defined as follows: $P'\in D_P$ iff there is a P-chain $P_o,\ldots,P_s$ such that $P'=P_s$. By L_P we denote the restriction of L to D_P .

For the sake of a brief repetition we introduce

Definition 33: a) A formula $A(\alpha^{i_1}_{u_1},\ldots,\alpha^{i_s}_{u_s})$ without free variables is true if $(\xi_1,\ldots,\xi_s)A(u_1*\xi_1,\ldots,u_s*\xi_s)$ is true. b) Let $B_1,\ldots,B_s,A$ be formulas without free variables. Then $\longrightarrow A$, $B_1,\ldots,B_s\longrightarrow A$ and $B_1,\ldots,B_s\longrightarrow$ are true iff A , $B_1\wedge B_2\wedge\ldots\ldots\wedge B_s\supset A$ and $B_1\wedge B_2\ldots\ldots\wedge B_s\supset 0=1$ respectively are true.

A basic property of L_P is described by the following
Theorem 32: Let P_o be an intuitionistic s.n.s. proof of some type whose endsequent S_o is either $\longrightarrow$ or else of the form $\longrightarrow A$, where A does not contain $\neg$, $\supset$. If L_{P_o} is wellfounded then S_o is true.

Proof: The proof is by transfinite induction over L_{P_o}. To this end we note: if P is in D_{P_o} then P is again an intuitionistic s.n.s. proof of type (m,j), $i \leq j$ (where P is of type (m,i)) whose endsequent is $\longrightarrow$ or has the form $\longrightarrow B$ where B does not contain $\neg$, $\supset$. Furthermore, it is clear that if $P \in D_{P_o}$ then L_P is also wellfounded. The transfinite induction essentially amounts to show the following: if $P \in D_{P_o}$, and if for all P' with $L(P,P')$ the endsequent S' of P' is true, then P has true endsequent S. Hence let us assume: a) $P \in D_{P_o}$, b) if $L(P,P')$ then P' has true endsequent S'. We distinguish between cases, subcases, subsubcases etc. Subcases and subsubcases are denoted by SC, SSC, etc. Case 1: P is saturated and does not admit preliminary reduction steps. SC1: P admits a strictly essential reduction step; let P' be the resulting proof and S' its endsequent. S' is either $\longrightarrow$ or $\longrightarrow A$ for some A. According to the inductive assumption S' is true; hence S' has to be $\longrightarrow A$ and so S is $\longrightarrow A$, hence true too. SC2: P does not admit an essential reduction step. Then P cannot have $\longrightarrow$ as endsequent, since this would imply that P coincide with its final part according to theorem 30; but from true saturated mathematical axioms we cannot derive $\longrightarrow$, using only cuts, interchanges inductions and conversions. Therefore the endsequent S of P must be $\longrightarrow A$ for some A, and a subformula reduction step must be applicable to P. We have to distinguish between cases according to the form of A. We content ourself by treating two of them; those left out are even easier to treat. SSC1: A is $(\xi)B(\alpha^{i}_{u}, \xi)$; for simplicity we assume that only one special function constant is present. The subformula reduction step transforms P into a proof P' of $\longrightarrow B(\alpha^i_u, \alpha^j_{<>})$ (for some $j \neq i$). Since $L(P,P')$ holds, $B(\alpha^i_u, \alpha^j_{<>})$ is true, that is, $(\eta, \xi)B(u*\eta, <>*\xi)$ is true, hence $(\eta, \xi)B(u*\eta, \xi)$ and so A are true. SSC2: A is $(x)B(\alpha^i_u, x)$. Then there is a list $P_o, P_1, \ldots$ of proofs such that 1) P_n is a proof of $\longrightarrow B(\alpha^i_u, n)$, 2) $L(P,P_n)$ holds. According to the inductive hypothesis $B(\alpha^i_u, n)$ is true for all n. That is $(\xi)B(u*\xi, n)$ is true for all n, hence $(x)(\xi)B(u*\xi, x)$ and so A is true. Case 2: P is saturated but admits preliminary reduction steps. Let $P_o, \ldots, P_N$ be a chain such that a) $P_o = P$, b) P_{i+1} follows from P_i by means of a preliminary reduction step, c) no preliminary reduction step is applicable to P_N. Obviously P_N is still saturated. If $L(P_N, P')$ then $L(P,P')$ as is easily verified. Hence $L(P_N, P')$ implies that P' has true endsequent. But then we can apply the reasoning presented under case 1 in order to conclude that P_N has true endsequent. But this implies that P and P_N have the same endsequent, hence the endsequent of P is true. Case 3: P is not saturated. Assume for simplicity that there is only one special function constant present in P, say α^i_u; in the more general case the reasoning remains exactly the same. If we replace α^i_u by α^i_{u*w}, we obtain a new proof, denoted by P_w, whose endsequent is S_w. Let τ_P be the prim. rec. continuity function associated with P according to lemma 9, the remark following it and definition 20. As before, we write $\tau_P(\xi(i)) \neq\!\!\!\neq 0$ as abbreviation for "$\tau_P(\xi(i)) \neq 0$ and $\tau_P(\xi(s)) = 0$ for all $s < i$". By definition, if $\tau_P(w) \neq\!\!\!\neq 0$, then P_w is saturated and $L(P,P_w)$. Since S_w is true according to the inductive hypothesis, it is not $\longrightarrow$. Hence S is not $\longrightarrow$ but has the form $\longrightarrow A(\alpha^i_u)$. Now: if $\tau_P(w) \neq\!\!\!\neq 0$, then $\longrightarrow A(\alpha^i_{u*w})$ is true, according to the inductive hypothesis. Hence $(\xi)A(u*w*\xi)$ is true whenever $\tau_P(w) \neq\!\!\!\neq 0$. From this one infers by barinduction over τ_P that $(\xi)A(u*\xi)$ is true; hence S is true.

B. The previous theorem gives rise to a certain subclass of s.n.s. proofs, the so called "graded proofs". This subclass is given by

Definition 34: a) An intuitionistic s.n.s. proof P is said to be "good" if its endsequent has the form $\longrightarrow A$ with A not containing $\neg$ nor $\supset$ and if in addition L_P is wellfounded. b) An intuitionistic or classical proof is said to be "graded" if all its side proofs are "good".

The following lemma is evident:

Lemma 13: A preliminary reduction step, the operation "omission of a cut" or a classical logical reduction step applied to a graded proof P yield a graded proof P'. An intuitionistic logical reduction step, an induction reduction, a T_1- or a T_2-reduction step applied to an intuitionistic graded proof P yield an intuitionistic proof P'.

The only case not covered by this lemma is that of a IV_N-reduction step whose role will become clearer below. In order to associate ordinals with graded proofs, we use some notation. If P is a good proof of $\longrightarrow W'(<_R)$, then $\|\subset_R\|$ is the ordinal associated with the partial ordering $\subset_R$, which is wellfounded according to the previous theorem; if a is in the domain of R, that is, if $R(a)$ (or what amounts to the same $t_R(a)=0$) is true, then $\|a\|_R$ denotes the ordinal associated with the restriction of $\subset_R$ to $\{x/\ x \subset_R a\}$. By Ω we denote the smallest ordinal ξ having the property: if P is a good proof of $\longrightarrow W'(<_R)$ then $\|\subset_R\| < \xi$. Ω is evidently denumberable. Now we can describe our ordinal assignement. Let P be a graded proof and S a sequent in it. With each such S we associate inductively an ordinal, to be denoted by $o(S)$. Case 1: S is an axiom of P. Then $o(S)=1$. Case 2: S is the conclusion of a conversion or a one-premiss structural rule, say S'/S. Then $o(S)=o(S')$. Case 3: S is the conclusion of a one-premiss logical inference S'/S. Then $o(S)=o(S')\#1$. Case 4: S is the conclusion of a two-premiss logical inference $S_1,S_2/S$. Then $o(S)=o(S_1)\#o(S_2\#1$. Case 5: S is conclusion of an induction S'/S. Then $o(S)=\omega_d(o(S')\omega)$ where $d=h(S')-h(S)$. Case 6: S is conclusion of a IV_N-inference S'/S. Then $o(S)=\omega_d((o(S')\#\omega^{\Omega+L})\,\omega^{\Omega+1})$ where $d=h(S')-h(S)$. Case 7: S is conclusion of a $T(P_1)$-inference S'/S, where P_1 is a proof of $\longrightarrow W'(<_R)$. Then $o(S)=\omega_d((o(S')\#\omega^{\lambda+1})\,\omega^{\lambda+1})$ where $d=h(S')-h(S)$ and $\lambda = \|\subset_R\|$. Case 8: S is the conclusion of a $T(P_1,a)$-inference S'/S. Then $o(S)=\omega_d((o(S')\#\omega^{\nu+1})\,\omega^{\nu+1})$ where $\nu = \|a\|_R$ ($t_R(a)=0$ and hence $R(a)$ are true) and $d=h(S')-h(S)$.

The ordinal of the endsequent is called the ordinal of P and denoted by $o(P)$. This assignement of ordinals has all the familiar properties of the assignements described in earlier chapters. We collect these properties by means of the following

Theorem 33: 1) The operation "omission of a cut" lowers the ordinal of a graded proof P. 2) Preliminary reduction steps do not increase the ordinal of a graded proof. 3) A classical logical reduction step lowers the ordinal of a graded proof. 4) An intuitionistic logical reduction step, applied to an intuitionistic graded proof P, lowers the ordinal of P. 5) An induction reduction, a T_1- or a T_2-reduction step, applied to an intuitionistic graded proof P, lowers the ordinal of P. 6) A subformula reduction step lowers the ordinal of an intuitionistic graded proof.

The proof of this theorem leads exactly to the same calculations as in earlier cases and is omitted. The case of a IV_N-reduction step is not covered by the above theorem since it is not clear whether a IV_N-reduction step transforms an intuitionistic graded proof always in an intuitionistic graded proof. However, the following can be said:

Theorem 34: Let P be an intuitionistic graded s.n.s. proof and assume that a IV_N-reduction step is applied to the critical IV_N-inference

$$IV_N \quad \frac{t_R(y)=0,\ (x) <_R y\, A(x),\ \Gamma \longrightarrow A(y)}{W'(<_R),\ t_R(y)=0,\ \Gamma \longrightarrow A(y)}$$

Let P_1 be the side proof determined by this inference. If P_1 is a good proof, then the IV_N-reduction step, determined by the above IV_N-inference, transforms P into an intuitionistic graded s.n.s. proof P' whose ordinal is smaller than that of P.

Proof: Let P_1 have the endsequent $\longrightarrow W'(<_Q)$ and put $\lambda = \| <_Q \|$. The reduction step looks as follows:

$$
T(P_1)\quad
\begin{array}{ll}
t_R(y)=0,\ (x) <_R y\, A(x),\ \Gamma \longrightarrow A(y) & \\
\hline
t_Q(y)=0,\ (x) <_R y\, A(x),\ \Gamma \longrightarrow A(y) & \text{conversion} \\
\hline
t_Q(q)=0,\ \Gamma \longrightarrow A(q) & \\
\hline
t_R(q)=0,\ \Gamma \longrightarrow A(q) & \text{conversion} \\
\hline
W'(<_R),\ t_R(q)=0,\ \Gamma \longrightarrow A(q) & \text{thinning}
\end{array}
$$

Let S' and S be premiss and conclusion of the IV_N-inference and α the ordinal of S' in P . The ordinal of S in P is by definition $\omega_d((\omega^\alpha \# \omega^{\Omega+1})\,\omega^{\Omega+1})$. Calculating the ordinal of S in P', we evidently obtain $\omega_d((\omega^\alpha \# \omega^{\lambda+1})\,\omega^{\lambda+1})$. Since $\lambda < \Omega$, the second ordinal is smaller than the first one what proves essentially the statement.

Below we have to use the full force of basic lemma II. There is a slightly sharpened version of basic lemma II, namely

<u>Basic lemma II_1:</u> Let P be an intuitionistic graded s.n.s. proof of type (m,j). Let $S_1,\dots,S_n$ be the uppermost sequents of the final part, listed from left to right; let S_i be $\Gamma_i \longrightarrow A_i$. Then the following holds: 1) for every $i<n$ there is an intuitionistic graded s.n.s. proof P_i of type (m,j) of $\longrightarrow A_i$, whose ordinal is smaller than that of P ; 2) for every $i \leqq n$, if B occurs in Γ_i , then there is an intuitionistic graded s.n.s. proof P' of type (m,j) of $\longrightarrow B$, whose ordinal is smaller than that of P.

<u>Proof:</u> The construction of P_i, P' respectively remains the same as in the proof of basic lemma II; the inequalities $o(P_i) < o(P)$ and $o(P') < o(P)$ follow from the fact that the operation "omission of a cut" is used in the construction of P_i and P' . An important special case of this sharpened version of basic lemma II is

<u>Corollary:</u> Let P be an intuitionistic graded s.n.s. proof of type (m,i) and S/S' a critical IV_N-inference in P . The side proof P_1 determined by this inference is again an intuitionistic graded s.n.s. proof of type (m,i) whose ordinal $o(P_1)$ is smaller than $o(P)$.

Proof: Follows immediately from basic lemma II.

5.4. The wellfoundedness proof

Theorem 35: If P is an intuitionistic graded s.n.s. proof then L_P is wellfounded.

Proof: We proceed by transfinite induction with respect to the ordinal $o(P)$ of P. Let P be an intuitionistic graded s.n.s. proof with $o(P)=\xi$ and assume that for all intuitionistic graded s.n.s. proofs P' with $o(P')=\lambda<\xi$ the relation $L_{P'}$ is wellfounded. We want to show that L_P is wellfounded and note in this connection that L_P is wellfounded iff for all P' with $L(P,P')$ $L_{P'}$ is wellfounded. Case A: We first prove the wellfoundedness of L_P under the assumption that P is saturated and does not admit preliminary reduction steps. If $L(P,P')$ then P' necessarily follows from P by means of a strictly essential reduction step or a subformula reduction step. The proof is accomplished in this case if we can show that for each such P' $L_{P'}$ is wellfounded in virtue of the inductive assumption. We distinguish two subcases. Subcase 1: P' follows from P by means of a subformula reduction step or a strictly essential reduction step other than a IV_N-reduction step. Then $o(P')<o(P)$ by theorem 33. In virtue of our inductive assumption $L_{P'}$ is wellfounded. Subcase 2: P' follows from P by means of a IV_N-reduction step. More precisely, let S/S' be a critical IV_N-inference in P and let the IV_N-reduction step in question be that one determined by this critical IV_N-inference. Let P_1 be the side proof determined by the critical IV_N-inference S/S'. In virtue of the corollary of basic lemma II_1, it follows that P_1 is an intuitionistic graded s.n.s. proof with ordinal $o(P_1)$ smaller than $o(P)$. From the inductive assumption it follows that L_{P_1} is wellfounded: hence P_1 is good. Theorem 34 now implies that[1] P' is again an intuitionistic graded s.n.s. proof, but with $o(P')<o(P)$. Hence $L_{P'}$ is wellfounded too in virtue of the inductive assumption. Subcase 1 and 2 together imply the wellfoundedness of L_P.

Case B: P is saturated but preliminary reduction steps can be applied to P. Let $L(P,P')$ hold. Then there is a chain $P_0,\ldots,P_N$ such that 1) $P=P_0$, 2) P_{i+1} follows from P_i by means of a preliminary reduction step, 3) no preliminary reduction step is applicable to P_N, 4) P_N is saturated and P' follows from P_N

by means of a strictly essential or a subformula reduction step. That is, as shown in case A, we have $o(P') < o(P_N)$. But $o(P_N) \leqq o(P)$ by theorem 33, hence $o(P') < o(P)$. That is, if $L(P,P')$ holds, then $L_{P'}$ is wellfounded in virtue of our inductive assumption; hence L_P is wellfounded. Case C: P is not saturated. If $L(P,P')$ holds, then P' is saturated by definition of L and $o(P)=o(P')$. By case B $L_{P'}$ is wellfounded. Hence L_P is wellfounded what concludes the proof.

Corollary: The relation L_P is wellfounded for every s.n.s. proof in ZTi/IV_N .

Proof: An s.n.s. proof in ZTi/IV_N is evidently an intuitionistic graded s.n.s. proof since it contains no side proofs at all.

5.5. Remarks on applications

From the last theorem, and in particular from its corollary, we could again reobtain easily theorems 23, 24 and 25 (but restricted of course to ZTi/IV_N) . However, as we will see in later chapters, the present method enables us to prove much more general results than theorems 23, 24 and 25. We will therefore postpone the discussion of applications to these later chapters.

CHAPTER VI:

A formally intuitonistic theory equivalent to classical transfinite induction with respect to recursive wellfounded trees with function parameters

In this chapter we apply a proof-theoretic treatment to the theory ZTi/V (or rather to a conservative extension of ZTi/V), which is very similar to that one presented in the last chapter. The method, however, is no more involved since ZTi/V includes two additional features: a) the formula $W'(<_R)$ which appears in the rule of transfinite induction characterizing ZTi/IV_N is now replaced by $W^o(<_R)$; b) function parameters are admitted.

6.1. Some preparations

A. Let $R(x)$ be a quantifierfree standard formula, that is, of the form $R_o(x) \wedge seq(x)$, and let $t_R(x)=0$, $x <_R y$ be the quantifier-free formulas associated with $R(x)$ and $x \sqsubset_R y$ according to theorem 2* and its corollary. Let $p_1(x),\ldots,p_n(x)$ be a list of prime formulas. Assume that x is the only free variable in $R(x)$ and $p_i(x)$, $i=1,\ldots,n$; let $\alpha_{u_1}^{i_1},\ldots,\alpha_{u_s}^{i_s}$ be the list of special function constants which occur in $R(x)$ or in at least one $p_i(x)$. In order to indicate this occurences we write sometimes more explicitely $R(\alpha_{u_1}^{i_1},\ldots,\alpha_{u_s}^{i_s},x)$, $t_R(\alpha_{u_1}^{i_1},\ldots,\alpha_{u_s}^{i_s},x)$, $p_i(\alpha_{u_1}^{i_1},\ldots,\alpha_{u_s}^{i_s},x)$ or $R(\vec{\alpha}_u,x)$, $t_R(\vec{\alpha}_u,x)$, $p_i(\vec{\alpha}_u,x)$. Therby we use the following notation: if $v_1,\ldots,v_s$ are sequence numbers and t a term, then we denote $R(\alpha_{u_1*v_1}^{i_1},\ldots\alpha_{u_s*v_s}^{i_s},t)$ more briefly by $R(\vec{\alpha}_{u*v},t)$ or even $R_v(t)$; similarly, with the p_i's and other formulas.

Now we associate with R and $p_1,\ldots,p_n$ a certain partial ordering, to be denoted by $\sqsubset$. The domain of $\sqsubset$, to be denoted by D consists of ordered pairs $\langle\langle v_1,\ldots,v_s\rangle ,d\rangle$ which satisfy the following conditions: a) $v_1,\ldots,v_s$ are sequence numbers all having the same length; b) d is a sequence number ; c) $R(\vec{\alpha}_{u*v},d)$ is saturated and true, or what amounts to the same, $t_R(\vec{\alpha}_{u*v},d)$ is saturated and its value is 0 ; d) for all $i \leq \text{length}(v_1)$ and all $k \leq n$ $p_k(\vec{\alpha}_{u*v},i)$ is either not saturated or else saturated and $|p_k(\vec{\alpha}_{u*v},i)| = 0$. Instead of $\langle\langle v_1,\ldots,v_s\rangle,d\rangle$, we write $\langle v_1,\ldots,v_s/d\rangle$. The relation $\sqsubset$, whose domain is by definition D , is now defined as follows:

$\langle v_1,\ldots,v_s/a\rangle \sqsubset \langle w_1,\ldots,w_s/b\rangle$ iff 1) each v_i is a proper extension of w_i , that is $v_i \subset_K w_i$ for $i=1,\ldots,s$; 2) a is a proper extension of b (that is $a \subset_K b$) ; 3) both $\langle v_1,\ldots,v_s/a\rangle$ and $\langle w_1,\ldots,w_s/b\rangle$ belong to D .

Notation: With R , $p_1,\ldots,p_n$ we associate the formula $(\xi)(Ex)(\neg\,\xi(x+1) \subset_R \xi(x) \vee p_1(x) \neq 0 \ldots \vee p_n(x) \neq 0)$ and denote it by $F[R,p_i;\vec{\alpha}_u]$. Then we have

Theorem 36: If $F[R,p_i;\vec{\alpha}_u]$ is true, then $\sqsubset$ is wellfounded.

Proof: In order to simplify the notation, we treat only the case where s=1 , that is where only one special function constant is present, say, α^1_u ; for simplicity, we assume $u=\langle\,\rangle$. We also assume n=1 , that is that $p_1(x)$ is the only member of the list $p_1,\ldots,p_n$; we write p in place of p_1 . By replacing α^1_u in $R(\alpha^1_u,x)$, $p(\alpha^1_u,x)$ and $x \subset_R y$ by η we get new formulas which we denote by $R(\eta,x)$, $p(\eta,x)$ and $x \subset^\eta_R y$. By assumption $(\eta)(\xi)(Ex)(\neg\,\xi(x+1) \subset^\eta_R \xi(x) \vee p(\eta,x) \neq 0)$ is true. Let g be an arbitrary number-theoretic function. We have to find an i such that $\neg\, g(i+1) \sqsubset g(i)$ holds. To this end, we introduce two functions f,h . We define f as follows: a) if for all $i \leq s+1$ $g(i) = \langle u_i/v_i\rangle \in D$ and $g(0) \sqsupset g(1) \sqsupset \ldots\ldots \sqsupset g(s+1)$

holds then $f(s)=a_s$ where a_s is the s'th component of u_{s+1}, which by necessity must have length $\geqq s+1$ and hence be of the form $u_{s+1} = \langle a_o,\ldots\ldots,a_s,\ldots\ldots\rangle$; b) if the assumption stated in a) does not hold, then $f(s)=0$. The function h is defined as follows: a) if for all $i \leqq s$ $g(i)= \langle u_i/v_i\rangle \in D$ and $g(0) \sqsupset g(1) \sqsupset \ldots\ldots \sqsupset g(s)$, then $h(s)=v_s$; b) if the assumption in a) does not hold, then $h(s)=0$. From our assumption it follows that there is an m such that I) $\neg h(m+1) \subset^f_R g(m) \vee p(f,m) \neq 0$ is true. Now we distinguish cases.

Case 1: $g(0) \sqsupset g(1) \sqsupset \ldots\ldots \sqsupset g(m+1)$ is false; then an i with $\neg g(i+1) \sqsubset g(i)$ $(i \leqq m)$ can effectively be found.

Case 2: $g(0) \sqsupset \ldots\ldots \sqsupset g(m+1)$ is true; put $g(i)= \langle u_i/v_i\rangle$ for $i \leqq m+1$. Then we can effectively determine an N so large that the following holds: 1) $N >$ length(u_{m+1}) ; 2) $R(\alpha^1_w,m)$ and $p(\alpha^1_w,m)$ are saturated where $w=\bar{f}(N)$. We claim: $g(0) \sqsupset g(1) \sqsupset \ldots\ldots \sqsupset g(N)$ is false. Assume the contrary and put $g(i)= \langle u_i/v_i\rangle$ for $i \leqq N$. Then necessarily $u_N \subset_K u_{m+1}$, $u_N \subseteq_K \bar{f}(N)$ and hence $\bar{f}(N) \subset_K u_{m+1}$. Moreover, $h(m)=v_m$, $h(m+1)=v_{m+1}$. Since I) is true, it follows that either $\neg v_{M+1} \subset_R v_m$ or $p(\alpha^1_w,m) \neq 0$ is true . Now necessarily $\langle u_N/v_N\rangle \in D$; this implies that $p(\alpha^1_{u_N},m)$ is either not saturated or saturated with value 0 . Since $u_N \subseteq w$, this yields a contradiction.

The case where more p_i's and α^i_u's are present is treated in exactly the same way.

Remark: The particular case where the p_i's are absent, that is, where the list $p_1,\ldots\ldots,p_n$ is empty, is, of course, contained in the definition of $\sqsubset$ and D : condition d) which occurs in the definition of D is then emptily satisfied. This particular case can also be subsumed under the general case by taking $n=1$ and for p any of the formulas $0=0$, $\alpha^1_{\langle\rangle}(0)= \alpha^1_{\langle\rangle}(0)$. The behaviour of $\sqsubset$ and D in this particular case is described by

<u>Corollary:</u> If $W(<_R)$ is true then $\sqsubset$ wellorders D.

<u>Proof:</u> This is a particular case of theorem 36 by putting $n=1$ and taking as p the formula $0=0$.

<u>Definition 35:</u> Let R, $p_1,\dots,p_n$ and $\sqsubset$, D be as in theorem 36. By D^* we mean the set of sequence numbers $u=\langle u_0,\dots,u_{s-1}\rangle$ which satisfy one of the following conditions: a) $u=\langle\ \rangle$; b) $u=\langle u_0\rangle$ and $u_0 \in D$; c) $s \geqq 2$, $u_i \in D$ for all $i<s$ and $u_0 \sqsupset u_1 \sqsupset \dots \sqsupset u_{s-1}$;
d) $s \geqq 2$, $u_i \in D$ for all $i<s$, $\neg u_{s-1} \sqsubset u_{s-2}$ and if $s \geqq 3$ then also $u_0 \sqsupset u_1 \sqsupset \dots \sqsupset u_{s-2}$. If $u=\langle u_0,\dots,u_{s-1}\rangle \in D^*$ according to a), b) or c), then u is called unsecured, if $u \in D^*$ according to d), then u is called immediately secured. By $\prec^*$ we denote the Kleene Brower linear ordering restricted to D^*.

<u>Theorem 37:</u> If $F[R,p_i;\vec{\alpha}_u]$ is true, then $\prec^*$ is wellfounded.

<u>Proof:</u> This is an immediate consequence of theorem 36 and the wellknown equivalence between the wellfoundedness of trees and the associated Kleene Brower linear ordering.

6.2. Conservative extensions of ZT/V and ZTi/V

<u>A.</u> The system ZT/V is obtained from ZT by addition of the following rule:

$$V \quad \frac{t_R(y)=0\ ,\ (x)<_R y A(x),\ \Gamma \longrightarrow \Delta, A(y)}{t_R(q)=0,\ W^o(<_R),\ \Gamma \longrightarrow \Delta, A(y)}$$

with q,y subject to the usual stipulations. Here $t_R(x)$ and $x <_R y$ are associated with $R(x)$ and $x \subset_R y$ in the way described in the proof of theorem 2*. $W^o(<_R)$ is an abbreviation for $(\alpha)\neg(x)\neg(\neg\alpha(x+1) <_R \alpha(x))$. Since $x <_R y$ is prime, the tertium non datur is available for it in ZTi and hence $W^o(<_R)$ is provable equivalent with $(\alpha)\neg(x)(\alpha(x+1)<_R\alpha(x))$; in order to avoid a new notation we use in this chapter $W^o(<_R)$ as an abbreviation for $(\alpha)\neg(x)(\alpha(x+1)<_R\alpha(x))$ instead for $(\alpha)\neg(x)\neg\neg(\alpha(x+1)<_R\alpha(x))$. The system ZTi/V is as usual obtained by restricting attention to those proofs which contain at most one formula in the succedent.

<u>B.</u> We now are going to define what we call intuitionistic proofs of type (m,i) by induction with respect to i. The definition is very similar to that one presented in the preceeding chapter.

<u>1.</u> Proofs in ZTi/V in which only formulas with at most m logical symbols occur are intuitionistic proofs of type (m,i) for all i.

<u>2.</u> Let P' be an intuitionistic proof of type (m,i) whose endsequent is S'. Let S be a sequent with at most one formula on the right of the arrow and assume that every formula in S contains at most m logical symbols. The tree

$$\begin{array}{c} P' \\ \vdots \\ S' \\ \hline S \end{array}$$

is an intuitionistic proof of type (m,i) if S'/S is an inference of the following type: structural, conversion, logical, induction, V-inference.

<u>3.</u> Let P_1, P_2 be intuitionistic proofs of type (m,i) with S_1, S_2 as endsequents, respectively. Let S be as in clause 2. The tree

$$\begin{array}{cc} P_1 & P_2 \\ \vdots & \vdots \\ S_1 & S_2 \\ \hline \multicolumn{2}{c}{S} \end{array}$$

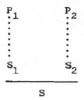

is an intuitionistic proof of type (m,i) if $S_1,S_2/S$ is an inference of the following type: cut, logical inference.

4. Let $R(x)$ be a quantifierfree standard formula (that is of the form $R_o(x)\wedge seq(x)$) and $p_1(x),\ldots,p_n(x)$ a list of terms; we assume that x is the only free variable which occurs in $R(x)$ and in the p_i's. Let $\alpha^{i_1}_{u_1},\ldots,\alpha^{i_s}_{u_s}$ be the list of those special function constants which occur in at least one of the expressions $R(x)$, $p_i(x)$. Here we use again the notation introduced at the beginning of section 6.1., part A. Let $v_1,\ldots,v_s$ be a list of sequence numbers all having the same length $\geqq 0$ and let P' be an intuitionistic proof of type $(m,i+1)$ of $t_{R_v}(y)=0,\ (x)\prec_{R_v}y\,A(x),\ \Gamma\longrightarrow A(y)$. Let P_1 be an intuitionistic proof of type (m,i) of $(x)p_1(x)=0,\ldots,(x)p_n(x)=0\longrightarrow W^o(\prec_R)$. The following tree is an intuitionistic proof of type $(m,i+1)$:

$$\cfrac{\begin{matrix}P'\\ \vdots\\ t_{R_v}(y)=0,\ (x)\prec_{R_v}y\,A(x),\ \Gamma\longrightarrow A(y)\end{matrix}}{(x)p\,(\vec{\alpha}_{u*v},x)=0,\ldots,(x)p_n(\vec{\alpha}_{u*v},x)=0,t_{R_v}(q)=0,\ \Gamma\longrightarrow A(q)}\;T(P_1)$$

where q and y are subject to the usual stipulations. The endsequent of this tree is said to follow from the premiss $t_{R_v}(y)=0,(x)\prec_{R_v}y\,A(x),\ \Gamma\longrightarrow A(y)$ by means of a $T(P_1)$-inference.

5. Let $R(x)$, $p_1(x),\ldots,p_n(x)$, $\alpha^{i_1}_{u_1},\ldots,\alpha^{i_s}_{u_s}$ be as before and let $\sqsubset$, D be the partial ordering and its domain associated with $R(x)$, $p_1(x),\ldots,p_n(x)$ according to section 6.1. Let $\prec^*$, D^* be the Kleene Brouwer ordering associated with $\sqsubset$, D according to definition 35. Let $a=\langle a_o,\ldots,a_{t-1}\rangle$ be an unsecured element of D^* and let a_{t-1} be $\langle v_1,\ldots,v_s/d\rangle$, in

particular. Let $w_1, \ldots\ldots, w_s$ be a list of sequence numbers, all having the same length and such that each w_i is a proper or improper extension of $v_i (w_i \subseteq_K v_i)$. Let t be a saturated term with value d . Let P' be an intuitionistic proof of type $(m,i+1)$ of $(x)p_1(x)=0, \ldots\ldots, (x)p_n(x)=0 \longrightarrow W^o(<_R)$. The following tree is an intuitionistic proof of type $(m,i+1)$:

$$
T(P_1,a)\ \frac{\begin{array}{c} P' \\ \vdots \\ y <_{R_w} t,\ (x) <_{R_w y} A(x),\ \Gamma \longrightarrow A(y) \end{array}}{(x)p_1(\vec{\alpha}_{u*w},x)=0, \ldots, (x)p_n(\vec{\alpha}_{u*w},x)=0,\ q <_{R_w} t,\ \Gamma \longrightarrow A(q)}
$$

where y,q are subject to the usual stipulations. The endsequent of the new tree is said to follow from the premiss $y <_{R_w} t,\ (x) <_{R_w y} A(x),\ \Gamma \longrightarrow A(y)$ by means of a $T(P_1,a)$-inference.

Remarks and definitions. The proof P_1 which appears in the clauses 4,5) above is called side proof of the $T(P_1)$- and $T(P_1,a)$-inference, respectively. If an intuitionistic proof P of some type contains a $T(P_1)$- or a $T(P_1,a)$-inference, then P_1 is said to be a side proof of P . The sequent number a in a $T(P_1,a)$-inference is called index of this inference. For simplicity, we did not include in the above clauses 1)-5) the notion of "node" of an intuitionistic proof of type (m,i) but this could of course be done in the same way as in the corresponding definition of the previous chapter. The main point to stress about nodes is the following: if P_1 is a side proof of an intuitionistic proof P of type (m,i), then we do not consider the nodes of P_1 as nodes of P .

Definition 36: A sequent S is said to be provable in ZTFi/V if there is an intuitionistic proof of type (m,j) (for some m,j) having S as endsequent.

There is a notion of classical proof of type (m,i) whose definition is given by clauses 1*-5*) which are obtained from clauses 1-5) by means of the following changes: a) in clauses 1-3) we allow S to contain more than one formula in the succedent; b) in clauses 4), 5) we allow premiss and conclusion of the $T(P_1)$- and $T(P_1,a)$-inference, respectively, to contain more than one formula in the succedent, that is to be of the form $\ldots\ldots \longrightarrow \Delta, A(y)$ and $\ldots\ldots \longrightarrow \Delta, A(q)$ respectively while the side proof P is still required to be intuitionistic. The classical system so obtained will be denoted by ZTF/V .

B. Again we have

Lemma 14: An intuitionistic proof of type (m,i) is also an intuitionistic proof of type (m',i') for $m \leqq m'$, $i \leqq i'$. Similarly, with classical proofs of type (m,i) .

Theorem 38: An intuitionistic proof of type (m,i) can be transformed effectively into an intuitionistic proof P' of type $(2m,0)$. Similarly, with classical proofs of type (m,i) .

Proof: The proof is essentially the same as the proof of theorem 14, that is, we proceed by induction over the proof tree P . Assume eg. that P contains a $T(P_1,a)$-inference, say

$$T(P_1,a)\ \frac{y <_{R_w} t,\ (x) <_{R_w y} A(x),\ \Gamma \longrightarrow A(y)}{(x)p_1(\vec{\alpha}_{u*w},x)=0,\ldots\ldots,q <_{R_w} t,\ \Gamma \longrightarrow A(q)}$$

(retaining thereby the notation used in clauses 4), 5)). P_1 is by definition an intuitionistic proof of type $(m,i-1)$ of $(x)p_1(\vec{\alpha}_{u*w},x)=0\ \ldots\ldots \longrightarrow W^o(<_{R_w})$. By induction, there is an intuitionistic proof P' of the premiss of the above $T(P_1,a)$-inference. By proceeding in exactly the same way as in the proof of theorem 14, case III, we obtain from P' a proof P" of $W^o(<_{R_w}),\ t_{R_w}(q)=0,\ \Gamma \longrightarrow q <_{R_w} t \supset A(q)$ which is intuitionistic of type $(2m,0)$. With the aid of P and with a little bit of intuitionistic predicate calculus, we can transform P" into an intuitionistic proof of type $(2m,0)$ of $(x)p_1(\vec{\alpha}_{u*w},x)=0,\ldots\ldots,q <_{R_w} t,\ \Gamma \longrightarrow A(q)$. Both for the classical and intuitionistic proofs of type (m,i) we can introduce the usual notions such as final part, normal proof, strictly normal proof, complexity of a cut, an induction, of a V-inference, of a $T(P_1)$- or a $T(P_1,a)$-inference. Similarly, we can define the notion of height of a sequent S in a proof P (denoted by $h(S)$) in the usual way, and the same holds for the notion of critical inference. Brief, the definitions of all these notions remain exactly the same as before. Basic lemmas I and II remain the same as before; however, a more general form of the basic lemma is needed below.

6.3. A generalisation of the basic lemma

Basic lemma III: Let P be a strictly normal intuitionistic proof of type (m,i). Assume that no thinning occurs in the final part. Let $G_1,\ldots\ldots,G_s \longrightarrow H$ be the endsequent. Let $S_1,\ldots\ldots,S_n$ be the uppermost sequents of the final part, listed from left to right; let S_j be $\Gamma_j \longrightarrow A_j$. Then: 1) for every $j<n$ there is a strictly normal intuitionistic proof P_j of type (m,i) whose endsequent is $G_1,\ldots\ldots,G_s \longrightarrow A_j$; 2) for every $j<n$, if B occurs in Γ_j and if B is not isomorphic with any $G_1,\ldots\ldots,G_s$, then there is a strictly normal intuitionistic proof P' of type (m,i) of $G_1,\ldots\ldots,G_s \longrightarrow B$.

Proof: Apart from minor variants the proof remains essentially the same as that of basic lemma II. a) We first prove 1). Since $j<n$, we must necessarily find a cut $S',S''/S$ in the final part with the property: 1) S' is equal to S_j or below S_j; 2) the cut formula F in S' is an image of A_j. Let $S',S''/S$ be more explicitely $\Sigma \longrightarrow F$; $F,\Pi \longrightarrow D/\Sigma,\Pi \longrightarrow D$. Let $P_{S'}$, $P_{S''}$, P_S be the subproofs of S', S'' and S in P respectively. We alter P as follows:

$$
\dfrac{\begin{array}{c} P_{S'} \\ \vdots \\ \Sigma \longrightarrow F \end{array} \qquad \dfrac{\begin{array}{c} P_{S''} \\ \vdots \\ F,\Pi \longrightarrow D \end{array}}{F,\Pi \longrightarrow F,D}\ \text{thinnings, interchange}}{\begin{array}{c} \Sigma,\Pi \longrightarrow F,D \\ \vdots \\ G_1,\ldots\ldots,G_s \longrightarrow F,H \end{array}}\ \text{cut}
$$

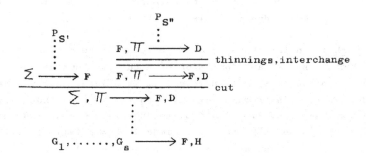

This new proof P^* is a classical proof of type (m,i). Clearly we can derive $\Sigma,\Pi \longrightarrow F,D$ from the left premiss of the cut indicated by thinning and interchange. That is, we can apply to P^* the operation "omission of a cut" and obtain a new proof P^{**} of type (m,i) having the following form:

$$\begin{array}{c}
P_{S'} \\
\vdots \\
\Sigma \longrightarrow F \\
\hline\hline
\Pi, \Sigma \longrightarrow F \\
\hline
\Pi, \Sigma \longrightarrow F, G \\
\vdots \\
G_1, \ldots\ldots, G_s \longrightarrow F, H
\end{array} \quad \text{thinning, interchanges}$$

P^{**} is clearly an almost intuitionistic proof in the sense of section 3.1., part A. According to lemma 9 (which remains invariably true in the present context) we can transform P^{**} by means of a series of preliminary reduction steps into an intuitionistic proof P' of type (m,i) of $G_1, \ldots\ldots, G_s \longrightarrow F$. By adding eventually a conversion if necessary, we finally obtain a strictly normal intuitionistic proof P_j of type (m,i) of $G_1, \ldots\ldots, G_s \longrightarrow A_j$.

b) In order to prove 2) it is sufficient to show: if B occurs in Γ_j and if B is not isomorphic with any of the formulas $G_1, \ldots\ldots, G_s$, then there is an A_k $(k<n)$ isomorphic with B . In virtue of the second half of the assumption, B has no image in the end-sequent. Hence there is a cut $S', S''/S$ with the property: 1) S'' is equal to S or below S ; 2) the cutformula F in S'' is an image of B . As in the proof of basic lemma I (chapter III, section 3.2.), we conclude that the cutformula F in S' is the image of some A_k, $k<n$. Hence B is isomorphic with A_k .

<u>Remarks:</u> In the above proof we have used the notions "preliminary reduction steps" and "omission of a cut" without having defined them in the present context. However, it is evident that the definition of these notions remain word by word the same as those given in chapter II, sections.2.2. and 2.6. Another remark concerns the proofs P_j and P' whose existence is claimed in basic lemma III. The content of the proof given above is that, as soon as A and B are given, we can construct the proofs P_j and P', respectively, in an effective way by applying to P certain preliminary reduction steps and the operation "omission of a cut". This gives rise to

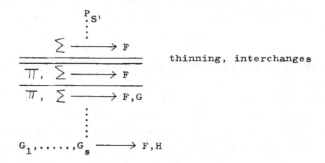

Definition 37: Let P be a strictly normal intuitionistic proof of type (m,i) whose endsequent is $G_1,\ldots,G_s \longrightarrow H$. Let $\Gamma_j \longrightarrow A_j$, $j=1,\ldots,n$ be the uppermost sequents of the final part, listed from left to right. The construction described in the proof of basic lemma III associates with every A_j $(j<n)$ a welldetermined strictly normal intuitionistic proof P_j of type (m,i) of $G_1,\ldots,G_s \longrightarrow A_j$; we call P_j the side proof determined by A_j. Similarly, a welldetermined strictly normal intuitionistic proof P' of $G_1,\ldots,G_s \longrightarrow B$ is associated with every $B \in \Gamma_j$, $(j \leq n)$ by means of the construction described in the proof of basic lemma III; we call P' the side proof determined by B.

6.4. Reduction steps

A. Let us first introduce reduction steps for classical proofs of type (m,i). The only kinds of reduction steps needed for our purposes are: a) preliminary reduction steps; b) elimination of forks, that is, logical reduction steps. Fork elimination in the present context will also be called "classical logical reduction step". Their definition remains the same as in all previous cases.

B. Next we introduce reduction steps for intuitionistic proofs of type (m,i). Apart from minor changes, they are essentially the same as those introduced in the last chapter for intuitionistic proofs of type (m,i). We have: a) preliminary reduction steps; b) intuitionistic logical reduction steps; c) induction reductions. The notion "substitution instance" is again given by definition 20; the definition of inessential reduction step, however, will slightly be modified below. Further reduction steps (V-, T_1-, T_2-reduction steps) will be introduced below. The definitions of the reduction steps a-c) remain invariably the same as in the previous chapters. An intuitionistic logical reduction step applied to an intuitionistic proof P of type (m,i) again splits up into a fork elimination, transforming P into an almost intuitionistic proof of type (m,i), plus a series of preliminary reduction steps transforming P' back into an intuitionistic proof P" of type (m,i), having the same endsequent as P. If P is strictly normal, then so is P". Since in most of the cases we have to do with intuitionistic proofs (of some type), we simply speak of logical reduction step instead of intuitionistic logical reduction step. The notion of substitution instance is, of course, again given by definition 20 in chapter IV. Now to the definition of T_1- , T_2- and V-reduction steps.

Notation: Below we use again the notation introduced in section 6.1. of this chapter, at the beginning of part A.

T_1-reduction steps. Let P be a saturated intuitionistic proof of type (m,i) containing a $T(P_1)$-inference, say

$$T(P_1)\quad \frac{t_{R_v}(y)=0,\ (x)\prec_{R_v} y A(x),\ \Gamma \longrightarrow A(y)}{(x)p_1(\vec{\alpha}_{u*v},x)=0,\ldots\ldots,t_{R_v}(q)=0,\ \Gamma \longrightarrow A(q)}$$

Here P_1 is an intuitionistic proof of type (m,i-1) of $(x)p_1(\vec{\alpha}_u,x)=0, \ldots\ldots, (x)p_n(\vec{\alpha}_u,x)=0 \longrightarrow W^o(\prec_R)$. By S' and S we denote premiss and conclusion of the above $T(P_1)$-inference. Let $\sqsubset$, D be the partial ordering and its domain associated with $R(\vec{\alpha}_u,x), p_1(\vec{\alpha}_u,x), \ldots\ldots, p_n(\vec{\alpha}_u,x)$ according to sect. 6.1.; let D* , $\prec^*$ be the Kleene Brouwer partial ordering associated with $\sqsubset$, D according to definition 35. Since P is saturated, both q and $t_R(q)$ are saturated. We distinguish three cases. Case 1: $|t_{R_v}(q)| \neq 0$. Then $t_{R_v}(q)=0 \longrightarrow$ is an axiom and the conclusion of the above $T(P_1)$-inference can be derived by means of thinnings and interchanges from this axiom. Let P_o be such a derivation. The reduction step in this case consists in replacing P_S by P_o . Case 2: $|t_{R_v}(q)|=0$ and $\langle v_1,\ldots,v_s/|q|\rangle \notin D$. Since $t_{R_v}(q)$ is saturated with value 0, it follows from the corollary of theorem 2* that $R(\vec{\alpha}_{u*v},q)$ is saturated and true. Since $\langle v_1,\ldots,v_s/|q|\rangle \notin D$, it follows from the definition of D that there is an $i \leq \text{length}(v_1)$ and a $k \leq n$ such that $p_k(\vec{\alpha}_{u*v},i)$ is saturated with value $\neq 0$. Hence $p_k(\vec{\alpha}_{u*v},i)=0 \longrightarrow$ is an axiom. Let P_o be the following proof:

$$\frac{p_k(\vec{\alpha}_{u*v},i)=0 \longrightarrow}{(x)p_k(\vec{\alpha}_{u*v},x)=0 \longrightarrow}$$
$$\overline{\overline{\quad S \quad}} \quad \text{thinnings, interchanges}$$

The T_1-reduction step in this case consists in replacing P_S by P_o.

Case 3: $|t_{R_v}(q)|=0$ and $\langle v_1,\ldots,v_s/|q|\rangle \in D$. By definition of D^* : $\langle\!\langle v_1,\ldots,v_s/|q|\rangle\!\rangle \in D^*$ and $a=\langle\!\langle v_1,\ldots,v_s/|q|\rangle\!\rangle$ is unsecured. The T_1-reduction step in this case consists in replacing P_S by the following derivation of S :

$$
\dfrac{
\dfrac{
\dfrac{
y <_{R_v} q \longrightarrow t_{R_v}(y)=0 \qquad \begin{matrix} P_{S'} \\ \vdots \\ S' \end{matrix}
}{
y <_{R_v} q\ ,\ (x) <_{R_v} y A(x),\ \Gamma \longrightarrow A(y)
}\ \text{cut}
}{
\dfrac{
\ldots\ldots (x)p_i(\vec{\alpha}_{u*v},x)=0,\ldots\ldots, s <_{R_v} q,\ \Gamma \longrightarrow A(q)
}{
\ldots\ldots (x)p_i(\vec{\alpha}_{u*v},x)=0,\ldots\ldots,\ \Gamma \longrightarrow (x) <_{R_v} q A(x)
}
}\ T(P_1,a)
\qquad
\begin{matrix} P^q_{S'} \\ \vdots \\ S'_q \end{matrix}
}{
\ldots\ldots (x)p_i(\vec{\alpha}_{u*v},x)=0,\ldots\ldots,\ t_{R_v}(q)=0,\ \Gamma \longrightarrow A(q)
}\ \text{cut}
$$

Here $P^q_{S'}$ denotes, as usual, the result which we obtain by replacing every (free) occurence of y in $P_{S'}$ by q ; S'_q is again the end-sequent of $P^q_{S'}$. We say that a T_1-reduction step has been applied to the particular $T(P_1)$-inference above.

T_2-reduction steps. Let $w_1,\ldots,w_s$ be a list of sequence numbers, all of the same length, such that each w_i is an extension of v_i ($w_i \subseteq_K v_i$), and let t be a saturated term. Let P be a strictly normal intuitionistic proof of type (m,i) which contains a critical $T(P_1,b)$-inference, say

$$
\dfrac{
y <_{R_w} t\ ,\ (x) <_{R_w} y A(x),\ \Gamma \longrightarrow A(y)
}{
\ldots\ldots (x)p_i(\vec{\alpha}_{u*w},x)=0,\ldots\ldots,\ q <_{R_w} t\ ,\ \Gamma \longrightarrow A(q)
}\ t(P_1,b)
$$

Here, P_1 is by definition an intuitionistic proof of type $(m,i-1)$ of $(x)p_1(x)=0,\ldots\ldots,(x)p_n(x)=0 \longrightarrow W^o(<_R)$. Since P is saturated, every constant term in the final part of P is saturated, $q <_{R_w} t$ is saturated, hence $R(\vec{\alpha}_{u*w},q)$ and $R(\vec{\alpha}_{u*w},t)$ and $q \subset_{R_w} t$ are saturated in virtue of the corollary of theorem 2*. Let b^w be $b_o,\ldots\ldots,b_{r-1}$ and let b_r, in particular, be $\langle v_1,\ldots\ldots,v_s/d\rangle$. By definition of $T(P_1,b)$-inference, b is an unsecured element of D^* , that is, $b_o \sqsupset b_1 \sqsupset \ldots\ldots \sqsupset b_{r-1}$;

moreover, $|t|=d$ and $\langle v_1,\ldots\ldots,v_s/d\rangle \in D$. By S' and S we denote again premiss and conclusion of the above $T(P_1,b)$-inference ; P_S is the subproof of S ; $P_{S'}$ is the subproof of S' . In order that a T_2-reduction step be applicable to the above $T(P_1,b)$-inference, we require that the following condition C. be satisfied: every w_i is a strict extension of v_i . In virtue of the definition of $T(P_1,b)$-inference, this amounts to require: $\mathrm{length}(v_1) < \mathrm{length}(w_1)$. We distinguish three cases. <u>Case 1:</u> $q <_{R_W} t$ is false. Then $q <_{R_W} t \longrightarrow$ is an axiom and we can derive S from $q <_{R_W} t \longrightarrow$ by means of thinnings and interchanges alone. Let P be such a derivation. The T_2-reduction step in this case consists in replacing P_S by P_o . <u>Case 2:</u> $q <_{R_W} t$ is true and $b^* = \langle b_o,\ldots\ldots,b_{r-1},b_r\rangle$ is not an unsecured element of D^* , where we have put $b_r = \langle w_1,\ldots\ldots,w_s/|q|\rangle$. Now $b_{r-1} \in D$ as noted above. Furthermore, $q <_{R_W} t$ is saturated and true, hence $|q| \subset_{R_W} |t|$ is saturated and true in virtue of the corollary to theorem 2* . If b_r would be in D then necessarily $b_r \sqsubset b_{r-1}$ in virtue of $w_i \subset_K v_i$ and the definition of D ; hence $\langle b_o,\ldots\ldots,b_r\rangle$ would be an unsecured element of D^* , contradicting the assumption. Hence we conclude $b_r \notin D$. But $R(\vec{\alpha}_{u*w},q)$ and hence $R(\vec{\alpha}_{u*w},q)$ are saturated and true as noted above. Looking at the definition of D, we see that the only reason for $\langle w_1,\ldots\ldots,w_s/|q|\rangle$ not to be an element in D is that there is a $k \leqq n$ and an $i \leqq \mathrm{length}(w_1)$ such that $p_k(\vec{\alpha}_{u*w},i)$ is saturated with value $\neq 0$. Hence $p_k(\vec{\alpha}_{u*w},i)=0 \longrightarrow$ is an axiom. Therefore the following derivation P_o of S can be found:

$$\forall\!\longrightarrow\ \cfrac{\cfrac{p_K(\vec{\alpha}_{u*w},i)=0 \longrightarrow}{(x)p_k(\vec{\alpha}_{u*w},x)=0 \longrightarrow}}{\overline{\overline{\quad S \quad}}}\ \text{thinnings, interchange}$$

The T_2-reduction step consists in replacing P_S by P_o .
<u>Case 3:</u> $q <_{R_W} t$ is true and $b^* = \langle b_o,\ldots\ldots,b_r\rangle$ is an unsecured element of D^* (with b_r as under case 2). The reduction step in this case consists in replacing P_S by the following derivation

$$
\begin{array}{lll}
\qquad\qquad\qquad\qquad\qquad\qquad\qquad\qquad P_{S'} & & \\
\qquad\qquad\qquad\qquad\qquad\qquad\qquad\qquad \vdots & & \\
y <_{R_w} q,\ q <_{R_w} t \longrightarrow y <_{R_w} t \qquad\qquad S' & & \\ \hline
y <_{R_w} q,\ q <_{R_w} t,\ (x) <_{R_w y} A(x),\ \Gamma \longrightarrow A(y) & \text{cut} & \\ \hline
\dots (x)p_i(\vec{\alpha}_{u*w}, x)=0, \dots\dots, s <_{R_w} t, q <_{R_w} t, \Gamma \longrightarrow A(s) & T(P_1, b^*) & P^q_{S'} \\ \hline\hline
 & & \vdots \\
\dots (x)p_i(\vec{\alpha}_{u*w}, x)=0, \dots\dots, q <_{R_w} t, \Gamma \longrightarrow (x) <_{R_w q} A(x) & & S'_q \\ \hline\hline
\qquad \dots\dots (x)p_i(\vec{\alpha}_{u*w}, x)=0, \dots\dots, q <_{R_w} t, \Gamma \longrightarrow A(q) & & \text{cut}
\end{array}
$$

The last double line indicates a cut combined with some interchanges and contractions. $P^q_{S'}$ and S'_q are again the results of replacing every (free) occurence of y in $P_{S'}$ and S' respectively by q. We say that a T_2-reduction step has been applied to the particular $T(P_1, b)$-inference.

<u>V-reduction steps.</u> Let P be a strictly normal intuitionistic proof of type (m,i). In order that a V-reduction step be applicable to P we require from the outset that the following condition D be satisfied: the endsequent of P has the form $(x)p_1(x)=0, \dots\dots, (x)p_n(x)=0 \longrightarrow A$ (A arbitrary). Let P have this property and assume that P contains a critical V-inference, say

$$
V \qquad \frac{t_R(y)=0,\ (x) <_{R y} A(x),\ \Gamma \longrightarrow A(y)}{W^o(<_R),\ t_R(q)=0,\ \Gamma \longrightarrow A(q)}
$$

Evidently $W^o(<_R)$ cannot have an isomorphic image in the endsequent in virtue of condition D. Therefore we can extract from P the side proof P_1 determined by $W^o(<_R)$ (def. 37, basic lemma III and the remark following it). P_1 is a strictly normal intuitionistic proof of type (m,i) whose endsequent is $(x)p_1(x)=0, \dots\dots, (x)p_n(x)=0 \longrightarrow W^o(<_R)$. Let S be the conclusion of the above V-inference and P_S its subproof. We replace P_S by the following derivation:

$$T(P_1)\frac{\dfrac{t_R(y)=0,\ (x)\underset{R}{<}{}_y A(x),\ \Gamma \longrightarrow A(y)}{\ldots(x)p_i(x)=0\ldots,t_R(y)=0,\Gamma \longrightarrow A(q)}}{\overline{W^{o}(<_R),t_R(q)=0,\Gamma,\ldots(x)p_i(x)=0,\ldots \longrightarrow A(q)}}\ \text{interchanges, thinning}$$

The resulting proof P^* is a strictly normal and intuitionistic proof of type $(m,i+1)$ and its endsequent looks as follows: $(x)p_1(x)=0,\ldots,(x)p_n(x)=0,(x)p_1(x)=0,\ldots,(x)p_n(x)=0 \longrightarrow A$. Now we apply to the endsequent of P^* a series of interchanges and contractions and finally obtain a proof P', which is strictly normal and intuitionistic of type $(m,i+1)$, whose endsequent is the same as that of P. We say that P' follows from P by means of a V-reduction step. We also say that the reduction step in question has been applied to the particular V-inference above.

C. Before proceeding further, let us quickly draw attention to the T_2-reduction steps. Let us for this purpose retain the notation used in the definition of T_2-reduction step. According to this definition a T_2-reduction step is applicable to the critical $T(P_1,b)$-inference only if each sequence number $w_1,\ldots,w_s$ is a strict extension of the corresponding sequence number $v_1,\ldots,v_s$. Now assume that the w_i's are not strict extensions of the v_i's; this implies, of course, $v_i=w_i$, $i=1,\ldots,s$. In this case we say that the $T(P_1,b)$-inference under consideration is incomplete; if each w_i is a strict extension of v_i, then we call the $T(P_1,b)$-inference complete. The $T(P_1,b)$-inference can, of course, be made complete by passing from P to a substitution instance P'. This suggests

Definition 38: A strictly normal proof is called strongly saturated if every constant term which occurs in the final part or in the premiss of a critical inference is saturated and if every critical $T(P_1,b)$-inference is complete.

Why we also require that every constant term which occurs in the premiss of a critical inference should be saturated will become clear below. With respect to the notion "strongly saturated" there is available a lemma which is the exact counterpart of lemma 9, namely

Lemma 14: We can effectively decide whether a proof P in ZTFi/V is strongly saturated or not. If it is not strongly saturated and if $\alpha^{i_1}_{u_1},\ldots,\alpha^{i_s}_{u_s}$ is a given listing of the distinct special function constants occuring in P, then we can find effectively a prim. rec. continuity function $\tau(x_1,\ldots,x_s)$ having the following property: if $\tau(v_1,\ldots,v_s)\neq 0$ and if P^* results from P by replacing every $\alpha^{i_k}_{u_k}$ by $\alpha^{i_k}_{u_k*v_k}$, then P^* is strongly saturated.

The proof of this lemma, like that of lemma 9, is an immediate consequence of the definition of term and saturated term and hence omitted.

Remark: With every strictly normal proof P in ZTFi/V which is not strongly saturated there is associated in an effective way a continuity function σ which is related to P in the way described by lemma 14; we denote this continuity function by σ_P and call it the continuity function strongly associated with P.

Definition 39: Let P be a strictly normal proof in ZTFi/V which is not strongly saturated, $\alpha^{i_1}_{u_1},\ldots,\alpha^{i_s}_{u_s}$ a listing of the special function constants which occur in P. Let σ_P be the continuity function strongly associated with P. Let $v_1,\ldots,v_s$ be a list of sequence numbers, all of the same length, and P^* the proof obtained from P by replacing every occurence of $\alpha^{i_1}_{u_1},\ldots,\alpha^{i_s}_{u_s}$ in P by $\alpha^{i_1}_{u_1*v_1},\ldots,\alpha^{i_s}_{u_s*v_s}$. P^* is said to follow from P by means of an inessential reduction step if the following holds:
a) $\sigma_P(v_1,\ldots,v_s)\neq 0$, b) if $w_1,\ldots,w_s$ is a list of sequence numbers such that $v_i \subset_K w_i$, $i=1,\ldots,s$ then $\sigma_P(w_1,\ldots,w_s)=0$.

D. A reduction step is called strictly essential, if it is a logical one, an induction reduction, a T_1-, T_2- or a V-reduction step.

Strictly essential reduction steps satisfy

Theorem 39: Let P be a strictly normal, strongly saturated intuitionistic proof of type (m,i) (for some (m,i)) whose endsequent has the form $(x)p_1(x)=0,\ldots\ldots,(x)p_n(x)=0 \longrightarrow A$ (where the p_i's or A or both may be absent). Assume the following: a) P does not coincide with its final part, b) no preliminary and no strictly essential reduction steps are applicable to P. Then the following is true: there is a critical logical inference whose principal formula has an image in the endsequent.

Proof: P cannot contain any critical induction inference, $T(P_1)$- , $T(P_1,b)$- or V-inference since in this case we could apply a corresponding reduction step to P, in contradiction with the assumption. No fork can occur in the final part of P since this would give rise to an intuitionistic fork elimination, contradicting the assumption. Hence we can proceed as in the proof of theorem 6.

E. Finally let us discuss the notion of subformula reduction step. To start with, let us fix necessary conditions which have to be satisfied by a proof P in order that a subformula reduction step may eventually be applicable to it. These conditions, summarily denoted by SFC, are 1) P has to be a strictly normal, strongly saturated intuitionistic proof of type (m,i), (for some (m,i)); 2) no preliminary and no strictly essential reduction step is applicable to P; 3) the endsequent of P must have the form $(x)p_1(x)=0,\ldots\ldots,(x)p_n(x)=0 \longrightarrow A$. According to the last theorem, there must be at least one critical logical inference in P, whose principal formula has an image in the endsequent. We distinguish two cases. Case 1: There is no critical inference in P which has an image in the antecedent of the endsequent of P. The critical inference provided by the above theorem must then by necessity be a logical inference which introduces a new logical symbol in the succedent, that is an inference of the following type: a) a functional quantification $\longrightarrow \forall$ or $\longrightarrow E$, b) a quantification $\longrightarrow \forall$ or $\longrightarrow E$ over individuals, c) a propositional inference $\longrightarrow \wedge$, $\longrightarrow \vee$, $\longrightarrow \supset$ or $\longrightarrow \neg$. That is, we are precisely in the situation considered in section 4.4. of chapter IV, part D. Hence we define the subformula reduction step in this case in precisely the same way as in section 4.4., part D, summarized by definition 21. Case 2: There is a critical inference whose principal formula has an image in the antecedent of the endsequent of P . This inference must necessarily have the form: $p(t)=0,\ \Gamma \longrightarrow B/(x)p(x)=0,\ \Gamma \longrightarrow B$, where $p(x)=0$ is isomorphic with one of the formulas $p_i(x)=0$. Since P is strictly normal, there is no free variable in the endsequent of P , and according to the definition of "normal" there is no free variable in $p(t)=0$. Since P is strongly saturated, both $p(t)$ and t are saturated. We distinguish two subcases. <u>Subcase 1:</u> $|p(t)| \neq 0$. Then by definition no subformula reduction step is applicable to P . <u>Subcase 2:</u> $|p(t)|=0$. Then $\longrightarrow p(t)$ is an axiom and we can replace the inference $p(t)=0,\ \Gamma \longrightarrow B/(x)p(x)=0,\ \Gamma \longrightarrow B$ by the following derivation:

$$\frac{\dfrac{\longrightarrow p(t)=0 \qquad p(t)=0,\ \Gamma \longrightarrow B}{\Gamma \longrightarrow B}\ \text{cut}}{(x)p(x)=0,\ \Gamma \longrightarrow B}\ \text{thinning}$$

The resulting proof P' is said to follow from P by means of a subformula reduction step. <u>Remark:</u> If P' is obtained from P by means of a subformula reduction step according to case 1 above, then it is, of course, possible that the endsequent of P' has no longer the particular form $(x)p_1(x)=0,\ldots\ldots,(x)p_n(x)=0 \longrightarrow A$; this may happen if the critical inference provided by theorem 39 is of type $\longrightarrow \supset$ or $\longrightarrow \neg$.

<u>F.</u> The list of reduction steps is completed. Let us summarize their properties. The properties of preliminary reduction steps are again given by theorem 4. A relation W can be introduced using definition 14 as it stands; theorem 5 remains invariably true in the present case. As we have seen, our attention is mostly restricted to proofs whose endsequents are of the particular form $(x)p_1(x)=0,\ldots\ldots,(x)p_n(x)=0 \longrightarrow A$. This gives rise to

<u>Definition 40:</u> A proof is said to have standard form if its endsequent has the form $(x)p_1(x)=0,\ldots\ldots,(x)p_n(x)=0 \longrightarrow A$. Thereby the p_i's or A or both may be absent. As before we use "s.n.s. proof" as abbreviation for strictly normal standard proof. Definitions 22 and 23 can be used without any change in order to introduce two relations R' and L' . The text of the definitions remains the same with one exception: "saturated" has to be replaced by

"strongly saturated". The notions "strictly essential reduction step", "subformula reduction steps", "inessential reduction step" have, of course, to be interpreted in the sense of the present chapter. The relations R' and L' are the counterparts of R and L, respectively, and have also the similar properties: theorem 22, part a) (with R' and L' in place of R and L) remains invariably true and its proof remains up to minor modifications the same. Again we simplify the notation by writing R and L in place of R' and L' ; no danger of confusion arises thereby. Subtrees L_P of L and its domains D_P can be introduced by using definition 32 as it stands. Finally, we call a formula $A(\alpha^{i_1}_{u_1},\ldots\ldots,\alpha^{i_s}_{u_s})$ as before true if $(\xi_1,\ldots,\xi_s)A(u_1*\xi_1,\ldots\ldots,u_s*\xi_s)$ is true in the usual sense (def. 33); $\alpha^{i_1}_{u_1},\ldots\ldots,\alpha^{i_s}_{u_s}$ is thereby the list of distinct special function constants which occur in $A(\alpha^{i_1}_{u_1},\ldots\ldots,\alpha^{i_s}_{u_s})$. Our goal is to prove that L_P is wellfounded for proofs P of a suitably large class. To this end we need a few definitions. In order to formulate them we use again the notation introduced at the beginning of part A of section 6.1. (this chapter). Let $R(\vec{\alpha}_u,x)$ be a quantifierfree formula, $p_i(\vec{\alpha}_u,x)$ $i=1,\ldots,n$ a list of terms and $\alpha^{i_1}_{u_1},\ldots\ldots,\alpha^{i_s}_{u_s}$ the list of those special function constants which occur in $R(\vec{\alpha}_u,x)$ or at least one $p_i(\vec{\alpha}_u,x)$. It is assumed that x is the only free variable in $R(\vec{\alpha}_u,x)$ and $p_i(\vec{\alpha}_u,x)$ $i=1,\ldots\ldots,n$, respectively. Let $v_1,\ldots\ldots,v_s$ be a list of sequence numbers, all having the same length; by $\vec{\alpha}_{u*v}$ we denote the list $\alpha^{i_1}_{u_1*v_1},\ldots\ldots,\alpha^{i_s}_{u_s*v_s}$. By $x <_{R_v} y$ we denote the prime formula associated with $x \subset_K y \wedge R(\vec{\alpha}_{u*v},x) \wedge R(\vec{\alpha}_{u*v},y)$ according to theorem 2*, and $x \subset_{R_v} y$ is used as an abbreviation for the latter formula.

<u>Definition 41:</u> An intuitionistic s.n.s. proof P of type (m,i) is said to be special if its endsequent has one of the forms listed below:

1) $(x)p_1(\vec{\alpha}_u,x)=0,\ldots\ldots,(x)p_n(\vec{\alpha}_u,x)=0 \longrightarrow W^o(<_R)$,

2) $(x)p_1(\vec{\alpha}_u,x),\dots,(x)p_n(\vec{\alpha}_u,x)=0 \longrightarrow \neg(x)\ \alpha^j_w(x+1) <_R \alpha^j_w(x)$

3) $(x)p_1(\vec{\alpha}_u,x)=0,\dots,(x)p_n(\vec{\alpha}_u,x)=0 \longrightarrow$

for some terms $p_i(\vec{\alpha}_u,x)$, $i=1,\dots,n$, some quantifierfree formula $R(\vec{\alpha}_u,x)$ and some special function constant α^j_w with j different from $i_1,\dots,i_s$. Thereby we allow the list $p_i(\vec{\alpha}_u,x)$, $i=1,\dots,n$, to be empty.

Lemma 15: If P is an intuitionistic s.n.s. proof which is special, if $L(P,P')$ holds then P' is also special.

Proof: The lemma is proved if we can show the following: if P^* is special and if P^{**} is obtained from P^* by means of a reduction step, then P^{**} is also special. Let S^* and S^{**} be the endsequents of P^* and P^{**} respectively and assume S^* to have form 1),2) or 3) in definition 41. Case 1: The reduction step is a preliminary one. Then we can derive S^* from S^{**} by means of thinnings and interchanges alone. Then S^{**} has clearly one of the forms 1),2) or 3) of definition 41. Case 2: The reduction step is an inessential one. Then S^{**} has the same form as S^* except that the list $\alpha^{i_1}_{u_1},\dots,\alpha^{i_s}_{u_s}$ is now replaced by a corresponding list $\alpha^{i_1}_{u_1*v_1},\dots,\alpha^{i_s}_{u_s*v_s}$ where the v_i's are sequence numbers all having the same length $\neq 0$.

Case 3: The reduction step is a strictly essential one. Then S^{**} is the same as S^* .

Case 4: The reduction step is a subformula reduction step. Then the following subcases arise: a) S^* has form 1) and S^{**} has form 1) or 2); b) S^* has form 2) and S^{**} has form 2) or 3); c) S^* has form 3) and S^{**} has form 3). In each of these cases S^{**} has form 1),2) or 3) listed in definition 41.

<u>Lemma 16:</u> Let P be an intuitionistic s.n.s. proof of type (m,i) which is special. Let P contain a critical V-inference, say

$$V \quad \frac{t_R(y)=0,\ (x) <_R y\, A(x),\ \Gamma \longrightarrow A(y)}{W^o(<_R),\ t_R(q)=0,\ \Gamma \longrightarrow A(q)}$$

The side proof P_1 determined by this inference according to Basic lemma III, the remark following it and definition 37 is again special.

<u>Proof:</u> This is immediate from Basic lemma III, the definition of side proof determined by a critical V-inference and the fact that P is special.

In order to state the main property of special proofs we need a further

<u>Definition 42:</u> Let $p_1(x),\dots,p_n(x)$ be terms, $R(x)$ a quantifier-free formula and x the only free variable occuring in all these expressions. Let $\alpha^{i_1}_{u_1},\dots,\alpha^{i_s}_{u_s}$ be the special function constants occuring in $R(x)$ or at least one $p_i(x)$. We allow the list $p_1,\dots,p_n$ to be empty and indicate this by putting $n=0$. Let S be any of the following sequents:

1) S_1: $(x)p_1(x)=0,\dots,(x)p_n(x)=0 \longrightarrow W^o(<_R)$;

2) S_2: $(x)p_1(x)=0,\dots,(x)p_n(x)=0 \longrightarrow \neg(x)\ \alpha^j_w(x+1) <_R \alpha^j_w(x)$;

3) S_3: $(x)p_1(x)=0,\dots,(x)p_n(x)=0 \longrightarrow$, (where j is different from $i_1,\dots,i_s$) . Consider the following formulas:

1) A_1: $(\xi)(Ex)(\neg\ \xi(x+1) \subset_R \xi(x) \vee p_1(x)\neq 0 \dots \vee p_n(x)\neq 0)$ (simply $(\xi)(Ex)(\neg\ \xi(x+1) \subset_R \xi(x)$ if $n=0$);

2) A_2: $(Ex)(\neg\ \alpha^j_w(x+1) \subset_R \alpha^j_w(x) \vee p_1(x)\neq 0 \vee \dots \vee p_n(x)\neq 0)$, (simply $(Ex)(\neg\ \alpha^j_w(x+1) \subset_R \alpha^j_w(x))$ if $n=0$);

3) A : $(Ex)(p_1(x)\neq 0 \vee \dots \vee p_n(x)\neq 0)$ (simply $0=1$ if $n=0$) .

The formula A is said to be induced by S if A is A_i when S is S_i . We say that S is strongly true if the induced formula is true.

Remark: 1) From a purely classical point of view the above definition is superfluous: if S is true under the usual interpretation, then its induced formula is necessarily true. From an intuitionistic point of view, however, the truth of S does not necessarily imply the truth of the induced formula. Although the considerations in the present chapter use the language of classical set theory, their presentation is as constructive as possible in view of the discussion presented in chapter X. Therefore we make the distinction between true and strongly true sequent.

Theorem 40: Let P_o be an intuitionistic s.n.s. proof in ZTFi/V (that is of some type (m,i)) which is special. Assume that L_{P_o} is wellfounded and let S_o be the endsequent of P_o . Then S_o is strongly true.

Proof: The proof is by transfinite induction with respect to L_{P_o} , that is, we prove: if $P \in D_{P_o}$ then its endsequent S is strongly true (P is again special in virtue of lemma 15). Hence, let $P \in D_{P_o}$ be given, and assume that for all P' , if $L(P,P')$ holds, then S' is strongly true, where S' is the endsequent of P' . With the aid of this hypothesis we have to show: S is strongly true. We distinguish between cases, within cases between subcases, within subcases between subsubcases etc. We abbreviate "subcase", "subsubcase" etc. by SC , SSC etc. Case 1: P is strongly saturated and does not admit preliminary reduction steps. SC1: P admits a strictly essential reduction step. Then $L(P,P')$ iff P' follows from P by application of a strictly essential reduction step. Take any such P' . The endsequent S' of P' is evidently the same as S . By the inductive assumption S' is strongly true, hence S is strongly true. SC2: P does not admit any strictly essential reduction step. In view of the special form of the endsequent S of P, it follows that P cannot coincide with its final part since this would clearly force S to be $\longrightarrow$; again $\longrightarrow$ is not provable from mathematical axioms using only interchanges, contractions, conversions and cuts. According to theorem 39, there is a critical logical inference whose principal formula has an image in the final part. We distinguish between subcases.
SSC1: There is no critical logical inference whose principal formula

has an image in the antecedent of the endsequent. Therefore, a well-defined subformula reduction step is applicable to P , transforming P into P' ; by definition L(P,P') holds. Let S' be the endsequent of P' . By necessity S is S_1 or S_2 in definition 42 for some terms $p_1, \ldots, p_n$, some quantifierfree formula R and some special function constant α_w^j, respectively. SSSC1: S is S_1 . The induced formula A is then given by
$(\xi)(ex)(\neg \xi(x+1) \subset_R \xi(x) \vee p_1(x) \neq 0 \vee \ldots \vee p_n(x) \neq 0)$. By necessity, S' is
$(x)p_1(x)=0, \ldots, (x)p_n(x)=0 \longrightarrow \neg(x)\ \alpha^j_{<>}(x+1) <_R \alpha^j_{<>}(x)$
for some j . The formula A' induced by S' looks as follows:
$(Ex)(\neg\ \alpha^j_{<>}(x+1) \subset_R \alpha^j_{<>}(x) \vee p_1(x) \neq 0 \vee \ldots \vee p_n(x) \neq 0)$. However, it is evident from definition 33 in chapter V that A is true iff A' is true. Since L(P,P') holds, S' is strongly true by the inductive assumption, that is A' , hence A, are true and so S is strongly true. SSSC2: S is S_2 . The formula A induced by S looks as follows:
$(Ex)(\neg\ \alpha^j_w(x+1) \subset_R \alpha^j_w(x) \vee p_1(x)=0 \vee \ldots \vee p_n(x)=0)$. Necessarily, S' is given by
$(x)p_1(x)=0, \ldots, (x)p_n(x)=0,\ (x)\ \alpha^j_w(x+1) <_R \alpha^j_w(x) \longrightarrow$. The formula A' induced by S' is obviously again A . S' is strongly true by the inductive assumption. It follows that A' and hence A are true; hence S is strongly true. SSC2: There is a critical logical inference whose principal formula has an image in the antecedent of S . Let $p(t)=0, \vdash \longrightarrow B / (x)p(x)=0, \vdash \longrightarrow B$ be this inference. $p(x)=0$ is necessarily isomorphic with some $p_i(x)=0$; let $i=1$ for simplicity. SSSC1: $p(t)$ (which is saturated) has value 0 . Then we can apply to P a subformula reduction step which transforms P into a proof P' whose endsequent S' is the same as that of P , that is, S . By the inductive hypothesis, S' is strongly true, hence S is strongly true. SSSC2: $p(t)$ has value $\neq 0$. Now $p_1(t)$ is saturated, too, and its value therefore also $\neq 0$. However, $p_1(t) \neq 0 \supset A_i$ (with A_i as in def. 42) are obviously all intuitionistically true formulas. Therefore S is strongly true, regardless whether S is S_1, S_2 or S_3 in def. 42. This exhausts the possibilities which might arise under the assumption of case 1.

Case 2: P is strongly saturated, but admits preliminary reduction steps. Let $P_0, \ldots, P_N$ be any chain such that a) P_0 is P ; b) P_{i+1} follows from P_i by means of a preliminary reduction step; c) P_N does not admit any preliminary reduction steps. Obviously

P_N is strongly saturated. If $L(P_N,P')$ then $L(P,P')$, as is easily verified; hence the endsequent S' of P' is strongly true in virtue of the inductive assumption about P . Therefore we can apply the considerations of case 1 to P_N and conclude that the endsequent S_N of P_N is strongly true. Now S can obviously be derived from S_N by means of thinnings and interchanges alone; from this one easily concludes that S is also strongly true.

Case 3: P is not strongly saturated. Let S be the endsequent of P and $\alpha_{u_1}^{i_1},\ldots\ldots,\alpha_{u_s}^{i_s}$ the list of special function constants occuring in P . Let A be the formula induced by S . The special function constants occuring in A are obviously contained in the list $\alpha_{u_1}^{i_1},\ldots\ldots,\alpha_{u_s}^{i_s}$. We indicate this by writing $A(\alpha_{u_1}^{i_1},\ldots\ldots,\alpha_{u_s}^{i_s})$ or $A(\vec{\alpha}_u)$, respectively. Replacement of $\alpha_{u_1}^{i_1},\ldots\ldots,\alpha_{u_s}^{i_s}$ by $\alpha_{u_1*w_1}^{i_1},\ldots\ldots,\alpha_{u_s*w_s}^{i_s}$, respectively, transforms P into another proof, to be denoted by $P_{w_1\ldots\ldots w_s}$; the endsequent of $P_{w_1\ldots\ldots w_s}$ is denoted by $S_{w_1\ldots\ldots w_s}$. According to lemma 14, the remark following it and definition 39, there is a prim. rec. continuity function σ_P with the property: if $\sigma_P(w_1,\ldots\ldots,w_s)\neq 0$ then $P_{w_1\ldots\ldots w_s}$ is strongly saturated. Let us call a list $\bar{\xi}_1(n),\ldots\ldots,\bar{\xi}_s(n)$ immediately secured with respect to σ_P if $\sigma_P(\bar{\xi}_1(n),\ldots\ldots,\bar{\xi}_s(n))\neq 0$ and $\sigma_P(\bar{\xi}_1(i),\ldots\ldots,\bar{\xi}_s(i))=0$ for $i<n$; the fact that $w_1,\ldots,w_s$ is immediately secured with respect to σ_P will be indicated by writing $\sigma_P(w_1,\ldots\ldots,w_s)\not\equiv 0$. It is evident that the formula A' induced by $S_{w_1\ldots\ldots w_s}$ is $A(\alpha_{u_1*w_1}^{i_1},\ldots\ldots,\alpha_{u_s*w_s}^{i_s})$. From the definition of inessential reduction step it follows that $L(P,P_{w_1\ldots w_s})$ holds whenever $\sigma_P(w_1,\ldots\ldots,w_s)\not\equiv 0$. Hence, using the inductive assumption, we have the following situation: if $\sigma_P(w_1,\ldots\ldots,w_s)\not\equiv 0$ then $A(\alpha_{u_1*w_1}^{i_1},\ldots\ldots,\alpha_{u_s*w_s}^{i_s})$ is true. Using bar induction with respect to the p.r. continuity function σ_P, one easily deduces the truth of $A(\alpha_{u_1}^{i_1},\ldots\ldots,\alpha_{u_s}^{i_s})$. Hence, S is strongly true.

There is an immediate and important corollary, namely

Corollary: Let P be an intuitionistic s.n.s. proof in $ZTFi/V$ whose endsequent S has the form $(x)p_1(x)=0,\ldots,(x)p_n(x)=0 \longrightarrow W^o(<_R)$ and let L_P be wellfounded. Then: a) $(\xi)(Ex)(\neg\,\xi(x+1) \subset_R \xi(x) \vee p_1(x)\neq 0 \vee \ldots \vee p_n(x)\neq 0)$ is true; b) the particular ordering $\sqsubset$ associated with the latter formula according to section 6.1., part A, is wellfounded; c) the Kleene Brouwer linear ordering $\prec^*$ associated with $\sqsubset$ according to def. 35 is a wellordering.

Proof: a) is a special case of the last theorem; b) follows from a) and theorem 36; c) is a consequence of the wellfoundedness of $\sqsubset$.

6.5. Ordinals

A. From now on we proceed in quite the same way as in the last chapter. First of all we introduce two classes of proofs by means of

Definition 43: a) An intuitionistic s.n.s. proof P (of some type (m,i)) is called "good" if it is special and if, moreover, L_P is wellfounded. b) An (intuitionistic or classical) s.n.s. proof P (of some type (m,i)) is said to be "graded" if all its side proofs are good.

Again we have the following evident

Lemma 17: A preliminary reduction step, the operation "omission of a cut" or a classical logical reduction step, applied to a graded proof, yield a graded proof P'. An intuitionistic logical reduction step, an induction reduction, a T_1- or T_2-reduction step, applied to an intuitionistic graded proof P, yield an intuitionistic graded proof P'.

In order to describe a certain ordinal assignement, we use again some suitable notation. Let P be a good proof of $(x)p_1(x)=0,\ldots\ldots,(x)p_n(x)=0 \longrightarrow W^o(<_R)$. The partial ordering $\sqsubset$ associated with $(\xi)(Ex)(\neg\,\xi(x+1) \subset_R \xi(x) \vee p_1(x)\neq 0 \vee \ldots\ldots \vee p_n(x)\neq 0)$ is then wellfounded according to the last corollary; and so is the Kleene-Brouwer ordering $\prec^*$ associated with $\sqsubset$ according to

def. 35. We denote the ordinal of $\prec^*$ by $\|\prec^*\|$. If, moreover, a is an element in the domain D* of $\prec^*$, then $\|a\|$ denotes the ordinal associated with the restriction of $\prec^*$ to $\{x/x \prec^* a\}$. By Ω we denote the smallest ordinal ξ with the property: if P is a good proof of
$(x)p_1(x)=0,\ldots\ldots,(x)p_n(x)=0 \longrightarrow W^o(<_R)$, then $\|\prec^*\| < \xi$.
Finally, if P is a proof of
$(x)p_1(x)=0,\ldots\ldots,(x)p_n(x)=0 \longrightarrow W^o(<_R)$, if $\sqsubset$ is the partial ordering associated with
$(\xi)(Ex)(\neg\, \xi(x+1) \subset_R \xi(x) \vee p_1(x)\neq 0 \vee \ldots\ldots \vee p_n(x)\neq 0)$, then we call $\sqsubset$ simply the partial ordering associated with P ; the Kleene-Brouwer ordering $\prec^*$ associated with $\sqsubset$ is also called the Kleene-Brouwer linear ordering associated with P .

Now to the description of the ordinal assignement announced above. Let P be a graded proof and S a sequent in it. With S we associate inductively an ordinal, denoted by o(S).

<u>Case 1:</u> S is an axiom (of P) . Then o(S)=1 .

<u>Case 2:</u> S is the conclusion of a one-premiss structural rule, or a conversion, say, S'/S . Then o(S)=o(S') .

<u>Case 3:</u> S is the conclusion of a one-premiss logical inference, say, S'/S, different from $A(t), \Gamma \longrightarrow \Delta / (x)A(x), \Gamma \longrightarrow \Delta$. Then o(S)=o(S')+1 .

<u>Case 4:</u> S is the conclusion of a one-premiss logical inference S'/S of the form $A(t), \Gamma \longrightarrow \Delta / (x)A(x), \Gamma \longrightarrow \Delta$. Then o(S)=o(S')+2 .

<u>Case 5:</u> S is the conclusion of a two-premiss logical inference, say, $S_1,S_2/S$. Then $o(S)=o(S_1) \# o(S_2) \# 1$.

<u>Case 6:</u> S is the conclusion of an induction S'/S . Then $o(S)=\omega_d(o(S')\,\omega)$ where d=h(S')-h(S) .

<u>Case 7:</u> S is the conclusion of a V-inference, say, S'/S . Then we put $o(S)=\omega_d((o(S') \# \omega^{\Omega+1})\,\omega^{\Omega+1})$ where d=h(S')-h(S) .

Case 8: S is the conclusion of a $T(P_1)$-inference, say, S'/S. Then we put $o(S) = \omega_d((o(S') \# \omega^{\lambda+1})\omega^{\lambda+1})$ where $d = h(S') - h(S)$ and $\lambda = \|\prec^*\|$, with $\prec^*$ the Kleene-Brouwer ordering associated with P_1.

Case 9: S is the conclusion of a $T(P_1, a)$-inference, say, S'/S. Then we put $o(S) = \omega_d((o(S') \# \omega^{\nu+1})\omega^{\nu+1})$ where $d = h(S') - h(S)$ and where ν is the ordinal associated with the restriction of $\prec^*$ to $\{x/x \prec^* a\}$ and where $\prec^*$ is the Kleene-Brouwer ordering associated with P_1.

The ordinal $o(P)$ of a graded proof is by definition the ordinal of its endsequent. We have

Theorem 41: Let P be a graded s.n.s. proof in ZTFi/V. 1) "Omission of a cut" lowers the ordinal of P; 2) preliminary reduction steps do not increase the ordinal of P; 3) a classical logical reduction step lowers the ordinal of P; 4) an intuitionistic logical reduction step lowers the ordinal of P; 5) an induction reduction, a T_1- or a T_2-reduction step lowers the ordinal of P; 6) a subformula reduction step lowers the ordinal of P (with P intuitionistic in clauses 4)-6)).

Proof: Verification of the clauses 1)-5) leads precisely to the same calculations and inequalities encountered before. In the verification of clause 6) one encounters just one case not treated up to now, namely: P is strongly saturated, no preliminary and no strictly essential reduction step is applicable to P, and P contains a critical inference $p(t)=0, \Gamma \longrightarrow A/(x)p(x)=0, \Gamma \longrightarrow A$ whose principal formula has an image in the endsequent and such that $p(t)$ has value 0. Let S' and S be premiss and conclusion of the above inference, P' the result of the subformula reduction step and $o(S') = \alpha$, $o(S) = \beta$. By definition $\beta = \alpha \# 2$. It is trivial to verify that the application of the subformula reduction step lowers the ordinal of S: it becomes $\alpha \# 1$. Hence $o(P')$ is smaller than $o(P)$.

We also have

Theorem 42: Let P be an intuitionistic graded s.n.s. proof and assume that a V-reduction step is applied to the critical V-inference

$$V \quad \frac{t_P(y)=0,\ (x) <_R y\, A(x),\ \Gamma \longrightarrow A(y)}{W^o(<_R),\ t_R(q)=0,\ \Gamma \longrightarrow A(q)}$$

Let P_1 be the side proof determined by $W^o(<_R)$. If P_1 is "good", then the V-reduction step determined by the above V-inference transforms P into an intuitionistic graded s.n.s. proof whose ordinal is smaller than that of P.

The proof is practically the same as that of theorem 34 and hence omitted.

<u>Basic lemma III$_1$:</u> Let P be an intuitionistic graded s.n.s. proof with endsequent $G_1,\dots,G_s \longrightarrow H$. Let $S_1,\dots,S_m$ be the uppermost sequents of the final part, listed from left to right; let S_j be $\Gamma_j \longrightarrow A_j$. Then: if B occurs in Γ_j, if P_1 is the side proof determined by B in S_j (according to basic lemma III, the remark following it and definition 37), then P_1 is a graded intuitionistic s.n.s. proof and $o(P_1) < o(P)$.

<u>Proof:</u> We proceed as in the proof of basic lemma III and use the fact that in the construction of P_1 we use the operation "omission of a cut".

Of main importance for us is

<u>Corollary:</u> Let P be a graded intuitionistic s.n.s. proof containing a critical V-inference

$$V \quad \frac{t_R(y)=0,\ (x) <_R y\, A(x),\ \Gamma \longrightarrow A(y)}{W^o(<_R),\ t_R(q)=0,\ \Gamma \longrightarrow A(q)}$$

The side proof P_1 determined by $W^o(<_R)$ is a graded intuitionistic s.n.s. proof and $o(P_1) < o(P)$.

<u>6.6. The wellfoundedness proof</u>

<u>Theorem 43:</u> If P is an intuitionistic graded s.n.s. proof, then L_P is wellfounded.

Proof: We proceed by transfinite induction with respect to the ordinal $o(P)$. Let P be an intuitionistic graded proof with $o(P)=\xi$; assume that for all intuitionistic graded proofs P' with $o(P')<o(P)$ the relation $L_{P'}$ is wellfounded. We have to show that L_P is wellfounded. Case 1: First we assume that P is strongly saturated and does not admit preliminary reduction steps. If $L(P,P')$, then there is necessarily a strictly essential reduction step or a subformula reduction step which transforms P into P'. We distinguish two subcases. Subcase 1: The reduction step in question is a subformula reduction step or a strictly essential reduction step other than a V-reduction step. Then $o(P')<o(P)$ according to theorem 41 and hence $L_{P'}$ is wellfounded. Subcase 2: P' follows from P by means of a V-reduction step. Let

$$\text{V}\qquad \frac{t_R(y)=0,\ (x)<_R{}^{y}A(x),\ \Gamma\longrightarrow A(y)}{W^o(<_R),\ t_R(q)=0,\ \Gamma\longrightarrow A(q)}$$

be the critical V-inference in P, to which the V-reduction step in question is applied. Let P_1 be the side proof determined by $W^o(<_R)$. According to the corollary to basic lemma III_1, P_1 is a graded intuitionistic s.n.s. proof whose ordinal $o(P_1)$ is smaller than that of P. By the inductive assumption, it follows that L_{P_1} is wellfounded; hence P_1 is "good". This combined with theorem 42 shows that P' is again a graded intuitionistic s.n.s. proof with ordinal $o(P')<o(P)$. Hence $L_{P'}$ is wellfounded. Combining subcase 1 with subcase 2, we conclude that $L(P,P')$ implies the wellfoundedness of $L_{P'}$. But L_P is wellfounded if and only if $L_{P'}$ is wellfounded for all P' with $L(P,P')$. Hence L_P is wellfounded. Case 2: P is strongly saturated but admits preliminary reduction steps. Proceeding as in the proof of theorem 35, case B, we conclude that $L(P,P')$ implies $o(P')<o(P)$, hence the wellfoundedness of $L_{P'}$. From this we again infer the wellfoundedness of L_P.
Case 3: P is not strongly saturated and admits preliminary reduction steps. If $L(P,P')$ then P' is by definition of L strongly saturated and is subject to case 2; since $o(P')=o(P)$ holds, we infer the wellfoundedness of $L_{P'}$. This in turn implies the wellfoundedness of L_P, concluding the proof of the theorem.

Corollary 1: The relation L_P is wellfounded for every s.n.s. proof P in ZTi/V.

<u>Proof:</u> Every such proof P is obviously a graded proof since it does not contain side proofs at all. Hence it is subject to the previous theorem.

In order to prove the last corollary we need

<u>Definition 44:</u> a) Let P be a strongly saturated intuitionistic s.n.s. proof which does not admit preliminary nor strictly essential reduction steps. A proof P' is said to follow from P by means of a weak subformula reduction step if P contains a critical inference $p(t)=o, \Gamma \longrightarrow A((x)p(x)=0, \Gamma \longrightarrow A$ with $p(t)$ true, and if P' follows from P by replacing this inference by

$$\cfrac{\cfrac{\longrightarrow p(t)=0 \quad p(t)=0, \Gamma \longrightarrow A}{\Gamma \longrightarrow A}\ \text{cut}}{(x)p(x)=0, \Gamma \longrightarrow A}$$

b) By L* we denote the relation which applies to P, P* (in signs L*(P,P*)) iff P, P' are intuitionistic s.n.s. proofs and if either R(P,P') holds, or if else there is a list $P_o, \dots, P_N$ of such proofs such that a) $P=P_o$, b) P is strongly saturated, c) P_{i+1} follows from P_i by means of a preliminary reduction step, d) no preliminary reduction step is applicable to P_N, e) P' follows from P_N by means of a weak subformula reduction step.

<u>Corollary 2:</u> Let P be an s.n.s. proof in ZTi/V whose endsequent S does not contain free variables nor special function constants.
a) If S is $(x)p_1(x)=0, \dots, (x)p_n(x)=0 \longrightarrow A \vee B$ (with the p_i's terms), then one effectively finds a proof P_1 in ZTi/V of either $(x)p_1(x)=0, \dots, (x)p_n(x)=0 \longrightarrow A$ or $(x)p_1(x)=0, \dots, (x)p_n(x)=0 \longrightarrow B$; b) if S is $(x)p_1(x)=0, \dots, (x)p_n(x)=0 \longrightarrow (E\xi)A(\xi)$, then one effectively finds a functor F without free variables and special function constants and a proof P in ZTi/V of $(x)p_1(x)=0, \dots, (x)p_n(x)=0 \longrightarrow A(F)$; c) similarly with (Ex) in place of $(E\xi)$ and a term t in place of F .

<u>Proof:</u> We consider e.g. b). P is clearly an intuitionistic graded s.n.s. proof, since no side proofs at all occur in P . Therefore L_P is wellfounded. Denote by L_P^* the restriction of L* to D_P . Since L* is a subrelation of L, it follows that L_P^* is wellfoun-

ded. Hence, we effectively find a chain $P_o,\dots,P_N$ with $P_o=P$ and such that a) $L^*(P_i,P_{i+1})$ holds for $i<N$, b) $(\forall X)\neg L^*(P_N,X)$. Obviously P_N is strongly saturated. By induction with respect to i one easily shows that the endsequent of P has the form

1) $(x)p_{\alpha_1}(x)=0,\dots\dots,(x)p_{\alpha_k}(x)=0 \longrightarrow (E\xi)A(\xi)$ or

$(x)p_{\alpha_1}(x)=0,\dots\dots,(x)p_{\alpha_k}(x)=0 \longrightarrow$. In case 1) k can be 0, in case 2) necessarily $k\neq 0$ since the last theorem implies consistency of ZTi/V . Let us apply in an arbitrary but fixed way preliminary reduction steps to P_N so as to obtain a proof P_N' which does not admit preliminary reduction steps. Evidently, P_N' is strongly saturated and does not admit strictly essential reduction steps since otherwise $L(P_N,P_N')$ would hold, contradicting the assumption. According to theorem 39, there is a critical logical inference whose principal formula has an image in the endsequent.

<u>Case 1:</u> The inference is $p(t)=0,\Gamma\longrightarrow C/(x)p(x)=0,\Gamma\longrightarrow C$. Then $p(t)=0$ is false by necessity. Otherwise we could apply a weak subformula reduction step to P_N', obtaining as result a proof P_N'' which would satisfy $L^*(P_N,P_N'')$, contradicting the assumption. Hence $p(t)=0$ is false, hence $p(t)=0\longrightarrow$ an axiom and $p(x)$ isomorphic with some $p_i(x)$. With the aid of an $\forall\longrightarrow$ inference, followed by conversions and interchanges, we can derive $(x)p_1(x)=0,\dots\dots,(x)p_n(x)=0 \longrightarrow A(F)$ for any functor F .

<u>Case 2:</u> There is no critical logical inference of the form $p(t)=0,\Gamma\longrightarrow C/(x)p(x)=0,\Gamma\longrightarrow C$ in P_N' . Then P_N' contains necessarily a critical logical inference of the form $\Gamma\longrightarrow A'(F)/\Gamma\longrightarrow(E\xi)A'(\xi)$ whose principal formula has an image in the endsequent. $A'(\xi)$ is necessarily isomorphic with $A(\xi)$. Without loss of generality we can assume that F does not contain free variables and special function constants: the first is a consequence of the fact that P_N' is an s.n.s. proof, the second can always be achieved by replacing eventually some special function constants by suitably chosen constants for prim. rec. functions. By application of a subformula reduction step to P_N' followed by a conversion, some thinnings and interchanges, we obtain a proof P_N'' in ZTFi/V of $(x)p_1(x)=0,\dots,(x)p_n(x)=0 \longrightarrow A(F)$. By means of theorem 38, we can transform P_N'' into a proof P' in ZTi/V of $(x)p_1(x)=0,\dots,(x)p_n(x)=0 \longrightarrow A(F)$, what concludes the proof.

<u>Remark:</u> In virtue of the equivalence of quantifierfree formulas with prime formulas, the last corollary remains true if we replace $p_1(x)=0,\dots\dots,p_n(x)=0$ by quantifierfree formulas $Q_1(x),\dots,Q_n(x)$, respectively.

CHAPTER VII:

A system containing barinduction with respect to decidable predicates

In this chapter we show that a reasoning very similar to that presented in chapters V, VI can be applied to the theory ZTi/I. There is, however, an essential difference between the methods presented in chapters V, VI and the method presented in this chapter: the former yield automatically the consistency of the theory to which they are applied, the latter, however, works only if we assume ab initio that ZTi/I is consistent. Hence let us assume throughout this chapter: ZTi/I is consistent.

7.1. The theory ZTi/I and a certain conservative extension

A. The theory ZT/I is obtained from ZT by additon of the following rule I:

$$\text{I.}\qquad \frac{R(y),\ (x)\subset_R y\, A(x),\ \Gamma \longrightarrow \Delta, A(y)}{W(\subset_R),\ R(q),\ \Gamma \longrightarrow \Delta, A(q)}$$

where y and q are subject to the usual stipulations. Here, R is an arbitrary standard formula, that is, a formula of the form $R_0(x) \wedge seq(x)$; no restrictions are thereby imposed on $R_0(x)$, that is, $R_0(x)$ can be any formula containing special function constants and free variables of any kind. $x \subset_R y$ and $W(\subset_R)$ are again abbreviations for $x \subset_K y \wedge R(x) \wedge R(y)$ and $(\xi)(Ex)\neg\, \xi(x+1) \subset_R \xi(x)$, respectively. ZTi/I is obtained from ZT/I by restricting attention to intuitionistic proofs.

B. Next, some notations. In part C below, R(x) denotes a standard formula whose special function constants are $\alpha^{i_1}_{u_1}, \ldots, \alpha^{i_s}_{u_s}$ and whose only free variable is x . In order to indicate the occurence of the $\alpha^{i_k}_{u_k}$'s, we write as before $R(\vec{\alpha}_u, x)$ or $R(\alpha^{i_1}_{u_1}, \ldots, \alpha^{i_s}_{u_s}, x)$. Replacement of $\alpha^{i_1}_{u_1}, \ldots, \alpha^{i_s}_{u_s}$ by $\alpha^{i_1}_{u_1 * v_1}, \ldots, \alpha^{i_s}_{u_s * v_s}$ transforms $R(\vec{\alpha}_u, x)$ into another formula which will be written as $R(\vec{\alpha}_{u*v}, x)$ or, more briefly, as $R_v(x)$ or even R_v . Of course $\alpha^{i_1}_{u_1}, \ldots, \alpha^{i_s}_{u_s}$ are precisely the special function constants which occur in $x \subset_R y$. In order to indicate their occurence in $x \subset_R y$, we sometimes also write $x \subset^u_R y$. Hence $x \subset^{\cdot}_R y$ and $x \subset^u_R y$ are both abbreviations of one and the same formula: $x \subset_K y \wedge R(\vec{\alpha}_u, x) \wedge R(\vec{\alpha}_u, y)$. If we replace $\alpha^{i_1}_{u_1}, \ldots, \alpha^{i_s}_{u_s}$ in $x \subset_R y$ by $\alpha^{i_1}_{u_1 * v_1}, \ldots, \alpha^{i_s}_{u_s * v_s}$, then we obtain a new formula which may be written as $x \subset_{R_v} y$. For convenience we denote this formula also by $x \subset^{u*v}_R y$. Hence, $x \subset_{R_v} y$ and $x \subset^{u*v}_R y$ both denote $x \subset_K y \wedge R(\vec{\alpha}_{u*v}, x) \wedge R(\vec{\alpha}_{u*v}, y)$.

C. We now introduce a conservative extension of ZTi/I which is related to ZTi/I in the same way as eg. $ZTEi/V_N$ is related to ZTi/V_N . This conservative extension is denoted by ZTGi/I and is obtained from ZTi/I by addition of two new rules $T(P_1)$ and $T(P_1, P_2)$ whose definition is given below. 1) Let $v_1, \ldots, v_s$ be a list of sequence numbers, such that $\text{length}(v_1) = \text{length}(v_i)$, $i = 1, \ldots, s$. Let P_o be a proof in ZTGi/I , whose endsequent S is $R(\vec{\alpha}_{u*v}, y), (x) \subset^{u*v}_R {}_y A(x), \Gamma \longrightarrow A(y)$; let P_1 be a proof in ZTGi/I , already at hand, whose endsequent is $\longrightarrow W(\subset^u_R)$. Then

$$T(P_1) \qquad \frac{\begin{array}{c} P_o \\ \vdots \\ S \end{array}}{R(\vec{\alpha}_{u*v}, q), \Gamma \longrightarrow A(q)}$$

is a proof in ZTGi/I ; we denote it by P . The inference

$$T(P_1) \qquad \frac{R(\vec{\alpha}_{u*v}, y), (x) \subset^{u*v}_R {}_y A(x), \Gamma \longrightarrow A(y)}{R(\vec{\alpha}_{u*v}, q), \Gamma \longrightarrow A(q)}$$

is called a $T(P_1)$-inference. P_1 is called side proof of this inference. P_1 is also said to be a side proof of P. 2) Let $v_1,\dots\dots,v_s$ and $w_1,\dots\dots,w_s$ be two lists of sequence numbers, denoted briefly by v and w ; assume $w_i \subseteq_K v_i$ for $i=1,\dots.,s$ and in addition $\mathrm{length}(v_1)=\mathrm{length}(v_i)$, $\mathrm{length}(w_1)=\mathrm{length}(w_i)$ for $i=1,\dots\dots,s$. Let P_o be a proof in ZTGi/I whose endsequent S is $y \subset_R^{u*w} t$, $(x) \subset_R^{u*w_y} A(x)$, $\Gamma \longrightarrow A(y)$; t is assumed to be saturated, $|t|=a$. Let P_1 be a proof in ZTGi/I already at hand of $\longrightarrow W(\subset_R^u)$ and P_2 another proof in ZTGi/I already at hand of $\longrightarrow R(\vec{\alpha}_{u*v}, t)$. Then

$$T(P_1,P_2)\ \frac{\begin{matrix} P_o \\ \vdots \\ S \end{matrix}}{q \subset_R^{u*w} t\ ,\ \Gamma \longrightarrow A(q)}$$

is a proof in ZTGi/I ; we denote it by P . The inference

$$T(P_1,P_2)\ \frac{y \subset_R^{u*w} t\ ,\ (x) \subset_R^{u*w_y} A(x),\ \Gamma \longrightarrow A(y)}{q \subset_R^{u*w} t\ ,\ \Gamma \longrightarrow A(q)}$$

is called a $T(P_1,P_2)$-inference. P_1 is called a side proof of this inference, P_2 is called the index proof of this inference. P_1 is again called side proof of P while P_2 is called an index proof of P .

<u>Remarks:</u> a) q and y in 2) and 3) above are subject to the usual stipulations. b) The description of ZTGi/I can, of course, be made more precise by associating inductively with every proof in ZTGi/I a type (m,i) in the same way as in chapters V, VI. c) If P_1 is a side proof of P , and if S is an occurence of a sequent in P_1 , then we do <u>not</u> consider S as an occurence of a sequent in P . Similarly, if P_1 is an index proof of P .

<u>D.</u> There is also a conservative extension ZTG/I of ZT/I whose definition is obtained from that of ZTGi/I by means of the following changes: a) in clause 1) in part B we permit P to be a proof in ZT/I ; b) in clauses 2) and 3) P_o is a proof in ZTG/I ; c) premiss and conclusion of a $T(P_1)$- or a $T(P_1,P_2)$-inference, respectively, are permitted to contain more than one formula in the succedent. The side proof P_1 and the index proof P_2, how-

ever, are still assumed to be proofs in ZTGi/I . The theory ZTG/I has been introduced for technical purposes only.

E. The main result about ZTG/I and ZTGi/I is given by

Theorem 44: a) ZTGi/I is a conservative extension of ZTi/I ; b) ZTG/I is a conservative extension of ZT/I.

The proof is essentially the same as that of theorem 38; one uses thereby the fact that types (m,i) can be associated with proofs in ZTGi/I and ZTG/I respectively.

F. For proofs P in ZTGi/I and ZTG/I, we can introduce the usual notions such as final part, complexity of a cut, of a I-inference, of a $T(P_1)$-inference, of a $T(P_1,P_2)$-inference, of a fork, etc. We use all these notions without any further comment; their definitions remain the same as before. A standard proof e.g. is again a proof whose endsequent has the form $\longrightarrow A$. Strictly normal standard proofs (s.n.s. proofs) will again be the objects with which we work most of the time. A further notion, which can be taken over without changes, is that of substitution instance; it is again given by definition 20, sect. 4.4., chapter IV.

7.2. Remarks about the basic lemma

A. The basic lemma will be used in the form given by basic lemma II (chapter III, sect. 3.2.). Let P be a proof in ZTGi/I , and $\Gamma \longrightarrow A$ an uppermost sequent in the final part of P and B a formula in Γ . The procedure described in the proof of basic lemma II associates with B a welldetermined proof P_1 in ZTGi/I of $\longrightarrow B$; we call P_1 the side proof determined by B in $\Gamma \longrightarrow A$. If, in particular, $\Gamma \longrightarrow A$ is the conclusion of a critical I-inference, say

$$\text{I.} \quad \frac{R(y),\ (x) \subset_{R^y} A(x),\ \Gamma_o \longrightarrow A(y)}{W(\subset_R),\ R(q),\ \Gamma_o \longrightarrow A(q)}$$

if B is $W(\subset_R)$, then P_1 is also called the side proof of $\longrightarrow W(\subset_R)$ determined by this particular I-inference.

7.3. Reduction steps for ZTGi/I

A. Now we introduce reduction steps for proofs P in ZTGi/I . Among these we have preliminary reduction steps, induction reductions and intuitionistic fork elimination (intuitionistic logical reduction steps). Their definitions remain the same as in all previous chapters. Next we have three kinds of reduction steps which are associated with I- , $T(P_1)$- and $T(P_1,P_2)$-inferences and which are called I-reduction steps, $T(P_1)$-reduction steps and $T(P_1,P_2)$-reduction steps respectively.

$T(P_1)$-reduction steps. Let P be a saturated s.n.s. proof in ZTGi/I , which contains a critical $T(P_1)$-inference S'/S , say

$$T(P_1)\qquad \frac{R(\vec{\alpha}_{u*v},y),\ (x)\subset^{u*v}_{R\;y}A(x),\ \Gamma\longrightarrow A(y)}{R(\vec{\alpha}_{u*v},q),\ \Gamma\longrightarrow A(q)}$$

where P_1 is by definition a proof of $\longrightarrow W(\subset^{u}_{R})$ in ZTGi/I. Let P be the side proof of $\longrightarrow R(\vec{\alpha}_{u*v},q)$, determined by $R(\vec{\alpha}_{u*v},q)$ according to basic lemma II. Let p_o be a cut free proof in ZTi which does not contain induction and whose endsequent is $y\subset^{u*v}_{R}q\longrightarrow R(\vec{\alpha}_{u*v},y)$. Let P_S be the subproof of S in P , $P_{S'}$ the subproof of S' in P and $P^q_{S'}$ the result of replacing every occurence of y in $P_{S'}$ by q ; let S'_q be the endsequent of $P^q_{S'}$. Then we can replace P_S by the following derivation P* :

$$\begin{array}{ll}
\begin{array}{c}P_o\\ \vdots\\ y\subset^{u*v}_{R}q\longrightarrow R(\vec{\alpha}_{u*v},q)\end{array}\qquad\qquad \begin{array}{c}P_{S'}\\ \vdots\\ S'\end{array} & \\
\hline
y\subset^{u*v}_{R}q\ ,\ (x)\subset^{u*v}_{R\;y}A(x),\ \Gamma\longrightarrow A(y) & \text{cut}\\
\hline
s\subset^{u*v}_{R}q,\ \Gamma\longrightarrow A(x) & T(P_1,P_2)\\
\hline\hline
\Gamma\longrightarrow (x)\subset^{u*v}_{R\;q}A(x)\qquad\qquad S'_q \quad (\Rightarrow\forall;\ P^q_{S'}) & \\
\hline\hline
R(\vec{\alpha}_{u*v},q),\ \Gamma\longrightarrow A(q) & \text{cut, interchanges}
\end{array}$$

The result of this replacement is a proof P' which is said to follow from P by means of a $T(P_1)$-reduction step. We say that the

$T(P_1)$-reduction step has been applied to the particular $T(P_1)$-inference above. We also say that the $T(P_1)$-inference is transformed by means of the $T(P_1)$-reduction step into the $T(P_1,P_2)$-inference, which appears in the last diagram.

<u>$T(P_1,P_2)$-reduction steps.</u> Let us retain the notation introduced in part B of sect. 7.1. and in the definition of $T(P_1,P_2)$-inference. In particular, $v_1,\ldots\ldots,v_s$ and $w_1,\ldots\ldots,w_s$ are two lists of sequence numbers such that $\mathrm{length}(v_1) = \mathrm{length}(v_i)$ and $\mathrm{length}(w_1) = \mathrm{length}(w_i)$, and such that $w_i \subseteq_K v_i$, $i=1,\ldots\ldots,s$. These two lists are again denoted by v and w respectively. Let P be a saturated s.n.s. proof which contains a critical $T(P_1,P_2)$-inference S'/S , say

$$T(P_1,P_2)\ \frac{y \subset_R^{u*w} t\ ,\ (x) \subset_R^{u*w}{}_y A(x),\ \Gamma \longrightarrow A(y)}{q \subset_R^{u*w} t\ ,\ \Gamma \longrightarrow A(q)}$$

Here P_1 is a proof in ZTGi/I of $\longrightarrow W(\subset_R^u)$ while P_2 is a proof of $\longrightarrow R(\vec{\alpha}_{u*v},t)$. Now to the $T(P_1,P_2)$-reduction step. First we note that the following sequents can be proved in ZTi without cuts and inductions: 1) $q \subset_R^{u*w} t \longrightarrow R(\vec{\alpha}_{u*w},q)$, 2) $y \subset_R^{u*w} q\ ,\ q \subset_R^{u*w} t \longrightarrow y \subset_R^{u*w} t$. Let $\hat{P}$ be such a proof of the first sequent and P_o be such a proof of the second sequent. Next we can extract according to basic lemma II the side proof $\tilde{P}$ determined by $q \subset_R^{u*w} t$ in S . By combining $\tilde{P}$ and $\hat{P}$ by means of a cut we obtain a proof P_2' in ZTGi/I of $\longrightarrow R(\vec{\alpha}_{u*w},q)$. Let again P_S and $P_{S'}$ be the subproofs of S and S' respectively. By $P_{S'}^q$ we denote the result of replacing every occurence of y in $P_{S'}$ by q : again S_q' denotes the end-sequent of $P_{S'}^q$. Then we replace P_S by the following derivation:

$$\dfrac{\dfrac{\begin{array}{c}P_o\\ \vdots\\ y \subset_R^{u*w} q\ ,\ q \subset_R^{u*w} t \longrightarrow y \subset_R^{u*w} t\end{array} \qquad \begin{array}{c}P_{S'}\\ \vdots\\ S'\end{array}}{y \subset_R^{u*w} q\ ,\ (x)\subset_R^{u*w}{}_y A(x),\ q \subset_R^{u*w} t,\ \Gamma \longrightarrow A(y)} \text{ cut, interchanges}}{\dfrac{\dfrac{s \subset_R^{u*w} q\ ,\ q \subset_R^{u*w} t,\ \Gamma \longrightarrow A(s)}{q \subset_R^{u*w} t,\ \Gamma \longrightarrow (x) \subset_R^{u*w} A(x)} \Rightarrow\forall \qquad \begin{array}{c}P^q_{S'}\\ \vdots\\ S'_q\end{array}}{q \subset_R^{u*w} t,\ \Gamma \longrightarrow A(q)} \text{ cut, interchanges, contractions}} \; T(P_1,P_2')$$

The result of this reduction is a proof P' which is said to follow from P by means of a $T(P_1,P_2)$-reduction step. We say that a $T(P_1,P_2)$-reduction step has been applied to the particular $T(P_1,P_2)$-inference above. The $T(P_1,P_2)$-inference, to which the reduction step is applied, is said to be transformed by the reduction step into the $T(P_1,P_2')$-inference, which appears in the last diagram.

<u>I-reduction steps.</u> Let P be a saturated s.n.s. proof in ZTGi/I, containing a critical I-inference, say

$$\text{I.} \qquad \frac{R(y),\ (x) \subset_R{}^y A(x),\ \Gamma \longrightarrow A(y)}{W(\subset_R),\ R(q),\ \Gamma \longrightarrow A(q)}$$

to be denoted by S'/S. Let P_1 be the side proof determined by $W(\subset_R)$ in S according to basic lemma II; its endsequent is $\longrightarrow W(\subset_R)$. Then we can alter P as follows:

$$\begin{array}{ll} & P_{S'} \\ & \vdots \\ T(P_1) & \dfrac{R(y),\ (x) \subset_R{}^y A(x),\ \Gamma \longrightarrow A(y)}{\dfrac{R(q),\ \Gamma \longrightarrow A(q)}{W(\subset_R),\ R(q),\ \Gamma \longrightarrow A(q)}} \\ \text{thinning} & \end{array}$$

The proof P' which is obtained from P by means of this alteration is said to follow from P by means of a I-reduction step. We say that a I-reduction step has been applied to the particular I-inference above. The I-inference is said to be transformed by the reduction step into the $T(P_1)$-inference, which appears in the last diagram.

B. What we actually need below are not the $T(P_1)$- and $T(P_1,P_2)$-reduction steps themselves, but slight variants of them, called strong $T(P_1)$- and strong $T(P_1,P_2)$-reduction steps. They are introduced by the following

Definition 45: Let P be a s.n.s. proof in ZTGi/I and $\alpha^{i_1}_{u_1}, \ldots\ldots, \alpha^{i_s}_{u_s}$ the special function constants which occur in P. Let $P_{v_1 \ldots\ldots v_s}$ be the result of replacing every occurence of $\alpha^{i_k}_{u_k}$ $(k=1,\ldots\ldots,s)$ by $\alpha^{i_k}_{u_k * v_k}$. An s.n.s. proof P' in ZTGi/I is said to follow from P by means of a strong $T(P_1)$- $(T(P_1,P_2)$-$)$ reduction step if there are sequence numbers $v_1,\ldots\ldots,v_s$ of length 1 such that P' follows from $P_{v_1 \ldots\ldots v_s}$ by means of a $T(P_1)$- $(T)P_1,P_2)$-$)$ reduction step.

C. For nonintuitionistic proofs P in ZTG/I, we merely need preliminary reduction steps (including "omission of a cut") and logical reduction steps (fork elimination) which are, of course, defined in the usual way. The only kind of nonintuitionistic proofs which will appear (implicitely) below are almost intuitionistic proofs in the sense of chapter III (sect. 3.1., pt. A). Such proofs appear in the proof of a theorem (a variant of theorems 33, 41) which states among others that an intuitionistic logical reduction step lowers the ordinal of the proof to which it is applied (with respect to an ordinal assignement to be defined below). Apart from this, nonintuitionistic proofs will not be encountered.

D. A reduction step is called strictly essential if it is a logical reduction step, an induction reduction, a I-reduction step, a strong $T(P_1)$-reduction step or a strong $T(P_1,P_2)$-reduction step. A saturated proof is as usual one all whose constant terms in the final part are saturated. The notion of inessential reduction step is again given by definition 20 (Chapter IV, sect. 4.4., pt. C). With respect to strictly essential reduction steps we have in analogy with theorem 39:

Theorem 45: Let P be a saturated s.n.s. proof in ZTGi/I which does not coincide with its final part and which does not admit preliminary nor strictly essential reduction steps. Then there exists a critical logical inference whose principal formula has an image in the endsequent.

Proof: The same as that of theorem 39.

E. The notion of subformula reduction step is introduced in the same way as in section 4.4. (part D) of chapter IV. In analogy with theorem 21 we have

Theorem 21*: If P is a saturated intuitionistic s.n.s. proof in ZTGi/I which does not coincide with its final part, and if P does not admit preliminary nor strictly essential reduction steps, then we can apply a subformula reduction step to P .

7.3a. Good proofs

A. In order to be able to introduce ordinals into our consideration, we introduce relations $\hat{R}$ and $\hat{L}$ whose definitions are given by definitions 22 and 23 in sect. 4.5. of chapter IV. $\hat{R}$ and $\hat{L}$ are counterparts of R and L and behave very similarly; in particular, they satisfy a slight variant of theorem 22, part a), which, however, will not be needed here. Without danger of confusion, we write R and L in place of $\hat{R}$ and $\hat{L}$. Using definition 32 in chapter V, sect. 5.3. as it stands, we can associate with every s.n.s. proof P in ZTGi/I the set D_P of proofs and the restriction L_P of L to D_P. With respect L_P and D_P, we have a theorem, which corresponds to theorem 32. In order to state it, we remind that R in $W(\subset^u_R)$ is a standard formula, whose only free variable is x and whose list of special function constants is given by $\alpha^{i_1}_{u_1}, \ldots, \alpha^{i_s}_{u_s}$.
$x \subset^u_R y$ is used as abbreviation for
$x \subset_K y \wedge R(\vec{\alpha}_u, x) \wedge R(\vec{\alpha}_u, y)$ and $W(\subset^u_R)$ is an abbreviation for $(\xi)(Ex)\neg\, \xi(x+1) \subset^u_R \xi(x)$. Now to the theorem.

Theorem 46: Let P be an s.n.s. proof in ZTGi/I of $\longrightarrow W(\subset^{u}_{R})$ and assume that L_P is wellfounded. Let $f_1,\ldots,f_s$ and g be numbertheoretic functions. Then we find an m and an n with $n+1 < m$ and a proof P' in D_P of $\longrightarrow \neg \xi_w(n+1) \subset^{u*v}_{R} \xi_w(n)$ where v denotes the system $v_i = \bar{f}_i(m)$, $i=1,\ldots,s$ of sequence numbers and where $w=\bar{g}(m)$.

Proof: The proof is essentially the same as that of theorem 24.

This gives rise to

Definition 46: An s.n.s. proof P in ZTGi/I is said to be a good proof if L_P is wellfounded.

Definition 46a According to theorem 46 we can associate with every good proof P , whose endsequent has the form $\longrightarrow W(\subset^{u}_{R})$, a continuity function τ^P having the following properties: if $f_1,\ldots,f_s$ and g are numbertheoretic functions, if moreover $\tau^P(\bar{f}_1(m),\ldots,\bar{f}_s(m),\bar{g}(m)) \neq 0$, then there is an n with $n+1 < m$ and a proof $P' \in D_P$ of $\longrightarrow \neg \xi_w(n+1) \subset^{u*v}_{R} \xi_w(n)$ where v and w have the same meaning as in theorem 46. τ^P is called the continuity function determined by P .

In connection with good proofs we again introduce the notion of graded proof.

Definition 47: An s.n.s. proof P in ZTGi/I or ZTG/I is said to be graded if all its side proofs are good.

Remark: We note that this definition imposes no condition on the index proofs of P . Lemma 13 in chapter V remains true in the present case as is evident to see.

7.4. Valuation of proofs

A. In order to be able to introduce ordinals into our considerations, we need an additional concept, that of valuation of a proof. We start with some preliminaries. By D^s we denote the set of ordered s+1-tuples of sequence numbers $\langle v_1,\ldots,v_s,v_{s+1}\rangle$ for which $\mathrm{length}(v_1)=\mathrm{length}(v_i)$, $i=1,\ldots,s+1$ holds. The partial ordering $\sqsubset^s$ of D^s is given as follows:

$<\bar{\alpha}_1(x),\ldots,\bar{\alpha}_s(x),\bar{\alpha}_{s+1}(x)> \sqsubset^s <\bar{\beta}_1(t),\ldots,\bar{\beta}_s(t),\bar{\beta}_{s+1}(t)>$ iff $t < x$ and $\bar{\alpha}_i(t)=\bar{\beta}_i(t)$ for $i=1,\ldots,s+1$.

Definition 48: Let P be a good proof of $\longrightarrow W(\subset^u_R)$ where R is the formula $R(\alpha^{i_1}_{u_1},\ldots,\alpha^{i_s}_{u_s},x)$. Let τ^P be the continuity function determined by P. An element $e=<v_1,\ldots,v_{s+1}> \in D^s$ is said to be unsecured with respect to P if $\tau^P(v_1,\ldots,v_{s+1})=0$ and secured otherwise.

B. In connection with the concept of unsecured element with respect to a good proof P, we use the following notation: 1) if P is a good proof of $\longrightarrow W(\subset^u_R)$ (with R denoting $R(\alpha^{i_1}_{u_1},\ldots,\alpha^{i_s}_{u_s},x)$), then $D^s(P)$ is the subset of D^s consisting of those elements $e \in D^s$, which are unsecured with respect to P; 2) the restriction of $\sqsubset^s$ to $D^s(P)$ is denoted by $\sqsubset^s_P$. Concerning $D^s(P)$, we have the following rather evident

Lemma 18: Let P be a good proof of $\longrightarrow W(\subset^u_R)$ (with R denoting $R(\alpha^{i_1}_{u_1},\ldots,\alpha^{i_s}_{u_s},x)$). The restriction $\sqsubset^s_P$ of $\sqsubset^s$ to $D^s(P)$ is wellfounded.

We omit the rather obvious proof.

C. Now to the concept of valuation. A valuation of a proof P in ZTG/I is a function (or an assignement) which associates with every $T(P_1,P_2)$-inference in P either a number e which satisfies a certain condition α) to be explained below, or else a pair of numbers e, e_1 which satisfy a certain condition β) to be explained below. In order to explain this concept more properly, let $v_1,\ldots,v_s$ and $w_1,\ldots,w_s$ be two lists of sequence numbers, denoted by v and w, respectively, such that

a) $\text{length}(v_1)=\text{length}(v_i)$, $i=1,\ldots,s$,

b) $length(w_1)=length(w_i)$, $i=1,\dots,s$,

c) $w_i \subseteq_K v_i$, $i=1,\dots,s$. Let P contain a $T(P_1,P_2)$-inference, say

$$T(P_1,P_2)\quad \frac{y \subset_R^{u*w} t,\ (x) \subset_R^{u*w_y} A(x),\ \Gamma \longrightarrow \Delta,A(y)}{q \subset_R^{u*w} t,\ \Gamma \longrightarrow \Delta,A(q)}$$

Here P is a proof in ZTGi/I of $\longrightarrow W(\subset_R^u)$ (with R as usual $R(\alpha_{u_1}^{i_1},\dots,\alpha_{u_s}^{i_s},x)$ containing no other free variable than x), while P_2 is a proof in ZTGi/I of $\longrightarrow R(\vec{\alpha}_{u*v},t)$. Let a valuation of P be given.

<u>Case 1:</u> The valuation associates with the above $T(P_1,P_2)$-inference a number e . Then e satisfies the following condition $\alpha)$:
a) e is of the form
$<\bar{\alpha}_1(x),\dots,\bar{\alpha}_s(x),\bar{\beta}(x)>$; b) $x=1$; c) $w_i \subseteq_K \bar{\alpha}_i(x)$, $i=1,\dots,s$; d) $\beta(0)=|t|$.

<u>Case 2:</u> The valuation associates with the above $T(P_1,P_2)$-inference a pair e , e_1 of numbers. Then e and e_1 satisfy the following condition $\beta)$: a) e has the form $<\bar{\alpha}_1(x),\dots,\bar{\alpha}_s(x),\bar{\beta}(x)>$ with $x\geq 2$; b) $\beta(x-1)=|t|$; c) e_1 has the form $\bar{\zeta}(x-1)$; d) if $i<x-1$, then there are sequence numbers $w_1',\dots,w_s'$, depending on i and all of the same length, such that $w_i \subseteq_K w_i'$, $i=1,\dots,s$ and such that $\zeta(i)$ is the Gödelnumber of a proof P_i in ZTGi/I of $\longrightarrow \beta(i+1) \subset_R^{u*w'} \beta(i)$ (where w' denotes the list $w_1',\dots,w_s'$) .

There are clearly proofs which do not admit a valuation: if eg. $w_i=<>$, $i=1,\dots,s$, then neither condition $\alpha)$ nor $\beta)$ can be satisfied. If, on the other hand, P does not contain $T(P_1,P_2)$-inferences at all, then it clearly admits a valuation, the so-called empty valuation. <u>Notation:</u> Valuations are denoted by symbols such as $\mathcal{V}$, $\mathcal{W}$, $\mathcal{V}_1$, $\mathcal{V}_2$ etc.. If S_1/S_2 is a $T(P_1,P_2)$-inference in P, then we denote the value of $\mathcal{V}$ for this inference by $\mathcal{V}(S_1/S_2)$.

<u>D.</u> Let P be an s.n.s. proof in ZTG/1 provided with a valuation $\mathcal{V}$. Let P' be a substitution instance of P or else be obtained from P by means of a reduction step. Then we can define on P' in a natural way a valuation $\mathcal{W}$ in terms of $\mathcal{V}$ which will be called the valuation induced by $\mathcal{V}$ on P' and denoted by $\mathcal{V}^*$. In order to define $\mathcal{V}^*$ it is useful to have three supplementary concepts at hand, that of <u>extension</u> of a $T(P_1,P_2)$-inference, of <u>data</u> and of <u>index</u> of a $T(P_1,P_2)$-inference or a $T(P_1)$-inference, respectively. Consider two $T(P_1,P_2)$-inferences, say

$$T(P_1,P_2)\quad \frac{y \subset_R^{u*w} t,\ (x) \subset_R^{u*w_y} A(x),\ \Gamma \longrightarrow \Delta, A(y)}{q \subset_R^{u*w} t,\ \Gamma \longrightarrow \Delta, A(q)}$$

and

$$T(P_1,P_2)\quad \frac{y \subset_R^{u*w'} t,\ (x) \subset_R^{u*w'_y} B(x),\ \Gamma' \longrightarrow \Delta', B(y)}{q' \subset_R^{u*w'} t,\ \Gamma' \longrightarrow \Delta', B(q')}$$

Here R denotes $R(\alpha_{u_1}^{i_1},\ldots\ldots,\alpha_{u_s}^{i_s},x)$, P_1 is a proof of $\longrightarrow W(\subset_R^u)$ and w and w' denote $w_1,\ldots,w_s$ and $w'_1,\ldots,w'_s$, respectively. The second inference is said to be an <u>extension</u> of the first if $w'_i \subseteq_K w_i$, $i=1,\ldots,s$; it is called a <u>strict extension</u> of the first if $w'_i \subset_K w_i$, $i=1,\ldots,s$. The formula $R(\alpha_{u_1}^{i_1},\ldots\ldots,\alpha_{u_s}^{i_s},x)$, the list $w_1,\ldots,w_s$ and the number $|t|$ are called the <u>data</u> of the first of the above $T(P_1,P_2)$-inferences and the term q is called the <u>index</u> of this inference. Similarly, if a $T(P_1)$-inference is given, say

$$T(P_1)\quad \frac{R(\vec{\alpha}_{u*v},y),\ (x) \subset_R^{u*v_y} A(x),\ \Gamma \longrightarrow \Delta, A(y)}{R(\vec{\alpha}_{u*v},q),\ \Gamma \longrightarrow \Delta, A(q)}$$

(with R,P_1 as before and v denoting $v_1,\ldots,v_s$), then $R(\vec{\alpha}_u,y)$ and $v_1,\ldots,v_s$ are the <u>data</u> of this inference, while the term q is called its <u>index</u>.

Now to the definition of γ^* . We distinguish cases according to the kind of reduction step which leads from P to P' .

<u>Case 1:</u> P' is a substitution instance of P . Then each $T(P_1,P_2)$-inference S/S' in P is transformed into a $T(P_1,P_2)$-inference S_1/S_1' in P' which is a strict extension of S/S' . Then we put $\gamma^*(S_1/S_1') = \gamma(S/S')$. γ^* , thus defined is certainly a valuation.

<u>Case 2:</u> P' is obtained from P by means of an inessential reduction step. This is a special case of case 1.

<u>Case 3:</u> P' follows from P by means of a subformula reduction step. Each $T(P_1,P_2)$-inference S/S' in P is transformed into a $T(P_1,P_2)$-inference S_1/S_1' in P' which is an extension of S/S' . We put $\gamma^*(S_1/S_1') = \gamma(S/S')$.

<u>Case 4:</u> P' is obtained from P by means of an induction reduction. This induction reduction transforms each $T(P_1,P_2)$-inference S/S' into n images S_i/S_i' , $i=1,\ldots,n$ (with n depending on S/S'), each of which is a $T(P_1,P_2)$-inference which is an extension of S/S'. We put $\gamma^*(S/S') = \gamma(S/S')$.

<u>Case 5:</u> P' is obtained from P by means of a classical fork elimination. Every $T(P_1,P_2)$-inference S/S' is transformed into at most three images S_i/S_i' , $i=1,2,3$, each of which is an extension of S/S' . We put $\gamma^*(S_i/S_i') = \gamma(S/S')$.

<u>Case 6:</u> P' follows from P by means of a preliminary reduction step or "omission of a cut". A $T(P_1,P_2)$-inference S/S' in P is either left unaffected by such a reduction step or else is cancelled out. We put $\gamma^*(S/S') = \gamma(S/S')$ if S/S' remains unaffected by the reduction step.

<u>Case 7:</u> P' follows from P by means of an intuitionistic fork elimination. This case can either be subsumed under case 5 followed by case 6, or else be treated directly in the same way as case 5.

<u>Case 8:</u> P' follows from P by means of a I-reduction step. Each $T(P_1,P_2)$-inference S/S' in P remains unaffected by this reduction step. Hence we put $\gamma^*(S/S') = \gamma(S/S')$.

<u>Case 9:</u> P' follows from P by means of a strong $T(P_1)$-reduction step, applied to the critical $T(P_1)$-inference S_o/S'_o in P . Each $T(\hat{P}_1,\hat{P}_2)$-inference S/S' in P , different from S_o/S'_o , is transformed by this reduction step into at most two images S_i/S'_i , i=1,2, each of which is an extension of S/S' . We put $\gamma^*(S/S') = \gamma(S/S')$ for such inferences. The $T(P_1)$-inference S_o/S'_o in P, however, is transformed by this reduction step into a $T(P_1,P_2)$-inference, say, S*/S**, and we have to define γ^* properly on S*/S** . Let $R(\vec{\alpha}_u,x)$ and $v_1,\ldots,v_s$ be the data of S_o/S'_o and q its index. According to the definition of strong $T(P_1)$-reduction step, the data of S*/S** are given by $R(\vec{\alpha}_u,x)$, $w_1,\ldots,w_s$ and $|q|$, where $w_i \subset_K v_i$ and where $\text{length}(w_i) = \text{length}(v_i)+1$, $i=1,\ldots,s$. Hence we find sequence numbers of length 1, say $\bar{\alpha}_1(1),\ldots, \bar{\alpha}_s(1), \bar{\beta}(1)$ such that $w_i \subseteq_K \bar{\alpha}_i(1)$, $i=1,\ldots,s$ and such that $\bar{\beta}(0) = |t|$. As value of γ^* for S*/S** we take $e = \langle \bar{\alpha}_1(1),\ldots, \bar{\alpha}_s(1), \bar{\beta}(1) \rangle$. Condition α) is obviously satisfied by e .

<u>Case 10:</u> P' follows from P by means of a strong $T(P_1,P_2)$-reduction step. Let S_o/S'_o be the critical $T(P_1,P_2)$-inference in P , to which the strong $T(P_1,P_2)$-reduction step is applied. If S/S' is a $T(\hat{P}_1,\hat{P}_2)$-inference in P other than S_o/S'_o , then S/S' is transformed by this reduction step into at most two $T(\hat{P}_1,\hat{P}_2)$-inferences S_1/S'_1 and S_2/S'_2 which are extensions of S/S' . We put $\gamma^*(S_i/S'_i) = \gamma(S/S')$. Now to S_o/S'_o . Let $R(\vec{\alpha}_u,x), v_1,\ldots,v_s$ and $|t|$ be the data of S_o/S'_o and q its index. The strong $T(P_1,P_2)$-reduction step transforms S_o/S'_o into another $T(P_1,P_2^*)$-inference S*/S**', whose data are given by $R(\vec{\alpha}_u,x)$, $w_1,\ldots,w_s$ and $|q|$ where a) $w_i \subseteq_K v_i$, $i=1,\ldots,s$; b) $\text{length}(w_i) = \text{length}(v_i)+1$, $i=1,\ldots,s$. <u>Subcase 1:</u> γ associates with S_o/S'_o a number e , say, $\langle \bar{\alpha}_1(1),\ldots, \bar{\alpha}_s(1), \bar{\beta}(1) \rangle$. By definition, $v_i \subseteq_K \bar{\alpha}_i(1)$ and $\bar{\beta}(0) = |t|$. Since w_i , $i=1,\ldots,s$ is a strict extension of v_i , we find sequence numbers $\bar{\alpha}_1(2),\ldots, \bar{\alpha}_s(2)$ which are extensions of $\bar{\alpha}_1(1),\ldots, \bar{\alpha}_s(1)$ and which satisfy $w_i \subset_K \bar{\alpha}_i(2)$, $i=1,\ldots s$. By defining $\beta(1) = |q|$, we obtain an extension $\bar{\beta}(2)$ of $\bar{\beta}(1)$.

Now we can extract from P by means of the basic lemma II an s.n.s. proof $\tilde{P}$ in ZTGi/I of $\longrightarrow |q| \subset_R^{u*v} |t|$. Let m be the Goedelnumber of this proof and put $\zeta(0)=m$. Then it is evident that $e' = \langle \bar{\alpha}_1(2), \ldots\ldots, \bar{\alpha}_s(2), \bar{\beta}(2) \rangle$ and $e'' = \zeta(1)$ satisfy condition β) . Thus we may define: The value of $\mathcal{V}^*$ for S*/S** is e',e" . The definition of $\mathcal{V}^*$ on P' is thus completed. Subcase 2: $\mathcal{V}$ associates with S_o/S_o' a pair of numbers, say, $\langle \bar{\alpha}_1(x), \ldots\ldots, \bar{\alpha}_s(x), \bar{\beta}(x) \rangle$ and $\zeta(z)$. According to condition β) , we have $x \geq 2$, $v_i \subseteq_K \bar{\alpha}_i(x)\ \beta(x-1) = |t|$ and $z=x-1$. For each $i=x-1$ there are in addition sequence numbers $v_1', \ldots\ldots, v_s'$ of equal length and an s.n.s. proof P_i in ZTGi/I of $\longrightarrow \beta(i) \subset_R^{u*v'} \beta(i+1)$ such that $v_i \subseteq_K v_i'$, $i=1,\ldots\ldots,s$ and such that $\zeta(i)$ is a Goedelnumber of P_i . Since w_i is a strict extension of v_i , $i=1,\ldots\ldots,s$, we find sequence numbers $\bar{\alpha}_1(x+1), \ldots\ldots, \bar{\alpha}_s(x+1)$ which are extensions of $\bar{\alpha}_1(x), \ldots\ldots, \bar{\alpha}_s(x)$ and which satisfy $w_i \subseteq_K \bar{\alpha}_i(x+1)$, $i=1,\ldots,s$. From P we can extract according to basic lemma II an s.n.s. proof $\tilde{P}$ in ZTGi/I of $\longrightarrow |q| \subset_R^{u*v} |t|$. Let m be the Goedelnumber of this proof and put $\zeta(x-1)=m$. Then it is clear that $e' = \langle \bar{\alpha}_1(x+1), \ldots\ldots, \bar{\alpha}_s(x+1), \bar{\beta}(x+1) \rangle$ and $e'' = \zeta(x)$ satisfy condition β) if we put $\bar{\beta}(x) = |q|$. Hence we define: the value of $\mathcal{V}^*$ for S*/S** is e',e" . $\mathcal{V}^*$ is thus fully defined on P' .

E. If P is a graded s.n.s. proof in ZTG/I, then there are certain valuations of P which are of particular interest.

Definition 49: Let P be a graded s.n.s. proof in ZTG/I and $\mathcal{V}$ a valuation of P . $\mathcal{V}$ is said to be compatible with P if for every $T(P_1,P_2)$-inference S/S' (whose data are assumed to be $R(\alpha_{u_1}^{i_1}, \ldots\ldots, \alpha_{u_s}^{i_s}, x)$, $w_1, \ldots\ldots, w_s$, $|t|$) the following holds: 1) if $\mathcal{V}(S/S')$ is $e = \langle \bar{\alpha}_1(1), \ldots\ldots, \bar{\alpha}_s(1), \bar{\beta}(1) \rangle$, then e is an unsecured element with respect to P_1 ; 2) if $\mathcal{V}(S/S')$ is $e = \langle \bar{\alpha}_1(x), \ldots\ldots, \bar{\alpha}_s(x), \bar{\beta}(x) \rangle$, $e' = \zeta(x-1)$, then e is an unsecured element with respect to P_1 .

Remark: Clause 1) of def. 49 is automatically satisfied according to our definition of "unsecured". Clause 1) has been included for convenience only.

<u>Lemma 19:</u> Let P be a graded s.n.s. proof and $\mathcal{V}$ a compatible valuation of P . Let P' be obtained from P by means of a preliminary reduction step, "elimination of a cut", (intuitionistic or classical) fork elimination, an inessential reduction step, an induction reduction or a subformula reduction step. The induced valuation $\mathcal{V}^*$ on P' is compatible with P' (which is still a graded proof) .

<u>Proof:</u> Is obvious from the definition of $\mathcal{V}^*$.

<u>Lemma 20:</u> Let P be a graded s.n.s. proof and $\mathcal{V}$ a compatible valuation of P . Let P' be obtained from P by means of a strong $T(P_1)$-inference or a strong $T(P_1,P_2)$-inference. The induced valuation $\mathcal{V}^*$ on P' is compatible with P' (which is still a graded proof).

<u>Proof:</u>

<u>Case 1:</u> P' follows from P by means of a strong $T(P_1)$-reduction step. Let S_o/S_o' be the $T(P_1)$-inference in P to which the reduction step is applied; let $R(\alpha_{u_1}^{i_1},\ldots\ldots, \alpha_{u_s}^{i_s},x)$, $v_1,\ldots\ldots,v_s$ and $|t|$ be the data of this inference. Let S_1/S_1' be the $T(P_1,P_2)$-inference into which S_o/S_o' is transformed by the reduction step. The lemma is essentially proved if we can show that $\mathcal{V}^*$ associates with S_1/S_1' an element $e= \langle \bar{\alpha}_1(x),\ldots, \bar{\alpha}_s(x), \beta(x)\rangle$ which is unsecured with respect to P_1 (where P_1 is by assumption a good proof). Now $\mathcal{V}^*$ associates with S_1/S_1' by definition an element e of the form $\langle \bar{\alpha}_1(1),\ldots., \bar{\alpha}_s(1), \bar{\beta}(1)\rangle$. But such an element is by definition unsecured with respect to P_1 , hence $\mathcal{V}^*$ is compatible.

<u>Case 2:</u> P' follows from P by means of a strong $T(P_1,P_2)$-reduction step. Let S_o/S_o' be the $T(P_1,P_2)$-inference to which the strong $T(P_1,P_2)$-reduction step is applied; let $R(\alpha_{u_1}^{i_1},\ldots., \alpha_{u_s}^{i_s},x)$, $v_1,\ldots\ldots,v_s$, $|t|$ be the data of this inference and q its index.

Let S_1/S_1' be the $T(P_1,\hat{P}_2)$-inference into which S_o/S_o' is transformed by the reduction step; let $R(\vec{\alpha}_u,x)$, $w_1,\ldots,w_s$ and $|q|$ its data. Assume eg. that $\mathcal{V}$ associates with S_o/S_o' the pair $e = \langle \bar{\alpha}_1(x),\ldots,\bar{\alpha}_s(x),\bar{\beta}(x)\rangle$, $e' = \bar{\xi}(x-1)$. By definition, the induced valuation $\mathcal{V}^*$ associates with S_1/S_1' a certain pair of the form $\langle \bar{\alpha}_1(x+1),\ldots,\bar{\alpha}_s(x+1),\bar{\beta}(x+1)\rangle$, $\bar{\xi}(x)$; here $\beta(x-1) = |t|$, $\beta(x) = |q|$ and $\xi(x)$ is a Gödelnumber of a proof P^* in ZTGi/I of $\longrightarrow |q| \subset_R^{u*v} |t|$. Now assume that $\mathcal{V}^*$ is not compatible with P'. This implies that $\langle \bar{\alpha}_1(x+1),\ldots,\bar{\alpha}_s(x+1),\bar{\beta}(x+1)\rangle$ is secured with respect to P_1. By definition there is an $n<x$ and a proof $\hat{P} \in D_{P_1}$ of $\longrightarrow \xi_{w'}(n+1) \subset_R^{u*v'} \xi_{w'}(n)$, where v' denotes the list $\bar{\alpha}_1(x+1),\ldots,\bar{\alpha}_s(x+1)$ and where $w' = \bar{\beta}(x+1)$. By means of a conversion we obtain from $\hat{P}$ a proof $\tilde{P}$ in ZTGi/I of $\longrightarrow \beta(n+1) \subset_R^{u*v'} \beta(n)$. On the other hand, $\xi(n)$ is the Gödelnumber of a proof P^{**} in ZTGi/1 of $\longrightarrow \beta(n+1) \subset_R^{u*v''} \beta(n)$ where v'' denotes a list of sequence numbers $v_1'',\ldots,v_s''$, all of equal length, satisfying $w_i \subseteq_K v_i''$, $i=1,\ldots,s$. From $\tilde{P}$ we obtain a substitution instance $\tilde{P}_1$ whose endsequent is $\longrightarrow \neg\, \beta(n+1) \subset_R^{u*w} \beta(n)$ and from P^{**} we obtain a substitution instance P_1^{**} whose endsequent is $\longrightarrow \beta(n+1) \subset_R^{u*w} \beta(n)$. But this implies that ZTGi/I is inconsistent and via theorem 44 that ZTi/I is inconsistent, contradicting the assumed consistency of ZTi/I. The case where $\mathcal{V}$ associates with S_o/S_o' a number $\langle \bar{\alpha}_1(1),\ldots,\bar{\alpha}_s(1),\bar{\beta}(1)\rangle$ can be treated in precisely the same way.

<u>F.</u> Let $P_o,\ldots,P_n,\ldots$ be a list of s.n.s. proofs in ZTGi/I, each of which is obtained from the previous one by means of a reduction step, including "omission of a cut". If $\mathcal{V}_o$ is a valuation of P_o, then we obtain valuations $\mathcal{V}_i$ of P_i by means

of the inductive definition $\mathcal{V}_{i+1} = \mathcal{V}_i{}^*$. In such a case we say that $\mathcal{V}_i$ is the valuation induced by $\mathcal{V}_o$ on P_i. As example, consider an s.n.s. proof P in ZTGi/I provided with a valuation $\mathcal{V}_o$ and let $\Gamma \longrightarrow A$ be an uppermost sequent in the final part of P (denoted by S). Let B be a formula in Γ and let $\hat{P}$ be the side proof determined by B in S according to basic lemma II. $\hat{P}$ can be derived from P by means of preliminary reduction steps and the operation "omission of a cut". Hence there is a chain $P_o, \ldots, P_N$ with $P_o = P$, $P_N = \hat{P}$ and such that P_{i+1} follows from P_i by means of a preliminary reduction step or an "omission of a cut". The valuation $\mathcal{V}_N$ induced on P_N (that is on $\hat{P}$) by $\mathcal{V}_o$ will be called the valuation induced by $\mathcal{V}_o$ on the side proof $\hat{P}$. The valuation which is induced on $\hat{P}$ by $\mathcal{V}_o$ can, of course, be described directly. Each $T(P_1,P_2)$-inference S/S' in P occurs either unaffected in $\hat{P}$ or else is omitted. The induced valuation $\hat{\mathcal{V}}$ on P is then nothing else than the restriction of $\mathcal{V}_o$ to those $T(P_1,P_2)$-inferences S/S' which are not cancelled out. We have the obvious

<u>Lemma 21:</u> Let P be a graded s.n.s. proof in ZTGi/I , provided with a compatible valuation $\mathcal{V}$ and $\Gamma \longrightarrow A$ (denoted by S) an uppermost sequent in the final part of P . Let B be a formula in Γ and $\hat{P}$ the side proof determined by B in S according to basic lemma II. The valuation $\hat{\mathcal{V}}$ induced by $\mathcal{V}$ on $\hat{P}$ is compatible with $\hat{P}$ (where $\hat{P}$ is, of course, a graded proof).

<u>G.</u> Lemmas 19 and 20 do not include the case of a I-reduction step, since it is not clear whether a I-reduction step transforms a graded proof into a graded proof. We have, however,

<u>Lemma 22:</u> Let P be a graded s.n.s. proof in ZTGi/I provided with a compatible valuation $\mathcal{V}$. Let S/S' be a critical I-inference in P , P_1 the side proof determined by S/S' . Let finally P' be obtained from P by means of a I-reduction step, applied to S/S' and $\mathcal{V}'$ the valuation induced by $\mathcal{V}$ on P' . If P is "good", then P' is graded and $\mathcal{V}'$ is compatible with P' .

The evident proof is omitted.

7.5. Ordinals

A. Let P be a good proof of $\longrightarrow W(\subset^u_R)$ with R denoting $R(\alpha^{i_1}_{u_1},\ldots,\alpha^{i_s}_{u_s},x)$. As noted earlier, the restriction of $\sqsubset^s$ to the set $D^s(P)$ (denoted by $\sqsubset^s_P$) of unsecured elements with respect to P is wellfounded. If e is such an element, then we can associate with e as usual its ordinal with respect to $\sqsubset^s_P$; we denote it by $\|e\|_P$. The ordinal associated with $\sqsubset^s_P$ will be denoted by $\|\sqsubset^s_P\|$.

B. Now let P be a graded s.n.s. proof in ZTG/I and $\mathcal{V}$ a compatible valuation of P. If S/S' is a $T(P_1,P_2)$-inference in P, then $\mathcal{V}$ associates with S/S' either a number e or else a pair of numbers e, e_1, satisfying conditions α) or β), respectively. In both cases e is by definition an unsecured element with respect to the good proof P_1. The ordinal $\|e\|_{P_1}$ will be called the ordinal associated by $\mathcal{V}$ with S/S' and will be denoted by $O_{\mathcal{V}}(S/S')$.

C. The set of proofs in ZTGi/I is denumerable and so is the set of good proofs. Hence there is a smallest denumerable ordinal ξ having the property: if P is a good proof of $\longrightarrow W(\subset^u_R)$ (with R for $R(\alpha^{i_1}_{u_1},\ldots,\alpha^{i_s}_{u_s},x)$) then $\|\sqsubset^s_P\|<\xi$. We denote this smallest ordinal by Ω.

D. Given a graded s.n.s. proof P in ZTG/I provided with a compatible valuation $\mathcal{V}$, we can associate with every sequent S in P a certain ordinal (depending on $\mathcal{V}$) which we denote by $O(\mathcal{V}/S)$ and whose inductive definition is given as follows:

1) if S is an axiom, then $O(\mathcal{V}/S)=1$; 2) if S is the conclusion of a conversion or a one-premiss structural rule S/S', then

$o(\mathcal{V}/S)=o(\mathcal{V}/S')$; 3) if S is the conclusion of a one-premiss logical inference S'/S, then $o(\mathcal{V}/S)=o(\mathcal{V}/S') \# 1$;
4) if S is the conclusion of a two-premiss logical inference $S_1,S_2/S$, then $o(\mathcal{V}/S)=o(\mathcal{V}/S_1) \# o(\mathcal{V}/S_2) \# 1$;
5) if S is the conclusion of a cut $S_1,S_2/S$, then $o(\mathcal{V}/S)=\omega_d(o(\mathcal{V}/S_1) \# o(\mathcal{V}/S_2))$ where $d=h(S_1)-h(S)$;
6) if S is the conclusion of an induction S'/S, then $o(\mathcal{V}/S)=\omega_d(o(\mathcal{V}/S')\omega)$ where $d=h(S')-h(S)$; 7) if S is the conclusion of a I-inference S'/S, then $o(\mathcal{V}/S)=\omega_d((o(\mathcal{V}/S') \# \omega^{\Omega+1})\omega^{\Omega+1})$ where $d=h(S')-h(S)$;
8) if S is the conclusion of a $T(P_1)$-inference S'/S, then $o(\mathcal{V}/S)=\omega_d((o(\mathcal{V}/S') \# \omega^{\lambda+1})\omega^{\lambda+1})$ where $d=h(S')-h(S)$ and $\lambda = \| \sqsubset^{s}_{P_1} \|$; 9) if S is the conclusion of a $T(P_1,P_2)$-inference S'/S, then $o(\mathcal{V}/S)=\omega_d((o(\mathcal{V}/S') \# \omega^{\zeta+1})\omega^{\zeta+1})$ where $d=h(S')-h(S)$ and $\zeta = o_{\mathcal{V}}(S'/S)$. The ordinal of the end-sequent of P is called the ordinal of P and will be denoted by $o_{\mathcal{V}}(P)$ (indicating its dependence on $\mathcal{V}$) .

E. With respect to this ordinal assignement we have the following

Theorem 47:
A. Let P be a graded s.n.s. proof in ZTG/I and $\mathcal{V}$ a compatible valuation of P . Let P' be obtained from P by means of a reduction step and $\mathcal{V}^*$ the valuation induced by $\mathcal{V}$ on P' . Then $o_{\mathcal{V}^*}(P') < o_{\mathcal{V}}(P)$ if the reduction step in question belongs to the following list: 1) "Omission of a cut", 2) a classical fork elimination, 3) an intuitionistic fork elimination, 4) an induction reduction, 5) a strong $T(P_1)$-reduction step, 6) a strong $T(P_1,P_2)$-reduction step.

B. If P' is a substitution instance of P or follows from P by means of a preliminary reduction step then $o_{\mathcal{V}^*}(P') \leq o_{\mathcal{V}}(P)$.

Proof: a)The proof of clauses 1)-6) and of the last part of the theorem leads to exactly the same inequalities as in earlier cases. The proof of 3), in particular, uses the fact that an intuitionistic fork elimination is composed by a classical fork elimination plus some preliminary reduction steps. Hence 3) is reduced as usual to 1),2) and part B. b) Next consider the case where P' follows from P by means of a $T(P_1)$-reduction step. Let S/S' be the $T(P_1)$-inference to which the reduction step is applied and let S_1/S_1' be the

$T(P_1,P_2)$-inference into which S/S' is transformed by means of the reduction step. By definition, $\mathcal{V}(S/S') = \|\sqsubset^{s}_{P_1}\|$ (for some suitable s) and $\mathcal{V}^*(S_1/S_1') = \|e\|_{P_1}$ where e is an element in the domain $D^s(P_1)$ of $\sqsubset^{s}_{P_1}$. By definition, $\|e\|_{P_1} < \|\sqsubset^{s}_{P_1}\|$. If we put $\|e\|_{P_1} = \lambda$, $\|\sqsubset^{s}_{P_1}\| = \xi$, then the proof of 5) leads again to the verification of the inequality

$\omega_d((\alpha \# m \# \omega^{\lambda+1})\omega^{\lambda+1} \# \alpha \# 2) < \omega_d((\alpha \# \omega^{\xi+1})\omega^{\xi+1})$ which in turn is a consequence of the inequality E :

$\omega_d((\alpha \# m \# \omega^{\gamma})\omega^{\gamma} \# \alpha \# n) < \omega_d((\alpha \# \omega^{\gamma+1})\omega^{\gamma+1})$ (for all γ, α and all finite m,n,d) which is proved in chapter II, sect. 2.5., part C. c) Finally, let P' be obtained from P by means of a $T(P_1,P_2)$-reduction step. Let S/S' be the $T(P_1,P_2)$-inference, to which the reduction step is applied, and let S_1/S_1' be the $T(P_1,\hat{P}_2)$-inference into which S/S' is transformed by the reduction step. Assume e.g. that $\mathcal{V}$ associates with S/S' the pair e, e_1 and that $\mathcal{V}^*$ associates with S_1/S_1' the pair e', e_1'. By definition of $\mathcal{V}^*$ it follows that $e' \sqsubset^{s} e$ holds. By assumption and according to lemma 20, it follows that $e' \sqsubset^{s}_{P_1} e$ holds. Hence, $\|e'\|_{P_1} < \|e\|_{P_1}$. The verification of 6) again amounts to the proof of

$\omega_d((\alpha \# m \# \omega^{\lambda+1})\omega^{\lambda+1} \# \alpha \# 2) < \omega_d((\alpha \# \omega^{\nu+1})\omega^{\nu+1})$ with $\|e'\|_{P_1} = \lambda$, $\|e\|_{P_1} = \nu$, which in turn is a consequence of the inequality E. The situation is precisely the same in the case where $\mathcal{V}$ associates with S/S' a single number e.

If P' follows from P by means of a I-reduction step then it is not clear whether P' is again a graded proof. However, we have

<u>Theorem 48:</u> Let P be a graded s.n.s. proof in ZTGi/I, provided with a compatible valuation $\mathcal{V}$. Let S/S' be a critical I-inference in P and assume that P' is obtained from P by means of a I-reduction step, applied to S/S'. Let S_1/S_1' be the $T(P_1)$-

inference into which S/S' is transformed by the reduction step, and let $\mathcal{V}^*$ be the valuation induced by $\mathcal{V}$ on P'. If the side proof P_1 of S/S' in P is good, then P' is graded, $\mathcal{V}^*$ is compatible with P' and $0_{\mathcal{V}^*}(P') < 0_{\mathcal{V}}(P)$.

<u>Proof:</u> That P' is graded and $\mathcal{V}^*$ compatible with P' is stated in lemma 22. By definition $o(\mathcal{V}/S') = \omega_d((\alpha \# \omega^{\Omega+1})\,\omega^{\Omega+1})$ where $\alpha = o(\mathcal{V}/S)$. Similarly, $o(\mathcal{V}^*/S_1') = \omega_d((\alpha \# \omega^{\zeta+1})\,\omega^{\zeta+1})$ where $\zeta = \|\sqsubset^{s}_{P_1}\|$. By definition, $\zeta < \Omega$. The proof of the theorem amounts to proving $o(\mathcal{V}^*/S_1') < o(\mathcal{V}/S)$ which, in turn, is a consequence of the strict monotonicity of $\omega_d((\alpha \# {}^{x+1})\,\omega^{x+1})$ as function of x.

<u>7.6. The wellfoundedness proof</u>

<u>A.</u> <u>Theorem 49:</u> Let P be a graded s.n.s. proof in ZTGi/I, provided with a compatible valuation $\mathcal{V}$. Then L_P is wellfounded.

<u>Proof:</u> We proceed by transfinite induction with respect to $0_{\mathcal{V}}(P)$. There are three subcases to be distinguished: A) P is saturated and does not admit preliminary reduction steps, B) P is saturated but preliminary reduction steps can be applied to P, C) P is not saturated and preliminary reduction steps can be applied to P. We content ourself with the proof of A). Cases B) and C) are easy consequences of case A) and can be treated in the same way as the corresponding cases B,C in, say, theorem 35. Case A) is proved if we can show: if $L(P,P')$ holds, then $L_{P'}$ is wellfounded. In view of the assumptions stated under case A), this is the same as to prove: if P' follows from P by means of a strictly essential reduction step or a subformula reduction step, then $L_{P'}$ is wellfounded. <u>Subcase 1:</u> Let P' be obtained from P by means of a strictly essential reduction step different from a I-reduction step or by means of a subformula reduction step. Let $\mathcal{V}^*$ be the valuation induced by $\mathcal{V}$ on P'. According to theorem 47 we have $0_{\mathcal{V}^*}(P') < 0_{\mathcal{V}}(P)$; hence $L_{P'}$ is wellfounded. <u>Subcase 2:</u> Let P' be obtained from P by means of a I-reduction step. Let S/S' be the critical I-inference in P to which the reduction step is applied and P_1 the side proof determined by S/S' (in P). According to its construction, described in basic lemma II, P_1 is derived from P by means of preliminary reduction steps, including the

operation "omission of a cut". Let $\hat{\mathcal{V}}$ be the valuation induced by $\mathcal{V}$ on P_1. According to theorem 47 it follows that $0\,\hat{\mathcal{V}}(P_1) < 0\,\mathcal{V}(P)$ holds. From the inductive assumption of our transfinite induction it follows that L_{P_1} is wellfounded, that is, that P_1 is good. The proof P' is therefore again a graded proof, and the valuation $\mathcal{V}^*$ induced by $\mathcal{V}$ on P' compatible with P', as follows from lemma 22. From theorem 48 we conclude that $0\,\mathcal{V}^*(P') < 0\,\mathcal{V}(P)$ holds; hence $L_{P'}$ is wellfounded. By combining subcase 1 with subcase 2 we infer the wellfoundedness of L_P. This proves case A and thus essentially the whole theorem.

An immediate consequence of the above theorem is

Corollary: If P is an s.n.s. proof in ZTi/I then L_P is wellfounded.

Proof: We can treat such a proof as graded proof provided with the empty valuation.

7.7. Remarks on applications

From the last theorem and its corollary we could again deduce theorems 23, 24, 25 (but restricted to ZTi/I). However, the method described in the last three chapters has a much wider range of applications and so we postpone the discussion of applications to the next chapters.

CHAPTER VIII:

Harrop formulas

In the present chapter we generalize the results obtained in chapters IV - VII by using some quite elementary combinatorial considerations which are intimately connected with basic lemmas I and II. The main applications of our methods, which we have obtained so far, are results of the form: "if $\longrightarrow A \vee B$ has been proved (in some suitable theory) then there is a proof of $\longrightarrow A$ or $\longrightarrow B$ ", etc.. Now we generalize these results and prove theorems of the following kind: "if $A_1,\ldots\ldots,A_s$ are formulas belonging to a certain class C of formulas (yet to be defined) and if $A_1,\ldots\ldots,A_s \longrightarrow A\vee B$ has been proved (in some suitable theory), then there is a proof of $A_1,\ldots\ldots,A_s \longrightarrow A$ or of $A_1,\ldots\ldots,A_s \longrightarrow B$" . The above-mentioned combinatorial arguments can be combined either with the methods described in chapter IV or else with the methods described in chapters V - VII. It turns out that the results obtained in the second case are much stronger than those obtained in the first case. This makes it evident that the methods described in chapters V - VII are more substantial than those described in chapter V; other arguments in favour of this statement will be given in the last chapter.

8.1. Intuitionistic number theory and Harrop formulas

A. To start with, let us introduce a class of formulas, called the class of Harrop formulas and denoted by M . The inductive definition of M is given by

Definition 50: a) prime formulas belong to M ; b) if A is in M, then $(x)A$ and $(\alpha)A$ are in M ; c) if A and B are in M then $A\wedge B$ is in M ; d) if A is in M and B is arbitrary, then $B\supset A$ is in M ; e) for arbitrary A , $\neg A$ is in M .

Remark: From now on we call a formula closed if it does not contain free variables nor special function constants.

In connection with the above definition we note the obvious

Lemma 23: 1) If $A\supset B$ is in M, then $B\in M$; 2) if $A\wedge B\in M$ then $A\in M$ and $B\in M$; 3) if $(\alpha)A(\alpha)\in M$, then $A(F)\in M$ for any functor F free for α in A ; 4) if $(x)A(x)\in M$ then

$A(t) \in M$ for every term t free for x in A. The first who recognized that the formulas of M play a certain role in the theory of intuitionistic systems was R. Harrop. In [2] he proved certain results for a Hilbert-type version of intuitionistic number theory. We formulate his result in terms of sentential calculus, using our version of intuitionistic number theory, namely ZTi . In this language Harrop's result can be stated as follows: a) if $A_1, \ldots, A_s$ are closed formulas in M and if A, B are closed formulas such that $ZTi \vdash A_1, \ldots, A_s \longrightarrow A \vee B$ holds, then $ZTi \vdash A_1, \ldots, A_s \longrightarrow A$ or $ZTi \vdash A_1, \ldots, A_s \longrightarrow B$; 2) if $ZTi \vdash A_1, \ldots, A_s \longrightarrow (E\xi)A(\xi)$ holds with $(E\xi)A(\xi)$, a closed formula, then $ZTi \vdash A_1, \ldots, A_s \longrightarrow A(F)$ for some functor F free for ξ in A ; c) similarly, with Ex in place of $E\xi$ and a term t in place of F. We will refer to this result henceforth as Harrop's result. In [8] we gave a proof of Harrop's result using the techniques which Gentzen introduced in [1] . In the meantime, however, it turned out that there is a much more elegant proof of this result which shows clearly the close relationship between Harrop formulas and Gentzen's reduction techniques. This proof will be given below.

B. In order to reformulate Harrop's result in such a way as to be easily accessible to Gentzen techniques, we need the following

Theorem 50: Let T be any of the theories considered so far, that is, any of ZT , ZTi , ZT/I , ZTi/I , ... or any of the conservative extensions ZTE/II_N , $ZTEi/II_N$, ZTEi/II , ZTE/II , ... etc. Let $A_1, \ldots, A_s$ be formulas without free variables. Then $T, \longrightarrow A_1, \ldots, \longrightarrow A_s \vdash \Gamma \longrightarrow \Delta$ iff $T \vdash A_1, \ldots, A_s, \Gamma \longrightarrow \Delta$.

Proof: The implication from right to left is obvious. Let P be a proof in $T, \longrightarrow A_1, \ldots, \longrightarrow A_s$ of $\Gamma \longrightarrow \Delta$. Then one proves by an almost trivial induction (starting with the axioms): if $\Gamma' \longrightarrow \Delta'$ is a sequent in P, then $T \vdash A_1, \ldots, A_s, \Gamma' \longrightarrow \Delta'$. The statement then follows by taking for $\Gamma' \longrightarrow \Delta'$ the endsequent of P .

This theorem allows us to reformulate Harrop's result in the following form

<u>Theorem 51:</u> Let $A_1,\ldots,A_s$ be closed Harrop formulas and A, B, $(E\xi)C(\xi)$ arbitrary closed formulas. a) If $ZTi, \longrightarrow A_1,\ldots, \longrightarrow A_s \vdash \longrightarrow A \vee B$, then $ZTi, \longrightarrow A_1,\ldots, \longrightarrow A_s \vdash \longrightarrow A$ or $ZTi, \longrightarrow A_1,\ldots, \longrightarrow A_s \vdash \longrightarrow B$; b) if $ZTi, \longrightarrow A_1,\ldots, \longrightarrow A_s \vdash \longrightarrow (E\xi)C(\xi)$, then there is a functor F free for ξ in $C(\xi)$ such that $ZTi, \longrightarrow A_1,\ldots, \longrightarrow A_s \vdash \longrightarrow C(F)$; c) similarly, with Ex and a term t in place of $E\xi$ and F. This in turn is a consequent of

<u>Theorem 52:</u> Let $A_1,\ldots,A_s$ be closed Harrop formulas such that $\longrightarrow A_1,\ldots, \longrightarrow A_s$, ZTi is consistent. Then a), b), c) of theorem 51 hold. If $\longrightarrow A_1,\ldots, \longrightarrow A_s$, ZTi is inconsistent, then $\longrightarrow$ is provable and so a), b), c) of theorem 51 hold trivially. So it remains only to consider the case where $\longrightarrow A_1,\ldots, \longrightarrow A_s$,ZTi is consistent. Here we make use of the tertium non datur, which could be avoided without difficulty; however, its use simplifies the considerations below.

<u>C.</u> Next, let T be any of the systems considered so far (e.g. ZTE/II) and T_i its intuitionistic version (that is, ZTEi/II). Let $A_1,\ldots,A_s$ be closed Harrop formulas. By $T(A_1,\ldots,A_s)$ we denote the system which we obtain by addition of $\longrightarrow A_1,\ldots, \longrightarrow A_s$ as new axioms to T, correspondingly by $Ti(A_1,\ldots,A_s)$ the system which we obtain by adding $\longrightarrow A_1,\ldots, \longrightarrow A_s$ as new axioms to Ti .

<u>Definition 51:</u> The Harrop hull $HTi(A_1,\ldots,A_s)$ of $Ti(A_1,\ldots,A_s)$ is obtained from $Ti(A_1,\ldots,A_s)$ by adding to it every sequent S as a new axiom which satisfies one of the following conditions: a) S is $\longrightarrow B$ and B is a Harrop formula such that $Ti(A_1,\ldots,A_s) \vdash \longrightarrow B$; b) S is $A \longrightarrow B$ and B is a Harrop formula such that $Ti(A_1,\ldots,A_s) \vdash \longrightarrow A \supset B$; c) S is $A \longrightarrow$ and $Ti(A_1,\ldots,A_s) \vdash \longrightarrow \neg A$. The Harrop hull $HT(A_1,\ldots,A_s)$ of $Ti(A_1,\ldots,A_s)$ is obtained from $T(A_1,\ldots,A_s)$ by addition of every sequent S which satisfies a), b) or c) above.

<u>Remark:</u> A sequent S which satisfies one of the conditions a), b) or c) above is called a Harrop axiom (with respect to $Ti(A_1,\ldots,A_s)$).

In connection with the above definition we note

Lemma 24: Let S be a Harrop axiom and assume that S is a sequent of the following list: 1) $\longrightarrow (\xi)B(\xi)$, 2) $\longrightarrow (x)B(x)$, 3) $\longrightarrow A \wedge B$, 4) $\longrightarrow A \supset B$, 5) $\longrightarrow \neg A$. If S is the i-th sequent in the above list then the i-th sequent in the list below is also a Harrop axiom: 1) $\longrightarrow B(F)$, F a functor free for ξ in B; 2) $\longrightarrow B(t)$, t a term free for x in B; 3) $\longrightarrow A$ and $\longrightarrow B$, 4) $A \longrightarrow B$, 5) $A \longrightarrow$.

Proof: S, having the form $\longrightarrow G$, can only be a Harrop axiom in virtue of clause a) of definition 51. In particular, G must be a Harrop formula. With the aid of this observation the statement immediately follows from the definition of Harrop formulas (in particular lemma 23) and from definition 51.

Systems of the form $HT(A_1,\ldots,A_s)$ will be called Harrop systems, those of the form $HTi(A_1,\ldots,A_s)$ are called intuitionistic Harrop systems. If e.g. Ti is $ZTEi/I1$, then $HTi(A_1,\ldots,A_s)$ is the theory obtained from $ZTEi/II$ by adding to it every sequent S as new axiom, which satisfies one of the clauses a), b) or c) in def.51. The following theorem is evident.

Theorem 53: a) $HTi(A_1,\ldots,A_s) \vdash S$ iff $Ti(A_1,\ldots,A_s \vdash S$, b) $HT(A_1,\ldots,A_s) \vdash S$ iff $T(A_1,\ldots,A_s) \vdash S$.

In other words, $HTi(A_1,\ldots,A_s)$ and $HT(A_1,\ldots,A_s)$ are conservative extensions of $Ti(A_1,\ldots,A_s)$ and $T(A_1,\ldots,A_s)$ respectively.

D. With respect to Harrop systems, we can introduce the notion of final part as usual: 1) the endsequent of a proof P is in its final part; 2) if S is in the final part of P and if S is the conclusion of a conversion or a structural inference, then the premiss(es) of this inference belong to the final part. An inference is called critical if it is neither a conversion nor a structural rule and if its conclusion belongs to the final part. Preliminary reduction steps and the operation "omission of a cut" can be introduced for proofs P with respect to Harrop systems in the usual way. An indispensable tool for the present section and the whole chapter is the basic lemma II, which in the present context reads as follows:

Basic lemma II_H: Let $HTi(A_1,\ldots,A_s)$ be an intuitionistic Harrop system and P a proof in it. Assume that the endsequent of P has the form $\longrightarrow A$. Let $S_1,\ldots,S_m$ be the uppermost sequents in the final part of P, listed from left to right; let S_i be $\Gamma_i \longrightarrow B_i$, $i=1,\ldots,m$. a) If $i < m$, then there is a proof P_i in $HTi(A_1,\ldots,A_s)$ of $\longrightarrow B_i$; b) if B occurs in Γ_j, then there exists a proof P' in $HTi(A_1,\ldots,A_s)$ of $\longrightarrow B$. The proofs P_i and the proof P' can be derived from P by means of preliminary reduction steps, including at least one "omission of a cut".

Proof: Word by word the same as that of basic lemma II.

Remark: The proof P' associated with B in S_j is welldetermined by B (and S_j) according to the construction described in the proof of basic lemma II. We call P' the side proof determined by B in S_j.

E. Now let $A_1,\ldots,A_s$ be arbitrary but fixed closed Harrop formulas. Throughout what follows we make the

Assumption: ZTi, $\longrightarrow A_1,\ldots,\longrightarrow A_s$ is consistent.

From theorem 53 we conclude

Lemma 25: $HZTi(A_1,\ldots,A_s)$ is consistent.

Notation: The theories $HZTi(A_1,\ldots,A_s)$ and $HZT(A_1,\ldots,A_s)$ will be denoted by HZi and HZ respectively.

For HZ and HZi we can introduce the whole complex of notions introduced in connection with ZT. That is, the following notions can be introduced without any changes in exactly the same way as before: 1) complexity of a cut; 2) of an induction; 3) height of a sequent in a proof; 3) fork I_1,I_2,I_3; 4) fork elimination (classical logical reduction step); 5) intuitionistic fork elimination (intuitionistic logical reduction step); 6) induction reduction; 7) saturated proof; 8) substitution instance; 9) inessential reduction step; 10) subformula reduction step; 11) preliminary reduction step; 12) strictly normal standard proof (s.n.s. proofs). To this list of concepts we add a new one, more precisely we introduce a new kind of reduction step, to be called "H-reduction step" with H indicating that the reduction step has something to do with Harrop formulas. Prior to the definition of H-reduction step we note an important lemma which connects the basic lemma II_H with the Harrop axioms.

Lemma 26: Let P be a standard proof in Zi (that is having an endsequent of the form $\longrightarrow C$). Let $A \longrightarrow B$ be a Harrop axiom in the final part of P . Then $\longrightarrow B$ is a Harrop axiom.

Proof: By assumption, $A \longrightarrow B$ is an uppermost sequent in the final part of P . Case a: $A \longrightarrow B$ is the rightmost one among the uppermost sequent in the final part of P . Then P , being a standard proof, has necessarily the endsequent $\longrightarrow B$. Hence, HZi $\vdash \longrightarrow B$ and so Zi $\vdash \longrightarrow B$ by theorem 53.
Case b: $A \longrightarrow B$ is not the rightmost one among the uppermost sequents in the final part of P . According to basic lemma II_H, there is a proof P' in HZi of $\longrightarrow B$; hence Zi $\vdash \longrightarrow B$ according to theorem 53.

On the other hand, it follows from the inspection of definition 51 that B is a Harrop formula. Hence, by combining this with cases a) and b), we obtain the lemma.

Now to the description of H-reduction step. Let P be an s.n.s. proof in HZi and S a Harrop axiom in the final part of P having the form $\Gamma \longrightarrow G$ where Γ contains at most one formula. Then we can apply to P a certain syntactical transformation, depending on the form of G . The specific form of this transformation is given by the clauses A-F below.

A) S is $\longrightarrow (\xi)B(\xi)$. By lemma 26 $\longrightarrow (\xi)B(\xi)$ and hence $\longrightarrow B(\alpha)$ are Harrop axioms. So we can replace S in P by the following derivation:

$$\cfrac{\longrightarrow B(\alpha)}{\Gamma \longrightarrow (\xi)B(\xi)} \qquad \longrightarrow \forall \text{ , eventually followed by a thinning,}$$

where α is a suitably chosen free variable.

B) S is $\longrightarrow (x)B(x)$. Then we proceed in the same way as under

A), but with a suitably chosen free individual variable y in place of α .

C) S is $\Gamma \longrightarrow A \wedge B$. By lemma 26 $\longrightarrow A \wedge B$ and hence $\longrightarrow A$ and $\longrightarrow B$ are Harrop axioms. Hence, we can replace S by the following derivation:

$$\frac{\longrightarrow A \qquad \longrightarrow B}{\Gamma \longrightarrow A \wedge B} \quad \longrightarrow \wedge$$

, eventually followed by a thinning.

D) S is $\Gamma \longrightarrow A \supset B$. By lemma 26 $\longrightarrow A \supset B$ is a Harrop axiom and by definition 51, clause b), $A \longrightarrow B$ is also a Harrop axiom. Hence we can replace S by the following derivation:

$$\frac{A \longrightarrow B}{\Gamma \longrightarrow A \supset B} \quad \longrightarrow \supset$$

, followed eventually by a thinning.

E) S is $\Gamma \longrightarrow \neg A$. By lemma 26 $\longrightarrow \neg A$ is a Harrop axiom and by definition 51, clause c), $A \longrightarrow$ is a Harrop axiom. Hence we can replace S by the following derivation:

$$\frac{A \longrightarrow}{\Gamma \longrightarrow \neg A} \quad \longrightarrow \neg$$

, eventually followed by a thinning.

F) S is $\Gamma \longrightarrow p=q$ and Γ not empty. Then $\longrightarrow p=q$ is a Harrop axiom and we can replace S by the following derivation:

$$\frac{\longrightarrow p=q}{\Gamma \longrightarrow p=q} \quad \text{thinning.}$$

The proof P' which one obtains by applying to P any of the transformations described under A) - F) is said to follow from P by means of an H-reduction step. We say that the H-reduction step is applied to the Harrop axiom S .

It is evident that there is no infinite chain of proofs $P_o, P_1, \ldots$ such that P_{i+1} follows from P_i by means of an H-reduction step. We even can find an upper bound N in terms of P_o with the property: if $P_o,\dots,P_s$ is a chain of proofs in HZi such that P_{i+1} follows from P_i by an H-reduction step and such that no such reduction step is applicable to P_s then $s \leqq N$. An important property of H-reduction steps is described by the following

Lemma 27: Let P be a saturated s.n.s. proof in HZi which does not admit any H-reduction step. Then every sequent S in the final part of P is either a true prime sequent or else a mathematical axiom $D \longrightarrow D'$, D isomorphic with D' .

Proof: Assume the lemma to be false. The sequent S which violates the lemma must then by necessity be a Harrop axiom. We show that a contradiction arises and distinguish cases according to which clause of definition 51 S is a Harrop axiom. Case a: S is $\longrightarrow B$ with B a Harrop formula such that $Zi \vdash \longrightarrow B$. If B were not a prime formula, then B would contain as outermost logical symbol either a propositional connective $\wedge$, $\neg$, $\supset$, or else a universal quantifier applied to a functional variable or an individual variable. In any case, we could apply an H-reduction step to S , contradicting the assumption. Hence B is a prime formula p=q and, since P is a saturated proof, both p and q are saturated. Since Zi is consistent by assumption, it follows from theorem 53 that $|p| = |q|$ holds; hence S is a true saturated prime formula, contradicting the assumption about S . Case b: S is $A \longrightarrow B$ and $Zi \vdash \longrightarrow A \supset B$. From lemma 26 it follows that $\longrightarrow B$ is a Harrop axiom, that is, $Zi \vdash \longrightarrow B$. As under a), it follows that B cannot contain a logical symbol. Hence B must be a saturated prime formula p=q . From lemma 26, the assumed consistency of Zi and theorem 53, we conclude that $|p| = |q|$ must hold, contradicting the assumption about S . Case c: S is $A \longrightarrow$ and $Zi \vdash \longrightarrow \neg A$ holds. Since S is an axiom in the final part of P, it is an uppermost sequent in the final part of P, and so we can infer from basic lemma II_H that there exists a proof P* in HZi of $\longrightarrow A$. Since HZi is a conservative extension of Zi, this contradicts $Zi \vdash \longrightarrow \neg A$ and the assumed consistency of Zi .

F. Now we associate with every formula A inductively a natural number, called its degree and denoted by d(A) . a) If A is prime, then d(A)=1 ; b) $d(A \wedge B)=d(A)+d(B)+1$;
c) $d(A \vee B)=d(A)+d(B)+1$; d) $d(A \supset B)=d(A)+d(B)+1$;
e) $d(\neg A)=d(A)+1$; f) $d((x)A(x))=d(A(x))+1$;

g) $d((\xi)A(\xi))=d(A(\alpha))+1$; h) $d((E\xi)A(\xi))=d(A(\alpha))+1$; i) $d((Ex)A(x))=d(A(x))+1$. After this, we associate with every sequent S in a proof P in HZ inductively an ordinal $<\varepsilon_0$. The inductive clauses in the definition of this ordinal assignement are invariably given by clauses 2) - 6) in section 2.4., part A of chapter II. Only clause 1) has to be replaced by another one, to be denoted by 1*). In order to state 1*) explicitely, let S be an axiom in P . Clause 1*) is then given as follows: 1) if S is $\Gamma \longrightarrow$, then $O(S)=1$; 2) if S is $\Gamma \longrightarrow B$, then $O(S)=d(B)$. As ordinal of P we take, as usual, the ordinal associated with its endsequent; it is denoted by $O(P)$. The reason for replacing 1) by 1*) is given by

<u>Theorem 54:</u> If P and P' are s.n.s. proofs in HZi such that P' follows from P by means of an H-reduction step, then $O(P') \leq O(P)$.

<u>Proof:</u> Let S be the Harrop axiom in P to which the H-reduction step is applied. We treat two representative cases; all other cases are equally trivial to treat.

<u>Case 1:</u> S is $\Gamma \longrightarrow A \supset B$. By definition $O(S)=d(A)+d(B)+1$. The H-reduction step amounts to replace in P the sequent S by the derivation

$$\frac{A \longrightarrow B}{\Gamma \longrightarrow A \supset B} \quad \longrightarrow \supset$$

eventually followed by a thinning,

where $A \longrightarrow B$ is again a Harrop axiom by lemmas 26 and 24. The theorem is essentially proved if we can show that the ordinal of $\Gamma \longrightarrow A \supset B$ in P' is not larger than the ordinal of $\Gamma \longrightarrow A \supset B$ in P . The first, however, is by definition $d(B) \# 1$, that is, $d(B)+1$, while the second is $d(A)+d(B)+1$, that is, larger than the first one.

<u>Case 2:</u> S is $\Gamma \longrightarrow \neg A$. The reduction step replaces S in P by the derivation

$$\frac{A \longrightarrow}{\Gamma \longrightarrow \neg A} \quad \longrightarrow \neg$$

, plus eventually a thinning.

The ordinal of S in P is d(A)+1 by assumption, the ordinal of S in P' is $1 \# 1$, that is, 2, hence not larger than the ordinal of S in P.

Concerning the other reduction steps, everything remains the same as in chapter II, that is, we have

Theorem 55: A) Preliminary reduction steps and inessential reduction steps do not increase the ordinal of the proof to which they are applied. B) Fork elimination (classical and intuitionistic), "omission of a cut" and induction reductions lower the ordinal of the proof to which they are applied. C) A subformula reduction step lowers the ordinal of the proof to which it is applied.

The proof is the same as usual and can be omitted. On the purely syntactical level we also have

Theorem 56: Let P be a saturated s.n.s. proof in HZi which does not admit preliminary reduction steps, H-reduction steps, induction reductions and fork elimination. If P does not coincide with its final part, then there is a critical logical inference whose principal formula has an image in the endsequent; hence a subformula reduction step is applicable to P.

Proof: Since no H-reduction step is applicable to P, it follows from lemma 27 that every axiom in the final part of P is either a true saturated prime sequent, or else of the form $D \longrightarrow D'$ with D, D' isomorphic. Since no preliminary reduction step is applicable to P, we conclude that only true prime sequents can occur as axioms in the final part of P. Since P is saturated and no induction reduction is applicable to P, it follows that P does not contain a critical induction inference. Hence, the only critical inferences in P are the logical ones. Now we proceed in exactly the same way as in the proof of theorem 2 in [8].

G. Now we come to the proof of theorem 51. In virtue of theorem 53, theorem 51 is proved if we can prove

Theorem 56: a) If A,B are closed formulas such that $HZi \vdash \longrightarrow A \vee B$ then either $HZi \vdash \longrightarrow A$, or else $HZi \vdash \longrightarrow B$. b) if $(E\xi)A(\xi)$ is a closed formula such that HZi $\vdash \longrightarrow (E\xi)A(\xi)$ then there is a functor F free for ξ in $A(\xi)$ such that HZi $\vdash \longrightarrow A(F)$ holds. c) Similarly as in b), but with $(Ex)A(x)$ in place of $(E\xi)A(\xi)$ and a term t in place of F .

Proof: We prove b) . The proofs of a),c) are practically the same. Hence let P be a proof in HZi of $\longrightarrow (E\xi)A(\xi)$. Without loss of generality we can assume that P is strictly normal and saturated (since $(E\xi)A(\xi)$ is closed). Let us call reduction chain every finite or infinite sequence of proofs $P_o, P_1, \dots$ having the following properties: 1) $P_o=P$; 2) each P_i is an s.n.s. proof in HZi ; 3) P_{i+1} follows from P_i by means of a preliminary reduction step, by an H-reduction step, an induction reduction or an intuitionistic fork elimination. Then it follows from our considerations above (in particular theorem 56) that no infinite reduction chain exists. Hence there exists a finite reduction chain $P_o, P_1, \dots, P_N$ having the property: no reduction step other than a subformula reduction step is applicable to P_N . By induction with respect to i , using thereby the consistency of HZi , one proves that P_i and hence P_N has the same endsequent as P , namely $\longrightarrow (E\xi)A(\xi)$. From theorem 56 we infer that a subformula reduction step is applicable to P_N . The result of this subformula reduction step must by necessity be a proof P' in HZi of $\longrightarrow A(F)$ for a certain functor F, free for ξ in $A(\xi)$ and determined by P_N . This proves b) of our theorem. Statements a) and c) are proved in the same way.

8.2. Harrop formulas and the theories ZTi/II_N and $ZTEi/II_N$

A. In this section, we consider only a special type of Harrop formulas, namely those given by the following

Definition 52: By MT we understand the set of those Harrop formulas which are classically true, whereby the truth of formulas containing special function constants is reduced to the truth of those without special function constants via definition 33.

If we restrict our attention to formulas belonging to MT , then we can extend the considerations of the previous section in an almost straightforward way to the theories ZTE/II_N and ZTE/II . It is the purpose of this section to extend the considerations of the previous section to the case where ZT and ZTi are replaced by ZTE/II_N and $ZTEi/II_N$, respectively, and where the class of Harrop formulas to be considered belongs to the subset MT of M .

<u>B.</u> For the time being , let $A_1, \ldots, A_s$ be arbitrary closed Harrop formulas. Then $ZT/II_N(A_1, \ldots, A_s)$ denotes by definition the theory obtained from ZT/II_N by addition of $\longrightarrow A_1, \ldots, \longrightarrow A_s$ as new axioms. Similarly, $ZTi/II_N(A_1, \ldots, A_s)$ is the theory obtained from ZTi/II_N by addition of $\longrightarrow A_1, \ldots, \longrightarrow A_s$ as new axioms; $ZTi/II_N(A_1, \ldots, A_s)$ is, of course, nothing else than the intuitionistic version of $ZT/II_N(A_1, \ldots, A_s)$.

From $ZTi/II_N(A_1, \ldots, A_s)$ we can pass to a conservative extension $ZTEi/II_N(A_1, \ldots, A_s)$ by addition of two new inference rules, Ti(P) and $Ti(P,P_1,m)$, which have been introduced in part B of section 4.1. of chapter IV. The formal definition of the rules Ti(P) and $Ti(P,P_1,m)$ remains the same as in part B of section 4.1., with the following exception: a) the side proof P in Ti(P) is now assumed to be a proof in $ZTi/II_N(A_1, \ldots, A_s)$; b) the side proofs P, P_1 in $Ti(P,P_1,m)$ are now assumed to be proofs in $ZTi/II_N(A_1, \ldots, A_s)$.

Similarly, we can introduce the conservative extension $ZTE/II_N(A_1, \ldots, A_s)$ of $ZT/II_N(A_1, \ldots, A_s)$ by adding to $ZT/II_N(A_1, \ldots, A_s)$ the two new rules Ti(P) and $Ti(P,P_1,m)$; again P, P_1 range now over proofs in $ZTi/II_N(A_1, \ldots, A_s)$. Corresponding to theorem 14 we have

<u>Theorem 57:</u> a) $ZTEi/II_N(A_1, \ldots, A_s)$ is a conservative extension of $ZTi/II_N A_1, \ldots, A_s)$; b) $ZTE/II_N(A_1, \ldots, A_s)$ is a conservative extension of $ZT/II_N(A_1, \ldots, A_s)$.

The proof of this theorem is a mere copy of the proof of theorem 14. By specializing definition 51 to the case where T and Ti are $ZTE/II_N(A_1, \ldots, A_s)$ and $ZTEi/II_N(A_1, \ldots, A_s)$, we obtain their respective Harrop hulls to be denoted by $HZTE/II_N(A_1, \ldots, A_s)$ and $HZTEi/II_N(A_1, \ldots, A_s)$, respectively. The notion of Harrop axiom (with respect to $ZTEi/II_N(A_1, \ldots, A_s)$ now) is again given by the remark following definition 51; lemmas 24 and 26 remain, of course, true in the present case. Clearly we have

<u>Theorem 58:</u> a) $HZTEi/II_N(A_1,\ldots,A_s)$ is a conservative extension of $ZTEi/II_N(A_1,\ldots,A_s)$ and hence of $ZTi/II_N(A_1,\ldots,A_s)$;
b) $HZTE/II_N(A_1,\ldots,A_s)$ is a conservative extension of $ZTE/II_N(A_1,\ldots,A_s)$ and hence of $ZT/II_N(A_1,\ldots,A_s)$.

It is also clear that $HZTEi/II_N(A_1,\ldots,A_s)$ is nothing else than the intuitionistic restriction of $HZTE/II_N(A_1,\ldots,A_s)$.

For proofs P in $HZTE/II_N(A_1,\ldots,A_s)$ we can, of course, introduce the notions "final part", "omission of a cut" and "preliminary reduction step" in exactly the same way as in all previous cases. Throughout this section we will use basic lemma II_H for the special case where $HTi(A_1,\ldots,A_s)$ is $HZTEi/II_N(A_1,\ldots,A_s)$. If in particular $\Gamma \longrightarrow A$ is an uppermost sequent in the final part of a proof P in $HZTEi/II_N(A_1,\ldots,A_s)$, if B is a formula in Γ, if P' is the welldetermined proof of $\longrightarrow B$, whose existence is stated in basic lemma II_H , then we call P' the side proof of $\longrightarrow B$ determined by B in $\Gamma \longrightarrow A$.

<u>C.</u> From now on $A_1,\ldots,A_s$ are fixed, closed Harrop formulas which satisfy the

<u>Assumption:</u> $A_1,\ldots,A_s$ are classically true.

In order to avoid the steady use of the clumsy notation $HZTEi/II_N(A_1,\ldots,A_s)$ and $HZTE/II_N(A_1,\ldots,A_s)$, we denote the first theory simply by HZEi , the second by HZE . The theories $ZTE/II_N(A_1,\ldots,A_s)$ and $ZTEi/II_N(A_1,\ldots,A_s)$ on the other hand will be denoted simply by ZEi and ZE .

Next, we can carry over without the slightest changes the whole body of concepts introduced in chapter IV for ZTE/II_N and $ZTEi/II_N$, respectively, to the present case. A list of concepts, which can be defined for proofs P in HZE and HZEi, respectively, using the same definitions as in chapter IV, is given in what follows:
1) complexity of a cut; 2) of an induction; 3) complexity of a II_N-inference; 4) of a $Ti(P)$-inference; 5) of a $Ti(P,P_1,m)$-inference; 6) height $h(S)$ of a sequent S in P ; 7) fork I_1,I_2,I_3 ; 8) fork elimination (classical and intuitionistic); 9) induction reduction; 10) canonical II_N-reduction step, 11) canonical Ti_1-reduction step; 12) canonical Ti_2-reduction

step; 13) saturated proof; 14) preliminary reduction step; 15) inessential reduction step; 16) subformula reduction step; 17) preliminary reduction step; 18) strictly normal standard proof (s.n.s. proof). With respect to the clauses 10), 11), 12) in the above list, we refer thereby to part B of section 4.4. (chapter IV) and in particular to theorem 17 and definition 18. To this list of concepts we add the notion of H-reduction step which has been defined in the previous section and whose definition remains invariably the same. It has exactly the same properties as before; lemma 27, in particular, remains invariably true and its proof remains the same. Finally, we can associate with every formula A its degree $d(A)$, whose inductive definition is again given by the inductive clauses stated at the beginning of part F in the last section.

D. Before associating ordinals with proofs P in HZE and HZEi, we have to make some remarks which are closely connected with part A of section 4.3. (chapter IV). To this end, consider a $Ti(P_1)$-inference S_1/S_2, where S_1, S_2 have the particular form described in part B of section 4.1. P_1 is by definition a proof in $ZTi/II_N(A_1,\dots,A_s)$ of a sequent of the form $\longrightarrow W^o(\subset_R)$, where R is a standard formula of the form $R_o(x) \wedge seq(x)$ containing no special function constants and whose only free variable is x. Since $A_1,\dots,A_s$ are by assumption classically true formulas, it follows that $W^o(\subset_R)$ is a classically true formula. In other words, the relation $\{\langle n,m\rangle / n \subset_K m \text{ and } R(n) \text{ and } R(m) \text{ true}\}$ is indeed wellfounded. The ordinal which is associated with this relation will be denoted by $\|\subset_R\|$.

Next, let there be given a $Ti(P_1,P_2,m)$-inference S_1/S_2, where S_1, S_2 have the particular form described in part B of section 4.1. By definition, P_1 is a proof in $ZTi/II_N(A_1,\dots,A_s)$ of a sequent having the form $\longrightarrow W^o(\subset_R)$, with R as above. The proof P_2 on the other hand is by definition a proof in $ZTi/II_N(A_1,\dots,A_s)$ whose endsequent has the form $\longrightarrow R(t)$, where t is a certain saturated term whose value $|t|$ is m. As before, we conclude that $R(t)$ and hence $R(m)$ are classically true formulas. This means that m belongs to the domain of definition of the wellfounded relation $\{\langle u,v\rangle / u \subset_K v \text{ and } R(u), R(v) \text{ classically true}\}$. Therefore, there is a welldefined ordinal associated with m as a member of the domain of definition of the relation $\{\langle u,v\rangle / u \subset_K v \text{ and } R(u), R(v) \text{ classically true}\}$. We denote this ordinal by $\|m\|_R$. Finally, we can introduce as in part B of section 4.3. the ordinal Ω which is the smallest among all ordinals ξ having the following property: if P is a proof in $ZTi/II_N(A_1,\ldots,A_s)$ of $\longrightarrow W^o(\subset_R)$, then $\|\subset_R\| < \xi$ (with R as above).

E. Now, if we are given a proof P in HZE we can associate inductively an ordinal $O(S)$ with every sequent S ocurring in P . The inductive clauses of this assignement are as follows: 1*) if S is an axiom of the form $\Gamma \longrightarrow$, then $O(S)=1$, if S is an axiom of the form $\Gamma \longrightarrow B$, then $O(S)=d(B)$; 2) if S is the conclusion of a conversion, a structural inference, an induction or a logical inference, then we proceed as in part A of section 2.4.; 3) if S is the conclusion of a II_N-inference S'/S, then $O(S)=\omega_d((O(S')\#\omega^{\Omega+1})\omega^{\Omega+1})$ where $d=h(S')-h(S)$; 4) if S is the conclusion of a $Ti(P_1)$-inference S'/S, then we put $O(S)=\omega_d((O(S')\#\omega^{\xi+1})\omega^{\xi+1})$ where $d=h(S')-h(S)$, and where P is a proof (in $ZTi/II_N(A_1,\ldots,A_s)$) of $\longrightarrow W^o(\subset_R)$ and $\lambda=\|\subset_R\|$; 5) if S is the conclusion of a $Ti(P_1,P_2,m)$-inference S'/S, then we put $O(S)=\omega_d((O(S')\#\omega^{\xi+1})\omega^{\xi+1})$ with $d=h(S')-h(S)$, where P_1 and P_2 are proofs (in $ZTi/II_N(A_1,\ldots,A_s)$) of $\longrightarrow W^o(\subset_R)$ and $\longrightarrow R(t)$ with $|t|=m$, respectively, and where $\xi=\|m\|_R$. As ordinal $O(P)$ of a proof, we take as usual the ordinal of its endsequent. The main property of this ordinal assignement is given by

Theorem 59: Let P and P' be two s.n.s. proofs in HZEi and let P' follow from P by means of an H-reduction step. Then $O(P')\leqq O(P)$.

Theorem 60: A) Preliminary reduction steps and inessential reduction steps do not increase the ordinal of the proof to which they are applied. B) A reduction step lowers the ordinal of the proof to which it is applied if it belongs to the following list: 1) fork elimination (classical or intuitionistic); 2) omission of a cut; 3) induction reduction; 4) canonical II_N-reduction step; 5) canonical Ti_1-reduction step; 6) canonical Ti_2-reduction step; 7) subformula reduction step.

The proof of theorem 59 is, of course, exactly the same as the proof of theorem 54 in the previous section; the proof of theorem 60, on the other hand, leads to precisely the same calculations and inequalities encountered in chapters II and III.

F. Before coming to the main result, we note that theorem 19 remains invariably true in the present case, that is, we have

Theorem 61: Let P be a saturated s.n.s. proof in HZEi and assume that P does not admit either preliminary reduction steps, H-reduction steps, fork eliminations, induction reductions, canonical II_N-reduction steps, canonical Ti_1-reduction steps or canonical Ti_2-reduction steps. If P does not coincide with its final part, then there is a critical logical inference whose principal formula has an image in the final part. Hence, a subformula reduction step is applicable to P in this case.

Proof: Since P is saturated and does not admit any induction reduction, there is obviously no critical induction inference in P. Similarly, there are no critical II_N-, $Ti(P_1)$- and $Ti(P_1,P_2,m)$-inferences in P since otherwise a corresponding reduction step would be applicable to P, contradicting the assumption. Since no H-reduction step is applicable to P, it follows from lemma 27 in the last section that every axiom in the final part of P is either a saturated prime sequent or else a logical axiom $D \longrightarrow D'$. Since no preliminary reduction step is applicable to P, we conclude that no logical axiom $D \longrightarrow D'$ occurs in the final part of P. Finally there is no fork I_1, I_2, I_3 in the final part of P since otherwise an intuitionistic fork elimination would be applicable to P, contradicting the assumption. Hence, by proceeding in the same way as in the proof of theorem 2 in [8], we conclude that there is a critical logical inference whose principal formula has an image in the final part of P.

G. Now we can state the main result:

Theorem 62: a) If A, B are closed formulas such that HZEi $\vdash \longrightarrow A \vee B$ holds, then either HZEi $\vdash \longrightarrow A$ or HZEi $\vdash \longrightarrow B$; b) if $(Ex)A(x)$ is a closed formula such that HZEi $\vdash \longrightarrow (Ex)A(x)$ holds, then there is a saturated term t such that HZEi $\vdash \longrightarrow A(t)$ holds; c) if $(E\xi)A(\xi)$ is a closed formula such that HZEi $\vdash \longrightarrow (E\xi)A(\xi)$ holds, then there is a functor F without free variables such that HZi $\vdash \longrightarrow A(F)$ holds.

<u>Proof:</u> The proof parallels the proof of theorem 56. Consider e.g. part c) and let P be a proof in HZEi of $\longrightarrow (E\xi)A(\xi)$, with $(E\xi)A(\xi)$ closed. Without restriction we can assume that P is an s.n.s. proof. A finite or infinite chain $P_o, P_1, \ldots$ of proofs in HZEi is called a reduction chain if the following holds:
1) $P_o = P$; 2) each P_i is an s.n.s. proof; 3) P_{i+1} follows from P_i by means of a preliminary reduction step, an H-reduction step, an intuitionistic fork elimination, an induction reduction, a canonical II_N-, Ti_1- or Ti_2-reduction step. Given any proof P* in HZEi, it is clear that we cannot apply indefinitely H-reduction steps and preliminary reduction steps to P* . From this observation and theorem 60, part B), it follows that infinite reduction chains do not exist. Let us call a reduction chain $P_o, P_1, \ldots, P_N$ terminating if no reduction step other than a subformula reduction step is applicable to P_N . Evidently, there exist terminating reduction chains. Let $P_o, P_1, \ldots, P_N$ be a fixed one. From the consistency of HZEi one infers by induction that each P, and in particular P_N, have $\longrightarrow (E\xi)A(\xi)$ as endsequent. From theorem 61 and the definition of terminating reduction chain, it follows that a subformula reduction step is applicable to P_N . The result of this subformula reduction step must necessarily be a proof P* in HZEi of $\longrightarrow A(F)$ for some functor F without free variables, determined by P_N .

Since HZEi is a conservative extension of $ZTi/II_N(A_1, \ldots, A_s)$ we can reformulate the above theorem in the following way:

<u>Theorem 63:</u> Let $A_1, \ldots, A_s$ be closed, classically true Harrop formulas and A,B, $(Ex)A(x)$, $(E\xi)A(\xi)$ arbitrary closed formulas.
a) If $ZTi/II_N(A_1, \ldots, A_s) \vdash \longrightarrow A \vee B$ then either $ZTi/II_N \vdash \longrightarrow A$ or $ZTi/II_N \vdash \longrightarrow B$; b) if $ZTi/II_N(A_1, \ldots, A_s) \vdash \longrightarrow (E\xi)A(\xi)$, then there exists a functor F without free variables such that $ZTi/II_N(A_1, \ldots, A_s) \vdash \longrightarrow A(F)$ holds; c) similarly with $(Ex)A(x)$ and a term t in place of $(E\xi)A(\xi)$ and F, respectively.

There is a special case of the last theorem which is of some interest. To this end let $B_1, \ldots, B_s$ be a list of closed formulas such that each B is an instance of the continuity axiom or of Church's thesis, which can be refuted in ZT/II_N . That is, for each i we have:

1) B_i is an instance of the continuity axiom or of Church's thesis; 2) $ZT/II_N \vdash \longrightarrow \neg B_i$. Then $B_1,\dots,B_s$ are obviously classically true formulas. This implies that theorem 65 applies to $ZTi/II_N(\neg B_1,\dots,\neg B_s)$:

<u>Corollary:</u> Let $B_1,\dots,B_s$ be closed formulas such that for each i the following holds: 1) B is an instance of the continuity axiom or of Church's thesis; 2) $ZT/II_N \vdash \longrightarrow \neg B$. Then a),b),c) of theorem 63 hold for $ZTi/II_N(\neg B_1,\dots,\neg B_s)$.

<u>H.</u> It causes no difficulties to reprove theorem 24 for $HZEi/II_N(A_1,\dots,A_s)$ with $A_1,\dots,A_s$ classically true Harrop formulas. The proof of this theorem remains essentially the same as the proof of theorem 24 in section 4.5. of chapter IV, provided with the necessary supplements due to the presence of Harrop axioms. We leave the proof to the reader.

8.3. Harrop formulas and the theories ZTi/II and ZTEi/II

<u>A.</u> The considerations of the previous section can be extended in a straightforward way to the case where ZTi/II_N and $ZTEi/II_N$ are replaced by ZTi/II and ZTEi/II, respectively. All that has to be done is to replace certain notions that are characteristic for $ZTEi/II_N$ by the corresponding notions belonging to ZTEi/II . So, II_N-, $Ti(P_1)$- and $Ti(P_1,P_2,m)$-inferences will be replaced by II-, $TI(P_1)$- and $TI(P_1,P_2,m)$-inferences, respectively. Similarly, we replace canonical II_N-, Ti_1- and Ti_2-reduction steps by canonical II-, TI_1- and TI_2-reduction steps, respectively. Finally we have to replace the ordinal assignement described in section 4.3. by the ordinal assignement described in section 4.6., part C. Apart from this, changes, definition and treatment of the theories $ZTi/II(A_1,\dots,A_s)$, $ZTEi/II(A_1,\dots,A_s)$ and $HZTEi/II(A_1,\dots,A_s)$, parallel definition and treatment of the theories $ZTi/II_N(A_1,\dots,A_s)$, $ZTEi/II_N(A_1,\dots,A_s)$ and $HZTEi/II_N(A_1,\dots,A_s)$, respectively. In particular, all concepts connected with Harrop formulas, such as Harrop axiom, Harrop hull, H-reduction step, remain the same as before. In order to avoid repetitions, we omit a detailed treatment of $ZTi/II(A_1,\dots,A_s)$ and $HZTEi/II(A_1,\dots,A_s)$ and content ourself by stating the main results which parallel those obtained for $ZTi/II_N(A_1,\dots,A_s)$:

Theorem 64: Let $A_1,\dots,A_s$ be closed, classically true Harrop formulas and A, B, $(E\xi)C(\xi)$, $(Ex)D(x)$ arbitrary closed formulas.
a) If $HZTEi/II(A_1,\dots,A_s) \vdash \longrightarrow A \vee B$, then $HZTEi/II(A_1,\dots,A_s) \vdash \longrightarrow A$ or $HZTEi/II(A_1,\dots,A_s) \vdash \longrightarrow B$;
b) if $HZTEi \vdash \longrightarrow (E\xi)C(\xi)$, then there is a functor without free variables F such that $HZTEi \vdash \longrightarrow C(F)$ holds; c) similarly with $(Ex)D(x)$ and a term t in place of $(E\xi)C(\xi)$ and F. Since $HZTEi/II(A_1,\dots,A_s)$ is a conservative extension of $ZTi/II(A_1,\dots,A_s)$, clauses a),b),c) apply to $ZTi/II(A_1,\dots,A_s)$ as well.

By specializing $A_1,\dots,A_s$ in an appropriate way we obtain a corollary to the last theorem which corresponds to the corollary to theorem 63, namely

Corollary: Let $B_1,\dots,B_s$ be closed formulas such that for each i the following holds: 1) B_i is an instance of the continuity axiom or of Church's thesis; 2) $ZT/II \vdash \longrightarrow \neg B_i$. Then a),b),c) of theorem 64 hold for $ZTi/II(\neg B_1,\dots,\neg B_s)$.

It would again cause no trouble to reprove theorem 24, but with $ZTi/II(A_1,\dots,A_s)$ in place of ZTi/II_N where $A_1,\dots,A_s$ are closed, classically true Harrop formulas. We omit the proof.

8.4. Harrop formulas and the theories ZTi/I and ZTGi/I

This is the most important section of this chapter. Its main purpose is to combine the considerations of the previous chapters with those of section 8.1. in order to obtain theorem 51, but with ZTi/I in place of ZTi.

A. To start with, let $A_1,\dots,A_s$ be arbitrary closed Harrop formulas. Then $ZTi/I(A_1,\dots,A_s)$ is the theory obtained from ZTi/I by addition of $\longrightarrow A_i$, $i=1,\dots,s$, as new axioms; $ZT/I(A_1,\dots,A_s)$ is obtained from ZT/I by addition of $\longrightarrow A_i$, $i=1,\dots,s$, as new axioms. $ZTi/I(A_1,\dots,A_s)$ is, of course, the intuitionistic restriction of $ZT/I(A_1,\dots,A_s)$. From $ZTi/I(A_1,\dots,A_s)$ we pass to a certain conservative extension, to be denoted by $ZTGi/I(A_1,\dots,A_s)$, by addition of two new rules $T(P_1)$ and $T(P_1,P_2)$. The formal definitions of $T(P_1)$ and $T(P_1,P_2)$

remain exactly the same as in the definitions of the rules $T(P_1)$ and $T(P_1,P_2)$, respectively, given in chapter VII, section 7.1., part A) (clauses 1), 2)) with the following exception: a) the proofs P_o and P_1 in the definition of $T(P_1)$ (clause 1), part A), sect. 7.1.) are now proofs already at hand in $ZTGi/I(A_1,\ldots.,A_s)$; b) the proofs P_o,P_1,P_2 in the definition of $T(P_1,P_2)$ (clause 2), part A), sect. 7.1.) are now proofs already at hand in $ZTGi/I(A_1,\ldots.,A_s)$. If we add to $ZT/I(A_1,\ldots.,A_s)$ and to $ZTi/I(A_1,\ldots.,A_s)$ the new rules $T(P_1)$ and $T(P_1,P_2)$, then we obtain correspondingly conservative extensions $ZTG/I(A_1,\ldots.,A_s)$ and $ZTGi/I(A_1,\ldots.,A_s)$, respectively. $ZTGi/I(A_1,\ldots.A_s)$ is, of course, nothing else than the intuitionistic restriction of $ZTG/I(A_1,\ldots,A_s)$. To sum up, we have

<u>Theorem 65:</u> a) $ZTG/I(A_1,\ldots.,A_s)$ is a conservative extension of $ZT/I(A_1,\ldots.,A_s)$; b) $ZTGi/I(A_1,\ldots.,A_s)$ is a conservative extension of $ZTi/I(A_1,\ldots.,A_s)$; c) $ZTGi/I(A_1,\ldots.,A_s)$ is the intuitionistic restriction of $ZTG/I(A_1,\ldots.,A_s)$. The proof of a), b) remains the same as the proof of theorem 14. From $ZTG/I(A_1,\ldots.,A_s)$ and $ZTGi/I(A_1,\ldots.,A_s)$ we can pass to their respective Harrop hulls $HZTG/I(A_1,\ldots.,A_s)$ and $HZTGi/I(A_1,\ldots.,A_s)$; the notion of Harrop axiom (with respect to $ZTGi/I(A_1,\ldots.,A_s)$) now remains, of course, the same as before. Lemma 24 remains true in the present case and we clearly have

<u>Theorem 66:</u> a) $HZTGi/I(A_1,\ldots.,A_s)$ is a conservative extension of $ZTGi/I(A_1,\ldots.,A_s)$ and hence of $ZTi/I(A_1,\ldots.,A_s)$; b) $HZTG/I(A_1,\ldots.,A_s)$ is a conservative extension of $ZTG/I(A_1,\ldots.,A_s)$ and so of $ZT/I(A_1,\ldots.,A_s)$.

<u>B.</u> From now on $A_1,\ldots.,A_s$ are arbitrary but fixed closed Harrop formulas which satisfy the following

<u>Assumption:</u> $ZTi/I(A_1,\ldots.,A_s)$ is consistent.

In order to avoid the lengthy notations $HZTGi/I(A_1,\ldots.,A_s)$, $HZTG/I(A_1,\ldots.,A_s)$, $ZTGi/I(A_1,\ldots.,A_s)$ and $ZTG/I(A_1,\ldots.,A_s)$, we replace them by HZGi, HZG, ZGi and ZG respectively. The next step consists in carrying over to HZGi and HZG certain notions and concepts, which have been introduced for ZTGi/I and ZTG/I . Among the simplest of these are the notions "final part", "preliminary reduction step" and "omission of a cut". In this connection we note that basic lemma II_H, formulated in section 8.1., remains invariably true in the present case if we take for $HTi(A_1,\dots,A_s)$ the theory HZGi . We also adopt the terminology introduced by the remark following basic lemma II_H: if B is a formula in $\Gamma \longrightarrow A$, if $\Gamma \longrightarrow A$ is an uppermost sequent in the final part of a proof P in HZGi , if P' is the proof of $\longrightarrow B$ whose existence is given by basic lemma II_H and whose construction is described in the proof of basic lemma II (chapter III, sect. 3.2.), then P' is called the side proof of $\longrightarrow B$, determined by B in $\Gamma \longrightarrow A$ according to basic lemma II_H . If, in particular, $\Gamma \longrightarrow A$ is the conclusion of a I-inference, S/S' , say

$$\text{I} \quad \frac{R(y),\ (x) \subset_R y A(x),\ \Gamma' \longrightarrow A(y)}{W(\subset_R),\ R(q),\ \Gamma' \longrightarrow A(q)} ,$$

if B is $W(\subset_R)$, then we call P' as before the side proof determined by this I-inference in P . Further notions which can be introduced for proofs P in HZGi . HZG in the same way as for proofs in ZTGi/I , ZTG/I are: 1) complexity of a cut; 2) an induction; 3) complexity of a I-inference; 4) complexity of a $T(P_1)$-inference; 5) complexity of a $T(P_1,P_2)$-inference; 6) height of a sequence S in P ; 7) fork I_1,I_2,I_3 ; 8) fork elimination (classical or intuitionistic); 9) induction reduction; 10) I-reduction step; 11) $T(P_1)$- and $T(P_1,P_2)$-reduction step; 12) strong $T(P_1)$- and strong $T(P_1,P_2)$-reduction step; 13) saturated proof; 14) substitution instance; 15) inessential reduction step; 16) subformula reduction step; 17) strictly normal standard proof; 18) side proof of a $T(P_1)$- or a $T(P_1,P_2)$-inference; 19) index proof of a $T(P_1,P_2)$-inference. All these notions are defined in precisely the same way as in chapter VII or in earlier chapters. To this list of notions, we add the concept of H-reduction step which has been introduced in section 8.1. and whose definition remains invariably the same. The notion of H-reduction step has the same properties as before; lemmas 26 and 27 in particular remain true and their proofs remain the same. The degree d(A) of a formula finally is defined in the same way as in part F of section 8.1.

C. Corresponding to theorem 56 in section 8.1. we have now

Theorem 67: Let P be an s.n.s. proof in HZGi which does not coincide with its final part. Assume that no reduction step of the following list is applicable to P : 1) preliminary reduction step, 2) intuitionistic fork elimination, 3) induction reduction, 4) I-reduction step, 5) strong $T(P_1)$-reduction step, 6) strong $T(P_1,P_2)$-reduction step, 7) H-reduction step. Then there is a critical logical inference in P whose principal formula has an image in the final part. Hence a subformula reduction step is applicable to P .

Proof: As in earlier cases, it follows that no critical I-inference, $T(P_1)$-inference and $T(P_1,P_2)$-inference occurs in P , since otherwise corresponding reduction steps could be applied to P ; for the same reason, there can be no critical induction in P . On the other hand, no H-reduction step and no preliminary reduction steps are applicable to P by assumption. Hence the final part of P contains only mathematical axioms (true saturated prime sequents), conversions, interchanges, contractions and cuts. Finally no fork can occur in P and so we can argue as in the proof of theorem 2 in [8] .

D. Our next aim is to introduce a suitable notion of "good" proof. For the sake of completeness, we discuss this notion in some detail and proceed thereby in a slightly different way than in chapters V and VII .

Definition 53: Let P be an s.n.s. proof in HZGi . A sequence (finite or infinite) $P_o,P_1,P_2,\ldots$ of s.n.s. proofs in HZGi is said to be a reduction chain of P if $P_o=P$, and if for each i P_{i+1} follows from P_i by means of a reduction step of the following list: 1) preliminary; 2) H-reduction step; 3) intuitionistic fork elimination ; 4) induction reduction; 5) I-reduction step; 6) strong $T(\hat{P}_1)$-reduction step; 7) strong $T(\hat{P}_1,\hat{P}_2)$-reduction step; 8) subformula reduction step; 9) inessential reduction step. A reduction chain is terminating if it is finite, say, $P_o,P_1,\ldots,P_N$, and if no reduction step listed above is applicable to P_N.

Definition 54: An s.n.s. proof P in HZGi is called "good" if every reduction chain of P is terminating.

Remarks on notation: In the theorem below we retain the notation used in connection with theorem 46 in section 7.3. of chapter VII; R in $W(\subset^u_R)$, in particular, is a standard formula whose only free variable is x and whose special function constants are $\alpha^{i_1}_{u_1}$, $\alpha^{i_2}_{u_2}$, ..., $\alpha^{i_s}_{u_s}$. More generally, we use throughout this section the notation introduced in part B of section 7.1., chapter VII.

The main property of good proofs is given by

Theorem 68: Let P be a good s.n.s. proof in HZGi of $\longrightarrow W(\subset^u_R)$. Let $f_1, ..., f_s, g$ be numbertheoretic functions. Then we find an m and an n with $n+1 < m$ and an s.n.s. proof P' in HZGi of $\longrightarrow \neg\, \xi_w(n+1) \subset^{u*v}_R \xi_w(n)$, where v denotes the system $v_i = \bar{f}_i(m)$, $i=1, ..., s$ of sequence numbers, and where $w = \bar{g}(m)$.

Proof: In order to save notation, we assume $s=1$, that is, just one special function constant, say, α^i_u, occurs in R and hence in $W(\subset^u_R)$. The upper index u in $\subset^u_R$ will then be identified with the lower index u in α^i_u. The function f_1 will be denoted by f. Now we proceed in steps.

1) From the definition of "good" proof it follows: if $P_o, P_1, ..., P_N$ is a reduction chain of P, then P_N is good.

2) Call a reduction chain $P_o, P_1, ..., P_N$ of P "short" if no P_{i+1} follows from P_i by means of a subformula reduction step. If $P_o, ..., P_N$ is a short reduction chain of P, then each P_i has an endsequent of the form $\longrightarrow W(\subset^{u*v_i}_R)$ where $v_{i+1} \subseteq_K v_i$. A short reduction chain is called compatible with f if v_i is an initial segment of f for all i. A short reduction chain of P is called terminating if there is no short reduction chain of P which extends the given one properly.

3) It is evident: there exist terminating short reduction chains of P which are compatible with f. Let $P_o,\ldots,P_N$ be any such reduction chain. According to its definition, no reduction step other than a subformula reduction step is applicable to P_N. Since the endsequent of P_N is $\longrightarrow W(\subset_R^{u*v_N})$, P_N does not coincide with its final part. By theorem 67 a subformula reduction step is applicable to P_N. The result is a proof P_{N+1} with endsequent $\longrightarrow (Ex)\neg\ \alpha^j_{<\ >}(x+1) \subset_R^{u*v_N} \alpha^j_{<\ >}(x)$. Here $i \neq j$ by definition of subformula reduction step.

4) P_{N+1} is good in virtue of 1). Consider a short reduction chain $P_{N+1}, P_{N+2},\ldots,P_M$ of P_{N+1}. Each of the P_i's has an endsequent of the form $\longrightarrow (Ex)\neg\ \alpha^j_{w_i}(x+1) \subset_R^{u*v_i} \alpha^j_{w_i}(x)$. Call such a reduction chain compatible with f,g if for each i v_i and w_i are initial segments of f and g respectively.

5) It is evident: there are short, terminating reduction chains of P_{N+1} which are compatible with f,g. Let $P_N,\ldots,P_M$ be any such chain. As before, we conclude that P_M admits a subformula reduction step. The result is a proof P* whose endsequent has the form $\longrightarrow \neg\ \alpha^j_{w_M}(t+1) \subset_R^{u*v_M} \alpha^j_{w_M}(t)$, where t is a constant term containing no other special function constants than

$\alpha^i_{u*v_M}$, $\alpha^j_{w_M}$.

6) Then it is obvious that we find an m so large that $T(\alpha^i_{u*\bar{f}(m)}, \alpha^j_{\bar{g}(m)})$ is saturated with value, say, n, such that: $\alpha)$ $n+1 < m$, $\beta)$ $\bar{f}(m) \subseteq_K u*v_M$, $\gamma)$ $\bar{g}(m) \subseteq_K w_M$. By substituting in P* $\alpha^i_{u*\bar{f}(m)}$ and $\alpha^j_{\bar{g}(m)}$ for $\alpha^i_{u*v_M}$ and $\alpha^j_{w_M}$ respectively and by adding eventually a conversion to the endsequent we finally obtain a proof P' which satisfies the conditions of the theorem.

Now we can associate with every good proof P of $\longrightarrow W(\subset^{u}_{R})$, exactly as we have done in sect. 7.3. of chapter VII, a continuity function $\tau(x_1,\ldots,x_s,y)$, having the properties: if $f_1,\ldots,f_s,g$ are numbertheoretic functions, if $\tau(\bar{f}_1(m),\ldots,\bar{f}_s(m),\bar{g}(m))\neq 0$, then there is an n with $n+1<m$ and a proof P' in HZGi of $\longrightarrow \neg \xi_w(n+1) \subset^{u*v}_{R} \xi_w(n)$, where v_i denotes the list $v_i=\bar{f}_i(m)$, $i=1,\ldots,s$, of sequence numbers and where $w=\bar{g}(m)$. We call τ the continuity function associated with P and denote it by τ^P. Actually, τ^P could be chosen recursive but we do not use this fact.

Definition 55: An s.n.s. proof in HZGi or HZG is called graded if all its side proofs are good.

E. The next tool which we need here is that of valuation. This concept is introduced in exactly the same way as in section 7.4. of the last chapter and has all the properties described there. So D^s is again the set of ordered $s+1$-tuples of sequence numbers, all having the same length, and $\langle v_1,\ldots,v_s,v_{s+1}\rangle \sqsubset^{s} \langle w_1,w_2,\ldots,w_s,w_{s+1}\rangle$ still holds iff $v_i \subseteq_K w_i$, $i=1,\ldots,s$ (where left and right arguments are elements of D). An element $v_1,\ldots,v_{s+1}$ from D is secured with respect to the good proof P iff $\tau^P(v_1,\ldots,v_{s+1})\neq 0$, unsecured otherwise. $D^s(P)$ is the set of those elements of D^s which are unsecured with respect to P and $\sqsubset^{s}_{P}$ is the restriction of $\sqsubset^{s}$ to $D^s(P)$. Clearly, $\sqsubset^{s}_{P}$ is wellfounded.

Now to the valuation. A valuation of a proof P in HZG is an assignement which associates with every $T(P_1,P_2)$-inference in P either a number e which satisfies a certain condition $\alpha)$, or else a pair of numbers e, e_1 which satisfies a certain condition $\beta)$. Condition $\alpha)$ in the present case is word by word the same as condition $\alpha)$ in part C of sect. 7.4. Condition $\beta)$ in the present case is the same as condition $\beta)$ in part C of sect. 7.4. with one exception: ZTGi/I in clause d) in the definition of $\beta)$, part C of sect. 7.4., has to be replaced by HZGi. In all other respects $\beta)$ in the present case is the same as $\beta)$ in C, 7.4. Valuations are again denoted by symbols such as $\mathcal{V}$, $\mathcal{W}$ etc; the value of $\mathcal{V}$ for an inference S/S' is written as $\mathcal{V}(S/S')$. From now on, we can treat valuations in exactly the same way as in sect. 7.4. In particular, we have the following three notions:
a) extension of a $T(P_1,P_2)$-inference; b) data of a $T(P_1,P_2)$-

inference; c) index of a $T(P_1,P_2)$-inference. Their definitions remain the same as in section 7.4. With these notions at hand, we can introduce the concept of induced valuation. That is, given a proof P in HZGi , a valuation $\mathcal{V}$ of P and a proof P' which follows from P by means of a reduction step, we can define on P' a valuation $\mathcal{V}^*$ in terms $\mathcal{V}$. This valuation is again called the valuation induced by $\mathcal{V}$ on P' . Its definition is described by cases 1 - 10 listed in part D of sect. 7.4. and an additional case 11 which takes into account H-reduction steps. Case 11: P' follows from P by means of an H-reduction step. Each $T(P_1,P_2)$-inference S/S' in P remains unaffected by this H-reduction step: we may therefore define $\mathcal{V}^*(S/S') = \mathcal{V}(S/S')$. $\mathcal{V}^*$ on P' is thus completely determined. If now P is a graded s.n.s. proof in HZG and $\mathcal{V}$ a valuation of P, then we call $\mathcal{V}$ compatible with P if the conditions in definition 49 (part D of sect. 7.4.) are satisfied. Lemma 19 is now replaced by the slightly modified

Lemma 19*: Let P be a graded s.n.s. proof and $\mathcal{V}$ a compatible valuation of P . Let P' be obtained from P by means of a reduction step from the following list: 1) preliminary; 2) omission of a cut; 3) H-reduction step; 4) intuitionistic or classical fork elimination; 5) induction reduction; 6) subformula reduction step . Then P' is still a graded proof and the induced valuation $\mathcal{V}^*$ is compatible with P' .

Lemma 20, on the other hand, remains true as it stands and its proof remains the same. Finally, let P be an s.n.s. proof in HZGi , provided with a valuation $\mathcal{V}$, let S/S' be a critical I-inference in P and P_1 the side proof determined by S/S' according to basic lemma II_H . Then we can define on P_1 a valuation $\mathcal{V}'$ in terms of $\mathcal{V}$ in exactly the same way as we have done it in part F of sect. 7.4. Without danger of confusion, we call $\mathcal{V}'$ the valuation induced by $\mathcal{V}$ on the side proof P_1 . Lemmas 21 and 22 about side proofs and induced valuation remain invariably true in the present case and their proofs remain the same.

F. Our next step consists in associating ordinals with graded proofs. More precisely, if P is a graded proof, then we associate with every sequent S in P a certain ordinal O(S) . The inductive definition of O(S) is exactly the same as in sect. 7.5., part D, that is, we use clauses 1) - 10) in section 7.5., part D, as they

stand in order to define $O(S)$. The notations $\|\sqsubset^{s}_{P_1}\|$, $\|e\|_{P_1}$ and $O_{\mathcal{V}}(S/S')$ retain thereby their meaning.
The properties of this ordinal assignement remain essentially the same as before. In place of theorem 47 we have the slightly modified

Theorem 47*: Let P be a graded s.n.s. proof in HZG and $\mathcal{V}$ a compatible valuation of P . Let P' be obtained from P by means of a reduction step and $\mathcal{V}^*$ the valuation induced by $\mathcal{V}$ on P'. Then $O_{\mathcal{V}^*}(P') < O_{\mathcal{V}}(P)$ if the reduction step in question is one of the following list: 1) omission of a cut; 2) classical fork elimination; 3) intuitionistic fork elimination; 4) induction reduction; 5) strong $T(P_1)$-reduction step; 6) strong $T(P_1,P_2)$-reduction step (with P intuitionistic in case of 3) - 6)). If P' is a substitution instance of P or follows from P by means of a preliminary of an H-reduction step, then $O_{\mathcal{V}^*}(P') \leqq O_{\mathcal{V}}(P)$.

Proof: The only new element which has to be taken into consideration is the case of H-reduction step, which can be treated in the same way as in the proof of theorem 54 in section 8.1. Apart from this, the proof of theorem 47* parallels that one of theorem 47.

Theorem 48 finally remains true as it stands and its proof remains the same.

G. Our final step consists in proving

Theorem 49*: If P is a graded proof in HZGi and $\mathcal{V}$ a compatible valuation of P, then P is good.

Proof: We proceed by transfinite induction with respect to $O_{\mathcal{V}}(P)$, that is, we assume: if P' is a graded proof in HZGi , and $\mathcal{V}'$ a compatible valuation of P' such that $O_{\mathcal{V}'}(P') < O_{\mathcal{V}}(P)$, then P' is good. We show that a contradiction follows from the assumption that P is not good. Hence let us make this assumption and let $P_o, P_1, \ldots$ be an infinite reduction chain of P . Then we clearly find an N with the following property: 1) if $i+1 \leqq N$, then P_{i+1} follows from P_i by means of a preliminary reduction step or an H-reduction step; 2) P_{N+1} follows from P_N by means of a reduction step which is neither an H-reduction step nor a preliminary reduction step. For $i \leqq N+1$ we define inductively valuations $\mathcal{V}_i$ on P_i as follows (part F, sect. 7.4.): $\mathcal{V}_{i+1} = \mathcal{V}_i^*$. From lemma 19* we conclude by induction that each P_i is still a graded proof and that $\mathcal{V}_i$ is compatible with P_i. From theorem 47* it follows that $o\,\mathcal{V}_{i+1}(P_{i+1}) \leqq o\,\mathcal{V}_i(P_i)$ holds in case $i < N$. Now we distinguish two subcases according to the kind of reduction step which leads from P_N to P_{N+1}. Subcase 1: The reduction step in question is a fork elimination, an induction reduction, a strong $T(P_1)$-reduction step, a strong $T(P_1,P_2)$-reduction step or a subformula reduction step. Then P_{N+1} is still a graded proof and $\mathcal{V}_{N+1}$ is compatible with P_{N+1} according to lemma 19* or 20, and $o\,\mathcal{V}_{N+1}(P_{N+1}) < o\,\mathcal{V}_N(P_N)$ according to theorem 47*. But then P_{N+1} is good according to our inductive assumption, contradicting the assumption that the reduction chain $P_o, P_1, \ldots, P_N, P_{N+1}, \ldots$ is infinite. Subcase 2: P_{N+1} follows from P_N by means of a I-reduction step. Let S/S' be the critical I-inference in P_N to which the reduction step is applied and $\hat{P}$ the side proof determined by S/S' in P_N. According to the construction described in the proof of basic lemma II (chapter III, section 3.2.), $\hat{P}$ is obtained from P_N with the aid of some preliminary reduction steps and at least one operation "omission of a cut". Let $\mathcal{V}'$ be the valuation induced by $\mathcal{V}_N$ on $\hat{P}$ (part F in section 7.4.). According to lemma 21 (still true now), $\hat{P}$ is graded and $\mathcal{V}'$ compatible with P. According to theorem 47*, $o\,\mathcal{V}'(\hat{P}) < o\,\mathcal{V}(P)$; hence $\hat{P}$ is good according to our inductive assumption. According to lemma 22 and theorem 48, P_{N+1} is graded, $\mathcal{V}_{N+1}$ is compatible with P_{N+1} and $o\,\mathcal{V}_{N+1}(P_{N+1}) < o\,\mathcal{V}_N(P_N)$. Hence P_{N+1} is good, again contradicting the assumption that the reduction chain $P_o, P_1, \ldots, P_N, P_{N+1}, \ldots$ is infinite.

From theorem 49* we obtain as an immediate consequence

Theorem 69: Let P be a graded proof in HZGi provided with a compatible valuation $\mathcal{V}$. Let $A, B, (E\xi)A(\xi), (Ex)A(x)$ be closed formulas. a) If P is a good proof in HZGi of $\longrightarrow A \vee B$, then HZGi $\vdash \longrightarrow A$ or HZGi $\vdash \longrightarrow B$. b) If P is a good proof of $\longrightarrow (E\xi)A(\xi)$, then there exists a closed functor F such that HZGi $\vdash \longrightarrow A(F)$. c) Similarly as in b), but with $(Ex)A(x)$ in place of $(E\xi)A(\xi)$ and a term t in place of F.

Proof: Consider eg. b). In virtue of theorem 49*, P is good. Hence we find a reduction chain $P_o, \ldots, P_N$ with the property:

a) no P_{i+1} follows from P_i by means of a subformula reduction step; b) no reduction step other than eventually a subformula reduction step is applicable to P_N . The endsequent of P_N is still $\longrightarrow (E\,\xi)A(\xi)$ and so P_N cannot coincide with its final part. According to theorem 67, we infer that there is a critical logical inference whose principal formula has an image in the final part and that a subformula reduction step is indeed applicable to P_N . The result of this subformula reduction step is by necessity a proof P^* whose endsequent is $\longrightarrow A(F)$ for some closed functor F . Clauses a) and c) are proved similarly.

From the last theorem we immediately get the main result:

Theorem 70: Let $A_1,\ldots,A_s$ be closed Harrop formulas such that $ZTi/I(A_1,\ldots,A_s)$ is consistent. Let A, B, $(E\,\xi)A(\xi)$, $(Ex)A(x)$ be closed formulas. Then we have: a) if $ZTi/I(A_1,\ldots,A_s) \vdash \longrightarrow A \vee B$ then $ZTi/I \vdash \longrightarrow A$ or $ZTi/I \vdash \longrightarrow B$; b) if $ZTi/I \vdash \longrightarrow (E\,\xi)A(\xi)$, then $ZTi/I \vdash \longrightarrow A(F)$ for some closed functor F ; c) similarly as in b), but with $(Ex)A(x)$ in place of $(E\,\xi)A(\xi)$ and with a term t in place of F .

Proof: Assume e.g. $ZTi/I(A_1,\ldots,A_s) \vdash \longrightarrow (E\,\xi)A(\xi)$. Then we obviously find an s.n.s. P proof of $\longrightarrow (E\,\xi)A(\xi)$. But with respect to $HZTGi/I(A_1,\ldots,A_s)$ (that is HZGi), P is clearly a graded proof: no $T(P_1)$- and $T(P_1,P_2)$-inferences occur in P . A compatible valuation of P is given by the empty valuation $\mathcal{V}_\phi$. Therefore we can apply the last theorem and conclude: $HZGi \vdash \longrightarrow A(F)$ for some constant functor F . Since HZGi is a conservative extension of $ZTi/I(A_1,\ldots,A_s)$, we obtain $ZTi/I(A_1,\ldots,A_s) \vdash \longrightarrow A(F)$, as stated by the theorem.

8.5. The theories ZTi/IV_N and ZTi/IV

A. The theories ZTi/IV_N and ZTi/IV are, of course, subtheories of ZTi/I . Despite this, we cannot specialize theorem 70 at once by replacing ZTi/I by ZTi/IV_N or ZTi/IV, respectively. The reasons are twofold: 1) from the consistency of e.g. $ZTi/IV(A_1,\ldots,A_s)$ we cannot necessarily infer the consistency of $ZTi/I(A_1,\ldots,A_s)$; 2) even if this is the case, and if e.g. $ZTi/IV(A_1,\ldots,A_s) \vdash \longrightarrow A \vee B$ holds, we can infer from theorem 70 only that either $ZTi/I(A_1,\ldots,A_s) \vdash \longrightarrow A$ or $ZTi/I(A_1,\ldots,A_s) \vdash \longrightarrow B$

holds. However, a closer inspection shows that if we restrict attention in the foregoing section to proofs P in $ZTi/IV_N(A_1,\dots,A_s)$ or $ZTi/IV(A_1,\dots,A_s)$, then we never have to take into account the larger theory $ZTi/I(A_1,\dots,A_s)$. By performing this inspection in some detail we would obtain theorem 70, but with ZTi/IV_N and ZTi/IV, respectively, in place of ZTi/I. We do not go into details but merely sum up the results which one obtains in this way:

<u>Theorem 71:</u> Let $A_1,\dots,A_s$ be closed Harrop formulas such that $ZTi/IV(A_1,\dots,A_s)$ is consistent. Let A, B, $(E\xi)A(\xi)$, $(Ex)A(x)$ be closed formulas. a) If $ZTi/IV(A_1,\dots,A_s) \vdash \longrightarrow A \vee B$ then $ZTi/IV(A_1,\dots,A_s) \vdash \longrightarrow A$ or $ZTi/IV(A_1,\dots,A_s) \vdash \longrightarrow B$. b) If $ZTi/IV(A_1,\dots,A_s) \vdash \longrightarrow (E\xi)A(\xi)$, then there is a constant functor F such that $ZTi/IV(A_1,\dots,A_s) \vdash \longrightarrow A(F)$ holds. c) Similarly, but with $(Ex)A(x)$ in place of $(E\xi)A(\xi)$ and a term t in place of F. Similarly, but with ZTi/IV_N in place of ZTi/IV.

There is a particular case of the last theorem which may be of some interest:

<u>Theorem 72:</u> Let $B_1,\dots,B_t$ be a list of closed formulas such that for each i the following holds: 1) B_i is an instance of the continuity axiom or of Church's thesis; 2) $ZT/IV \vdash \longrightarrow \neg B_i$. Let $C_1,\dots,C_q$ be a list of closed formulas such that for each i the following holds: α) C_i is an instance of the axiom of choice; β) $ZT/IV \vdash C_{i+1} \longrightarrow C_i$; γ) no $\longrightarrow C_i$ is provable from ZT/IV. Then clauses a), b), c) of the last theorem apply to $ZTi/IV(\neg B_1,\dots,\neg B_t, \neg C_1,\dots,\neg C_q)$.

<u>Proof:</u> All we have to do is to show that $ZTi/IV(\neg B_1,\dots,\neg B_t, \neg C_1,\dots,\neg C_q)$ (to be denoted for brevity by T) is consistent. To this end, assume the contrary. Then $T \vdash \longrightarrow$ follows, or what amounts to the same: $ZTi/IV \vdash \neg B_1,\dots,\neg B_t, \neg C_1,\dots,\neg C_q \longrightarrow$. Since $ZT/IV \vdash \longrightarrow \neg B_i$ by assumption, we obtain $ZT/IV \vdash \neg C_1,\dots,\neg C_q \longrightarrow$, that is $ZT/IV \vdash \longrightarrow C_1,\dots,C_q$. On the other hand, we have $ZT/IV \vdash C_{i+1} \longrightarrow C_i$ by assumption. Hence by application of a series of cuts and contractions we finally obtain: $ZT/IV \vdash \longrightarrow C_1$. However, this contradicts γ).

Expressed in an inexact way, the last theorem says: if we add to ZTi/IV the negation of the continuity axiom, of Church's thesis and of the axiom of choice, then we obtain a theory which still satisfies a), b), c) of theorem 71. The preceeding theorem is of course only of interest because there are formulas $B_1, B_2, \ldots, C_1, C_2, \ldots$ which satisfy 1), 2) and α), β), γ) ; thereby we tacitly use the fact that there exists an infinite list $C_1, C_2, \ldots$ of instances of the axiom of choice, such that $ZT(C_1, \ldots, C_n, \ldots)$ is as strong as classical analysis and such that $ZT \vdash C_{i+1} \longrightarrow C_i$ holds. Whether theorem 72 holds if we replace ZTi/IV by ZTi/I is not clear to the author.

CHAPTER IX:

The Markov principle

This chapter contains the main applications of the results contained in the preceeding chapter, namely a proof of the fact that the Markov principle (or at least a particular form of the Markov principle) is not derivable in a certain large class of intuitionistic formal theories. Since no new proof theoretic techniques will come into application, it is notationally somewhat simpler for us to consider Hilbert-type systems in place of Gentzen-type systems.

9.1. The Markov principle

A. We remember that according to our notation introduced in chapter I, ZH is the Hilbert-type version of the Gentzen-type system ZT of number theory, ZHi is the intuitionistic restriction of ZH and at the same time the Hilbert type version of ZTi . Briefly, ZHi is a Hilbert-type version of intuitionistic number theory, based on the language L . Since some Goedel type diagonal argument will be used below, it is advisable to make the distinction between natural numbers and the terms 0,0',0",.... which represent them in ZH : if n is a natural number, we denote the term 0 by $\bar{n}$ and call it the numeral of n . We also need

Definition 56: A theory T is said to be primitive recursive if it is primitive recursively axiomatizable, that is, if the set of its axioms can be chosen in a primitive recursive way.

Assumption: Throughout this chapter we assume that the assignement which associates with every term t a continuity function τ related with t is that one described in part L of section 1.4., chapter I. As mentioned there, we have then

Theorem 73: ZTi and hence ZHi are primitive recursive.

B. We distinguish between two kinds of Markov principle , the weak Markov principle, denoted by MP_0 , and the strong Markov principle, denoted by MP . The weak Markov principle is a certain axiom schema. A particular instance of MP_0 is given by a formula of the following type: $\neg(x)\neg R(x) \supset (Ey)R(y)$, where R(x) is a prime formula without special function constants and whose only free variable is x. A particular instance of MP on the other hand is given by $(x)\neg(y)\neg R(x,y) \supset (x)(Ey)R(x,y)$, where $R(x,y)$ is a prime formula without special function constants and with only x,y free. We say that MP_o (or MP) is not provable in a certain theory if a particular instance of MP_o (or MP) is not provable in this theory. Our main objective is to prove that MP_o and MP are not provable in a certain large class of intuitionistic theories.

C. Before proceeding further, we note a relation between MP_o and MP:

Lemma 28: MP_o can be derived from MP within ZHi.

Proof: Assume $\neg(y)\neg R(y)$. Let $R(x,y)$ be a prime formula such that $\neg R(x,y) \equiv \neg(R(y) \vee x \neq x)$ is provable in ZHi. Then $\neg R(x,y) \equiv \neg R(y)$ and $(x)\neg(y)\neg R(x,y) \equiv \neg(y)\neg R(y)$ are provable in ZHi. By application of MP to $(x)\neg(y)\neg R(x,y)$ we get $(x)(Ey)R(x,y)$. However, $(x)(Ey)R(x,y)$ is provable equivalent to $(Ey)R(y)$, that is, MP_o holds.

Sometimes we simply say that Markov's principle is not derivable, meaning that MP_o and hence MP is not derivable.

9.2. Markov principle and weak Harrop property

A. Definition 55: Let T be any extension of ZHi. We say that T has the weak Harrop property if T is consistent and if the following holds: if $R(x)$ and $Q(x)$ are prime formulas without free variables other than x and without special function constants, if $\neg(x)\neg R(x),\ T \vdash (Ez)Q(z)$, then there is an n such that $Q(\bar{n})$ is true.

Theorem 74: Let T be a primitive recursive extension of ZHi, which has the weak Harrop property. Then MP_o (and hence MP) is not provable in T.

Proof: Since T is a primitive recursive extension of ZHi and since ZHi contains the whole formalism of primitive recursive function theory, we find according to Goedel and Rosser a prime formula $R(y)$ such that the following holds: 1) $R(y)$ does not contain special function constants or free variables other than y ; 2) $(y) \neg R(y)$ is undecidable with respect to T ; 3) $(y) \neg R(y)$ is true. Clearly, $T, \neg (y) \neg R(y)$ is consistent. Otherwise, $T \vdash \neg\neg(y) \neg R(y)$ would hold. But $ZHi \vdash \neg\neg(y) \neg R(y) \equiv (y) \neg R(y)$ holds, since $(y) \neg R(y)$ is a formula without $\vee$ and E . Hence $T \vdash (y) \neg R(y)$ would follow, contradicting the undecidability of $(y) \neg R(y)$. Now assume $T \vdash MP_o$. Then $T, \neg (y) \neg R(y) \vdash (Ey)R(y)$. Since T has the weak Harrop property, it follows that there is an n such that $R(\bar{n})$ is true. This contradicts the fact that $(y) \neg R(y)$ is true. Hence $T \vdash MP_o$ is false.

Actually, if we inspect the proof of theorem 74, then we see that we have proved the following variant of theorem 74:

Theorem 74*: Let T be a primitive recursive extension of ZHi which has the weak Harrop property. Then we find a prime formula $R(x)$ whose only free variable is x , such that the following holds: $(Ey)R(y)$ is not provable from $\neg (y) \neg R(y), T$.

9.3. The Markov principle and some particular intuitionistic theories

A. In what follows we will apply theorem 74 to some particular intuitionistic theories. Since most of our results have been obtained in the frame of sentential calculus, we will rephrase them in the terminology of Hilbert-type systems. First we will pass from the Gentzen-type systems ZTi/V and ZTi/I to the corresponding Hilbert-type systems. To this end, consider the following formula:

$$W(\subset_R) \supset . \{(y)(R(y) \supset .(x) \subset_{R^y} A(x) \supset A(y)) \supset (z)(R(z) \supset A(z)) \} .$$

This formula is denoted by $T_o^*(R,A)$. The universal closure of $T_o^*(R,A)$ (that is, the formula obtained by universal quantification over all free variables which occur in $T_o^*(R,A)$) is denoted by $T_o(R,A)$. We also need formulas of the following type:

$$W(\subset_R) \supset . \{ (y)((x) \subset_{R^y} A(x) \supset A(y)) \supset (z)A(z) \} .$$

Such formulas are denoted by $T^*(R,A)$ and their universal closure by $T(R,A)$. Finally, we cite the axiom of barinduction such as stated in [5] in the form 26.3a:

$$\{(a)(seq(a) \supset .R(a) \vee \neg R(a)) \wedge (\alpha)(Ex)R(\bar{\alpha}(x)) \wedge (a)(seq(a) \wedge \wedge R(a). \supset A(a)) \wedge (a)(seq(a) \wedge (s)A(a*2^{s+1}). \supset A(a)) \}. \supset A(1) .$$

We denote it by $BI^*(R,A)$ and its universal closure by $BI(R,A)$.

<u>Definition 58:</u> 1) By $ZHti_o$ we denote the theory which we obtain by adding to ZHi all formulas $T_o(R,A)$ without special function constants as new axioms. 2) By ZHti we denote the theory which we obtain by adding to ZHi all formulas $T(R,A)$ without special function constants. 3) $ZHti_o^*$ is like ZHti, but R in $T_o(R,A)$ is required to be a bounded formula. 4) ZHti* is like ZHti, but R in $T(R,A)$ is required to be a bounded formula. 5) ZHBi is obtained by adding to ZHi all formulas $B(R,A)$ without special function constants. 6) ZHBi* is like ZHBi, but the R in $B(R,A)$ is required to be a bounded formula.

<u>Notation:</u> Let T be any of the theories listed in definition 58. The theory which we obtain by adding to T the formulas $A_1,\ldots,A_s$ as new axioms is denoted by $T(A_1,\ldots,A_s)$. We remind that, if T is a Gentzen-type theory, then $T(A_1,\ldots,A_s)$ denotes the theory obtained by adding $\longrightarrow A_1,\ldots,\longrightarrow A_s$ as new axiom to T. Closed formulas are again formulas without free variables and special function constants.

<u>Theorem 75:</u> Let $A_1,\ldots,A_s$ be closed formulas.
1) $ZTi/I(A_1,\ldots,A_s) \vdash \longrightarrow B$ iff $ZHti_o(A_1,\ldots,A_s) \vdash B$.
2) $ZTi/IV(A_1,\ldots,A_s) \vdash \longrightarrow B$ iff $ZHti_o^*(A_1,\ldots,A_s) \vdash B$.
3) $ZHti_o(A_1,\ldots,A_s) \vdash B$ iff $ZHti(A_1,\ldots,A_s) \vdash B$ and $ZHti_o^*(A_1,\ldots,A_s) \vdash B$ iff $ZHti^*(A_1,\ldots,A_s) \vdash B$.
4) If $ZHti(A_1,\ldots,A_s) \vdash B$ then $ZHBi(A_1,\ldots,A_s) \vdash B$.
5) $ZHti^*(A_1,\ldots,A_s) \vdash B$ iff $ZHBi^*(A_1,\ldots,A_s) \vdash B$.
6) The theories $ZHti_o(A_1,\ldots,A_s)$, $ZHTi^*(A_1,\ldots,A_s)$, $ZHti(A_1,\ldots,A_s)$, $ZHti^*(A_1,\ldots,A_s)$, $ZHBi(A_1,\ldots,A_s)$ and $ZHBi^*(A_1,\ldots,A_s)$ are all primitive recursive.

The proof of theorem 75 is completely routine and hence omitted; 6) in particular is an immediate consequence of theorem 73. Theorem 75 permits us to rephrase the results obtained in the preceeding chapter for ZTi/I and ZTi/IV in terms of their Hilbert-type versions ZHti and ZHti*, respectively, or what amounts to the same (in virtue of 3),4) of theorem 75) in terms of ZHti and ZHti*, respectively. That is, we have

<u>Theorem 76:</u> Let $A_1,\ldots,A_s$ be a list of closed Harrop formulas. Let T be any of the theories ZHti or ZHti* respectively. Let $A, B, (E\xi)A(\xi), (Ex)A(x)$ be closed formulas. If $T(A_1,\ldots,A_s)$

is consistent, then the following holds: a) if $T(A_1,\ldots,A_s) \vdash A \vee B$, then $T(A_1,\ldots,A_s) \vdash A$ or $T(A_1,\ldots,A_s) \vdash B$;
b) if $T(A_1,\ldots,A_s) \vdash (E\,\xi)A(\xi)$, then $T(A_1,\ldots,A_s) \vdash A(F)$ for some constant functor F ; c) if $T(A_1,\ldots,A_s) \vdash (Ex)A(x)$, then $T(A_1,\ldots,A_s) \vdash A(t)$ for some constant term t, and hence $T(A_1,\ldots,A_s) \vdash A(\bar{n})$ for some n .

The proof is an immediate consequence of theorems 70, 71 and theorem 75. From theorem 76 we infer

__Theorem 77:__ Let $A_1,\ldots,A_s$ be closed Harrop formulas. Let T be any of the theories ZHti and ZHti*, respectively. If $T(A_1,\ldots,A_s)$ is consistent, then it has the weak Harrop property.

__Proof:__ Let $R(x)$ and $Q(z)$ be prime formulas without special function constants and whose only free variables are x and z, respectively. Assume that $T(A_1,\ldots,A_s,(x)\neg(y)\neg R(y))$ is consistent and that $T(A_1,\ldots,A_s,(x)\neg(y)\neg R(y)) \vdash (Ez)Q(z)$ holds. Now we apply the last theorem, but with $A_1,\ldots,A_s,(x)\neg(y)\neg R(y)$ in place of $A_1,\ldots,A_s$ and infer that there is a number n such that $T(A_1,\ldots,A_s,(x)\neg(y)\neg R(y)) \vdash Q(\bar{n})$ holds. Now $Q(z)$ is numeralwise decidable in ZHi , that is, ZHi $\vdash Q(\bar{m})$ iff $Q(\bar{m})$ is true. If $Q(\bar{n})$ would be false, then ZHi $\vdash \neg Q(\bar{n})$ and hence $T(A_n,\ldots,A_s,(x)\neg(y)\neg R(y)) \vdash \neg Q(\bar{n})$, contradicting the assumed consistency of $T(A_1,\ldots,A_s,(x)\neg(y)\neg R(y))$. Hence $Q(\bar{n})$ is true what proves the weak Harrop property of $T(A_i,\ldots,A_s)$.

From the last theorem and theorem 74, we obtain immediately the main result of this chapter, namely

__Theorem 78:__ Let T be any of the theories ZHti of ZHti*, respectively. Let $A_1,\ldots,A_s$ be closed Harrop formulas. If $T(A_1,\ldots,A_s)$ is consistent, then Markov's principle is not derivable from $T(A_1,\ldots,A_s)$.

__Theorem 79:__ There are three primitive recursive lists of closed formulas $A_1,A_2,\ldots$, $B_1,B_2,\ldots$, $C_1,C_2,\ldots$ having the following properties: 1) each A_i is an instance of Church's thesis;
2) each B_i is an instance of the continuity axiom; 3) each C_i is an instance of the axiom of choice; 4) Markov's principle is not provable from ZHti$(A_1,A_2,\ldots,B_1,B_2,\ldots)$; 5) Markov's principle

is not provable from $ZHti^*(A_1, A_2, \ldots, B_1, B_2, \ldots, C_1, C_2, \ldots)$.

The proof of the theorem is via theorem 77, proceeding thereby essentially as in the case of corollary of theorem 63 and of theorem 72. From the last theorem and from theorem 75, we obtain

Corollary: There are primitive recursive lists of formulas $A_1, A_2, \ldots, B_1, B_2, \ldots, C_1, C_2, \ldots$ having properties 1) - 3) of theorem 79, and in addition the following properties: 4*) Markov's principle is not provable from $ZHBi(A_1, A_2, \ldots, B_1, B_2, \ldots)$; 5*) Markov's principle is not provable from $ZHBi^*(A_1, A_2, \ldots, B_1, B_2, \ldots, C_1, C_2, \ldots)$. The result obtained in the corollary can be stated in an imprecise way as follows: 1) if we add to the intuitionistic theory of barinduction for decidable formulas the negation of the axioms of continuity and of Church's thesis, then we cannot derive Markov's principle from the theory so obtained; 2) if we add to the intuitionistic theory of barinduction for quantifierfree formulas the negation of the axiom of choice, of continuity and of Church's thesis, then we cannot derive Markov's principle from the theory so obtained.

9.4. Markov principle and the theory of Kleene-Vesley

A. The reader might have wondered why up to now we did not say anything about the axiom of choice and the axiom of continuity. The reason is that our methods (at least, in the form in which we have presented them) do not extend to the case where the axiom of choice or the continuity axiom is present. In order to see this, let ZTiAC be intuitionistic number theory plus all instances of the axiom of choice. If Gentzen's proof-theoretic methods could be extended without modifications to ZTiAC, then we could prove among others the following statement S : If $ZTiAC \vdash \longrightarrow (E\xi)A(\xi)$, then $ZTiAC \vdash \longrightarrow A(F)$ for some constant functor F (where $(E\xi)A(\xi)$ is a closed formula). From this, however, we could derive a contradiction. In order to see this, let $T(z,x,y)$ be Kleene's T-predicate. Assume $ZTiAC \vdash \longrightarrow (E\xi)(x)T(e,x,\xi(x))$. Then, in virtue of the statement S, it follows that there is a constant functor F such that $ZTiAC \vdash \longrightarrow (x)T(e,x,F(x))$. However, all functors of ZTiAC represent primitive recursive functions. Therefore it follows that the recursive function $\{e\}(x)$ is primitive recursive. On the other hand, it is easy to find an e such that $\{e\}(x)$ is not primitive recursive and such that ZTiAC $\vdash \longrightarrow (E\xi)(x)T(e,x,\xi(x))$ holds; hence a contradiction is obtained. The difficulty is, of course, the same in the case of stronger theories such as the system of Kleene-Vesley, which will be denoted by KV .

B. Although Gentzen's methods are not directly applicable to KV, there are other methods (indirect methods) which permit us to infer that Markov's principle is not derivable from KV . All these methods are based on the fact that KV is interpretable in ZHti* . A detailed description lies outside the scope of this monograph; we content ourself with a few indications. One of these methods (the only one which we are going to consider) is based on work of Kreisel and Troelstra [7] and on work of Troelstra which is going to be published. In [7] , two theories CS and IDK are introduced. The first of these includes KV as a subsystem while the second is both a subsystem of CS and of classical analysis. CS contains a constant K , representing roughly speaking the species of recursive functions, variables for choice sequences and variables for constructive functions, together with suitable axioms. IDK is obtained from CS by dropping everything which refers to choice sequences. The major result concerning IDK and CS is the following: with every closed formula A from CS we can associate a formula A* from IDK (that is, one not containing variables for choice sequences), such that CS A iff IDK A* . If, in particular, A is itself a formula whithout choice variables, then A is A* . For formulas without choice variables, we can introduce a certain realizability notion which essentially coincides with that one introduced in [4] . In work which will appear, Troelstra proves the following statement S_1: if $A_1,\dots,A_s,B$ are closed formulas from IDK, and if IDK, $A_1,\dots,A_s \vdash B$ holds, then B is realizable whenever $A_1,\dots,A_s$ are realizable. This notion of realizability can be formalized within the language L which we have used throughout this work and there are closed formulas $\widetilde{R}_n$ with the property: if A is a closed formula from IDK with at most n logical symbols, then $\widetilde{R}_n([A])$ expresses intuitively that A is realizable where $[A]$ is the Goedelnumber of A . Although the author has not worked out the details, he believes that the following statement S_2 is provable: if $A_1,\dots,A_s,B$ are closed formulas from IDK each containing at most n logical symbols, if $A_1,\dots,A_s$,IDK $\vdash$ B holds,

then ZHti* $\vdash \breve{R}_n([A_1]) \wedge \ldots \wedge \breve{R}_n([A_s]) \supset R([B])$ holds. The following statement S_3, on the other hand, is easy to verify: if R(x) is a prime formula containing only x free and without special function constants, then, for n sufficiently large,
ZHti* $\vdash \breve{R}_n([\neg (y) \neg R(y)]) \equiv \neg (y) \neg R(y)$ and
ZHti* $\vdash \breve{R}_n([(Ey)R(y)]) \equiv (Ey)R(y)$ holds. From this, one can deduce the following statement S_4 : Let R(y) be the prime formula mentioned in theorem 74* (with ZHti* for T); then (Ey)R(y) is not derivable from $\neg (y) \neg R(y)$, CS .

<u>Proof:</u> Assume the contrary. Then CS $\vdash \neg (y) \neg R(y) \supset (Ey)R(y)$ and hence IDK $\vdash (\neg (y) \neg R(y) \supset (Ey)R(y))^*$ in virtue of the main result of Troelstra-Kreisel. Since A is A* if A does not contain variables for choice sequences, we infer IDK $\vdash \neg (y) \neg R(y) \supset (Ey)R(y)$. According to statement S_2, this implies
ZHti* $\vdash \breve{R}_n([\neg (y) \neg R(y)]) \supset \breve{R}_n([(Ey)R(y)])$. With the aid of statement S_3, finally we get ZHti* $\vdash \neg (y) \neg R(y) \supset (Ey)R(y)$, that is, ZHti*, $\neg (y) \neg R(y) \vdash (Ey)R(y)$, contradicting the combination of theorem 74* and theorem 78.

<u>C.</u> There are other ways of interpreting KV in ZHti* ; either of these could be used to prove statement S along the lines sketched above. We hope that this indications suffice to make clear that, at least with respect to the Markov principle, axiom of continuity and axiom of choice can be reduced to the theories treated in this monograph, although in an indirect way and at the expense of a considerable amount of work.

CHAPTER X:

Relative consistency proof of ZTN with respect to ZTi/I^*_N

Our arguments presented in chapters II - IX are essentially classical, that is, we looked at the proof theory of intuitionistic systems from a classical point of view. To be sure, we were careful not to use the law of excluded middle when it was not necessary; but ordinals were handled in a completely abstract and unconstructive way. It is the purpose of the present chapter to show that the reasoning presented in chapter VI can be reproduced in the theory ZTi/I^*_N (see chapter I for the definition of ZTi/I^*_N). This means that the consistency of ZTi/V can be reduced (in a primitive recursive way, in principle) to the consistency of ZTi/I^*_N. On the other hand, it is easily seen that ZT/V, that is, ZTi/V plus law of excluded middle, can be reduced in a primitive recursive way to ZTi/V: if $ZT/V \vdash A$, then $ZTi/V \vdash A^o$. Thus we obtain a consistency proof for ZT/V relative to ZTi/I^*_N. Actually, we do not formalize the theory presented in chapter VI in ZTi/I^*_N in the proper sense of the word. Our reasoning will be intuitive, but such that it will become clear that our arguments can be reproduced without difficulty in ZTi/I^*_N. For notational simplicity, we present our formalisation in the Hilbert type version of ZTi/I^*_N, that is, in the theory which we obtain from intuitionistic numbertheory ZHi by addition of all the axioms of the form $W(\subset_R) \supset .(y)((x)\subset_{R}{}_{y} A(x) \supset A(y)) \supset (z)(R(z) \supset A(z))$, with A a formula from the set W_N (sect. 1.5., def. 3) and R a bounded formula without function parameters (sect. 14, part K). Thus, if we say below that a formula B is provable in ZTi/I^*_N, we mean that $\longrightarrow B$ is provable in ZTi/I^*_N, or equivalently that B is provable in the Hilbert-type version of ZTi/I^*_N.

10.1. Preliminary remarks

A. Our task, to reduce the consistency of ZTi/V to that of ZTi/I^*_N, is, of course, accomplished if we can reduce the consistency of ZTFi/V to that of ZTi/I^*_N, where ZTFi/V is that particular conservative extension of ZTi/V which has been introduced in chapter VI. Denote by $ZTFi/V_n$ that subsystem of ZTFi/V which we obtain by considering those proofs in ZTFi/V only, which do not contain formulas with more than n logical symbols. Since ZTi/I^*_N is a subtheory of ZTFi/V, it is clear that we cannot reproduce the arguments of chapter VI as a whole in ZTi/I^*_N ; this would contradict Goedel's second incompleteness theorem. However, the arguments presented in chapter VI can be relativised to ZTi/V_n . This suggests that we try to prove in ZTi/I^*_N for each fixed n that ZTi/V_n is consistent, using thereby the methods of chapter VI, but now restricted to ZTi/V_n . That this can be done, will be shown in the following section.

B. Before proceeding further, we briefly recapitulate the definition of ZTi/I^*_N . To this end we remind that, according to definition 3, we denote by W_N the set of formulas which can be built up from Π^1_1-formulas without free-function variables by means of propositional combinations and quantifications over number variables. By ZTi/I^*_N we denote the theory obtained from ZTi by addition of the following rule of inference:

$$I^*_N \qquad \frac{R(y)\ ,\ (x) \subset_R y A(x)\ ,\ \Gamma \longrightarrow A(y)}{R(y)\ ,\ W(\subset_R)\ ,\ \Gamma \longrightarrow A(q)}$$

where R is a bounded formula without function parameters and where A belongs to W_N .

C. In this chapter we are not interested in the proof theory of ZTi/I^*_N ; we rather want to know what portion of chapter VI can be formalized within ZTi/I^*_N . It is therefore not necessary to take special function constants into account, as far as ZTi/I^*_N is concerned. Hence we will restrict ourself throughout this chapter to that portion of ZTi/I^*_N which does not contain special function constants; that is, we tacitly assume that the terms, formulas, sequents and proofs of ZTi/I^*_N with which we are concerned do not contain special function constants. Special function constants, however, reappear as soon as we are concerned with the proof theory of ZTi/V ; then they are objects about which we speak within ZTi/I^*_N .

10.2. Remarks about transfinite induction in ZTi/I^*_N

A. In ZTi/I^*_N we can perform transfinite induction only with respect to wellorderings of the form $\subset_R$ (that is $x \subset_K y \wedge R(x) \wedge R(y)$) where R is a bounded formula without function parameters (see part K of section 1.4., chapter I). It is not absolutely necessary but useful to know that in ZTi/I_N^* we can perform barinductions with respect to wellfounded trees $R(\overline{\alpha}(x))$ where $R(y)$ is recursive in the intuitionistic sense. More precisely, we have the following

<u>Theorem 80:</u> Let $D(x,y)$ and $\hat{D}(x,y)$ be two formulas not containing function parameters. Denote by $H_1,\dots,H_5$ consecutively the following formulas:
1) $(x)(\neg(Ey)D(x,y) \equiv (Ez)\hat{D}(x,z))$; 2) $(x)(\neg(Ey)D(x,y) \vee (Ey)D(x,y))$;
3) $(\alpha)(Ex,y)D(\overline{\alpha}(x),y)$; 4) $(\alpha,x)((z)A(\overline{\alpha}(x)*z) \supset A(\overline{\alpha}(x))$;
5) $(\alpha,x)((Ey)D(\overline{\alpha}(x),y) \supset A(\overline{\alpha}(x)))$. The formula A is thereby supposed to be in W_N. Then we can prove in ZTi/I_N^* the following implication: $H_1 \wedge \dots\dots \wedge H_5 \supset A(\langle\ \rangle)$.

<u>Remarks:</u> Clauses 1) and 2) express that $(Ey)D(x,y)$ is recursive in the intuitionistic sense ([4] , p.. 284). Since ZTi contains all primitive recursive functions, we can express every recursive enumerable set in the form $(Ey)D(x,y)$, with D a bounded formula. Although the proof of theorem 80 is not completely straightforward, it does not present any difficulties and therefore we omit it.

Wit the aid of theorem 80, other forms of transfinite induction can be proved in ZTi/I_N^*. In order to list them, let us introduce

<u>Definition 59:</u> A formula $A(x_1,\dots,x_s)$ is called intuitionistically recursive with respect to the intuitionistic system T if $x_1,\dots,x_s$ are its only free variables and if the following holds:
1) $A(x_1,\dots,x_s)$ has the form $(Ey)R(x_1,\dots,x_s,y)$ where R is a bounded formula; 2) there is a bounded formula $Q(x_1,\dots,x_s,z)$ such that $T \vdash \neg(Ey)R(x_1,\dots,x_s,y) \equiv (Ez)Q(x_1,\dots,x_s,z)$;
3) $T \vdash A(x_1,\dots,x_s) \vee \neg A(x_1,\dots,x_s)$.

<u>Remark:</u> We are mostly interested in the case where T is ZTi/I_N^*.

Now let $L(x,y)$, $D(x)$ and $R(x)$ be intuitionistically recursive formulas such that $L(x,y) \supset D(x) \wedge D(y)$ is provable in ZTi/I_N^*. Denote by $W(L)$ the formula $(\alpha)(Ex)\neg L(\alpha(x+1), \alpha(x))$. Then the following formulas are provable in ZTi/I_N^*, provided A belongs to W_N :

1) $W(\subset_R)\supset\{(y)(R(y)\wedge(x)(x\subset_R y\supset A(x)).\supset A(y))\supset(z)(R(z)\supset A(z))\}$

2) $W(\subset_R)\supset\{(y)((x)(x\subset_R y\supset A(x))\supset A(y))\supset(z)A(z)\}$

3) $W(L)\supset\{(y)((x)L(x,y)\supset A(x))\supset A(y))\supset(z)A(z)\}$

4) $W(L)\supset\{(y)(D(y)\wedge(x)(L(x,y)\supset A(x)).\supset A(y))\supset(z)(D(z)\supset A(z))\}$.

Formulas 1), 2) are special cases of 3), 4). Formulas 3), 4) follow from theorem 80 by means of standard devices such as presented in [3].

B. In order to apply theorem 80 and its implications successfully, it is important to know that certain particular sets and relations are indeed intuitionistically recursive. In many cases this is a consequence of the following well-known

Theorem 81: Let $A(x,y)$ be a quantifierfree formula. a) If $\vdash (x)(Ey)A(x,y)$ in classical number theory, then $\vdash (x)(Ey)A(x,y)$ in intuitionistic number theory. b) If $\vdash (x,y)A(x,y)$ in classical number theory, then $\vdash (x,y)A(x,y)$ in intuitionistic number theory.

From this theorem we infer the following

Theorem 82: If $A(x,y)$ and $B(x,y)$ are quantifierfree formulas and if $\vdash (x)((y)A(x,y)\equiv(Ez)B(x,z))$ in classical number theory, then $\vdash (x)((y)A(x,y)\equiv(Ez)B(x,z))$ in intuitionistic number theory.

Proof: a) In order to prove the theorem, we list four formulas which can be proved in intuitionistic predicate calculus and whose proof we leave to the reader: 1) $(Ez)(A\vee B(z))\supset(A\vee(Ez)B(z))$;
2) $(y)(U\supset V(y))\supset(U\supset(y)V(y))$;
3) $(Ey)(\neg A(y)\vee B)\supset((y)A(y)\supset B)$;
4) $(z)(U(z)\supset V)\supset((Ez)U(z)\supset V)$. In 2) and 3) y is not in U and B, respectively, in 1) and 4) z is not in A and V, respectively. b) Next we prove that $(y)A(x,y)\supset(Ez)B(x,z)$ can be proved intuitionistically. To this end we write $\vdash_c$ and $\vdash_i$ in order to indicate provability in classical and intuitionistic numbertheory, respectively. From $\vdash_c(y)A(x,y)\supset(Ez)B(x,z)$ we infer $\vdash_c(Ey)(Ez)(\neg A(x,y)\vee B(x,z))$ and thus from theorem 81 $\vdash_i(Ey)(Ez)(\neg A(x,y)\vee B(x,z))$. From formula 1) listed under a) we get $\vdash_i(Ey)(\neg A(x,y)\vee(Ez)B(x,z)$ and from formula 3) $\vdash_i(y)A(x,y)\supset(Ez)B(x,z)$. c) Now to the converse: $\vdash_i(Ez)B(x,z)\supset(y)A(x,y)$. From $\vdash_c(Ez)B(x,z)\supset(y)A(x,y)$ we infer $\vdash_c(y,z)(\neg B(x,z)\vee A(x,y))$, that is

$\vdash_i (y,z)(\neg B(x,z) \vee A(x,y))$ by theorem 81 and hence

$\vdash_i (y,z)(B(x,z) \supset A(x,y))$ (since A,B quantifierfree obey the law of excluded middle). From 4) in a) we infer

$\vdash_i (y)((Ez)B(x,z) \supset A(x,y))$ and from 2) in a) finally

$\vdash_i (Ez)B(x,z) \supset (y)A(x,y)$.

Corollary: If $A(x,y)$ and $B(x,z)$ are quantifierfree and if $\vdash_c (Ey)A(x,y) \equiv (z)B(x,z)$ holds, then:

a) $\vdash_i (Ey)A(x,y) \equiv (z)B(x,z)$, b) $\vdash_i (y)\neg A(x,y) \equiv (Ez)\neg B(x,z)$,

c) $\vdash_i (Ey)A(x,y) \vee \neg (Ey)A(x,y)$.

Proof: Part a) follows directly from theorem 82. Part b) follows from theorem 82 and the classical consequence $\vdash_c (y) \neg A(x,y) \equiv (Ez) \neg B(x,z)$. Now to part c). According to IM , p. 166, we have: I) $\vdash_i \neg (Ey)A(x,y) \equiv (y) \neg A(x,y)$. Next we have $\vdash_c (Ey)A(x,y) \vee \neg (Ey)A(x,y)$, that is, $\vdash_c (Ey)A(x,y) \vee (Ez) \neg B(x,z)$ in virtue of our assumption and therefore $\vdash_c (Ey)(Ez)(A(x,y) \vee \neg B(x,z))$. By applying formula 1) listed under a) in the proof of theorem 82, we infer from the last statement: $\vdash_i (Ey)A(x,y) \vee (Ez) \neg B(x,z)$. From b), already proved, we get therefore: $\vdash_i (Ey)A(x,y) \vee (y) \neg A(x,y)$. Using finally I), we obtain $\vdash_i (Ey)A(x,y) \vee \neg (Ey)A(x,y)$, what proves c) .

Thus, if a predicate can be proved to be recursive in classical numbertheory, then it can be proved to be recursive in intuitionistic number theory.

10.3. Syntax of ZTi/V

A. In the system ZTi/I_N^* we can speak about the syntax of ZTi/V ; one uses thereby a suitable Goedelnumbering of the symbols of ZTi/V , its terms, formulas, sequents and proofs. As noted at the beginning, we do not give a complete formalisation of the content of chapter VI in ZTi/I_N^* . We rather prefer to rephrase the arguments of chapter VI in a constructive, but intuitive way such that it will be evident that everything can be reproduced via Goedelnumbering in ZTi/I_N^* .

B. Chapter VI splits essentially into two parts: a rather elementary part presented in sections 6.1. - 6.4. and a nonelementary part, contained in sections 6.5. - 6.6. Formalizing the content of sections 6.1. - 6.4. requires obviously quite extensive routine work; however, this can be done in principle without difficulties. Among the more subtle parts in sections 6.1. - 6.4. for which it is not completely evident that they can be formalized in ZTi/I_N^* is perhaps theorem 40. Let us just outline how this can be done. First, it is clear that the relations R and L between proofs in ZTi/V introduced in part F of section 6.4., chapter VI, can be proved to be recursive in classical numbertheory. Using the corollary to theorem 82, it follows that both R and L can be proved to be recursive in intuitionistic numbertheory. If L is provably intuitionistic recursive, then so is L_P for every proof P in ZTi/V. Now consider the proof of theorem 40 as presented in section 6.4. In this proof, we assume that for a certain P_o , L_{P_o} is wellfounded. By transfinite induction over L_{P_o} we prove: if $P \in D_{P_o}$, then the endsequent of P is strongly true. In virtue of theorem 80 and its implications, this transfinite induction is accessible to ZTi/I_N^* if we can show that the statement "the endsequent of the special proof P is strongly true" is represented by a formula A(x) belonging to W_N (with x running over Goedelnumbers of special proofs). This, however, is an immediate consequence of the definitions of "special proof" and "strongly true" as given by definitions 41 and 42 in 6.4. Thus, there is in principle no obstacle to proving the Goedelized versions of sections 6.1. up to 6.4. in ZTi/I_N^* .

10.4. Ordinals

A. The main obstacle to a straightforward formalization of chapter VI within ZTi/I_N^* is obviously section 6.5. There we introduce ordinals, some of which are apparently nonconstructive. The most important among these nonconstructive ordinals is obviously Ω , whose definition is given at the beginning of part A of section 6.5. It is the purpose of this and the next sections to show that, despite the nonconstructive character of the ordinals introduced in 6.5., there is a way of handling them within ZTi/I_N^* .

B. Let L(x,y) be a formula containing no other free variables than x,y . We write xLy instead of L(x,y) . Assume that we have already proved xLy, yLz $\longrightarrow$ xLz . Even if we have good reasons to expect that xLy is wellfounded classically in virtue of its definition, we can hardly expect to prove $(\alpha)(Ex)(\neg\alpha(x+1)L\alpha(x))$ in ZTi/I_N^* because xLy might be highly undecidable. However, we can eventually hope to prove the following or a similar version of transfinite induction: $(y)((x)(xLy \supset A(x)) \supset A(y)) \supset (z)A(z)$. We will show that this is the case for certain particular formulas L .

Assumption A: In what follows, $P(z,x,y)$ and $G(z,x)$ are two intuitionistically recursive formulas and we assume that $P(z,x,y) \supset G(z,x) \wedge G(z,y)$ is provable in ZTi/I_N^* . We write $x \prec_z y$ and $x \in G_z$ in place of $P(z,x,y)$ and $G(z,x)$, respectively. By $W(\prec_z)$ we understand the formula $(\alpha)(Ex)(\neg\alpha(x+1) \prec_z \alpha(x))$ while $Progr_x(\prec_z, A(x))$ and $TI_x(\prec_z, A(x))$ are abbreviations for $(y)(y \in G_z \supset .(x)(x \prec_z y \supset A(x)) \supset A(y))$ and $W(\prec_z) \supset .Progr_x(\prec_z, A(x)) \supset (y)(y \in G_z \supset A(y))$. Finally, $F(z)$ is an arbitrary formula for which $F(z) \supset W(\prec_z)$ is provable in ZTi/I_N^* ; we sometimes write $x \in F$ instead of $F(x)$.

Notation: By $\langle x,y \rangle$ we denote the pairing function $\frac{1}{2}((x+y)^2 + 3x+y)$ which maps N^2 in a one-one way onto N (N the set of natural numbers).

Now we are going to define a relation $L(x,y)$, a family of relations $L(z,x,y)$ depending on the parameter z and their respective domains $D(x)$ and $D(z,x)$; we write xLy , $xL^z y$, $x \in D$ and $x \in D_z$ in place of $L(x,y)$, $L(z,x,y)$, $D(x)$ and $D(z,x)$, respectively. Their definition is as follows:

1) $\langle e,x \rangle L \langle e',y \rangle \equiv e \in F \wedge e' \in F \wedge x \in G_e \wedge y \in G_{e'} \wedge$
$\wedge (e < e' . \vee . (e=e' \wedge x \prec_e y))$,

2) $\langle e,x \rangle \in D \equiv e \in F \wedge x \in G_e$,

3) $\langle e,x \rangle L^z \langle e',y \rangle \equiv e \leqq z \wedge e' \leqq z \wedge \langle e,x \rangle L \langle e',y \rangle$,

4) $\langle e,x \rangle \in D_z \equiv \langle e,x \rangle \in D \wedge e \leqq z$.

By $Progr_x(L,A(x))$ and $Progr_x(L^z,A(x))$ we denote the formulas $(y)(y \in D \supset .(x)(xLy \supset A(x)) \supset A(y))$ and $(y)(y \in D_z \supset .(x)(xL\ y \supset A(x)) \supset A(y))$, respectively. $TI_x(L,A(x))$ and $TI_x(L^z,A(x))$, finally, are abbreviations for $Progr_x(L,A(x)) \supset (s)(s \in D \supset A(s))$ and $Progr_x(L^z,A(x)) \supset (s)(s \in D_z \supset A(s))$, respectively. Our aim is to prove

Lemma 29: Assume $A \in W_N$. Then: a) $ZTi/I_N^* \vdash TI_x(L^z, A(x))$, b) $ZTi/I_N^* \vdash TI_x(L, A(x))$. The lemma will be proved by first proving a) by induction over z and then by proving b) with the aid of a). We proceed in steps. a) First we claim $ZTi/I_N^* \vdash TI_x(L^o, A(x))$. To this end assume $Progr_x(L^o, A(x))$. Let $\langle u,v\rangle \in D_o$ be arbitrary. From the definition of D_o and D we infer:

$\langle u,v\rangle \in D_o \equiv u=0 \wedge 0 \in F \wedge v \in G_o$. Thus we have to prove $A(\langle 0,v\rangle)$. From the definition of L^z, on the other hand, we immediately infer: $\langle e,x\rangle\ L^o\ \langle e',y\rangle \equiv e=0 \wedge e=e' \wedge x \prec_o y$. Hence $Progr_x(L^o, A(x))$ is provably equivalent to

$(v)(0 \in F \wedge v \in G_o \supset .(x)(x \prec_o v \supset A(\langle 0,x\rangle)) \supset A(\langle 0,v\rangle))$, that is, to $0 \in F \supset .(v)(v \in G_o \supset .(x)(x \prec_o v \supset A(\langle 0,x\rangle)) \supset A(\langle 0,v\rangle))$. As noted, it follows from the assumption $\langle u,v\rangle \in D_o$ that $u=0$ and $0 \in F$ holds. Hence $Progr_x(L^o, A(x))$ is equivalent to

$(v)(v \in G_o \supset .(x)(x \prec_o v \supset A(\langle 0,x\rangle)) \supset A(\langle 0,v\rangle))$, that is, to $Progr_x(\prec_o, A(\langle 0,x\rangle))$. From our assumption A we infer $W(\prec_o)$, and since $A \in W_N$ so $A(\langle 0,x\rangle) \in W_N$. Hence we can derive in ZTi/I_N^* the formula

$W(\prec_o) \supset .Progr_x(\prec_o, A(\langle 0,x\rangle)) \supset (x)(x \in G_o \supset A(\langle 0,x\rangle))$. From this, $W(\prec_o)$ and $Progr_x(\prec_o, A(\langle 0,x\rangle))$ we immediately get $(x)(x \in G_o \supset A(\langle 0,x\rangle))$, that is, in particular, $A(\langle 0,v\rangle)$.

b) Next we want to show: $ZTi/I_N^* \vdash TI_x(L^z, A(x)) \supset TI_x(L^{z+1}, A(x))$. To this end assume $TI_x(L^n, A(x))$ and $Progr_x(L^{n+1}, A(x))$. Our aim is to prove $A(\langle u,v\rangle)$ for all $\langle u,v\rangle$ in D_{n+1}. To begin with, we list some equivalences and implications which immediately follow from the definition of L, L^z, D and D_z:

α) $\langle u,v\rangle \in D^{n+1} \equiv \langle u,v\rangle \in D^n \vee (u=n+1 \wedge n+1 \in F \wedge v \in D_{n+1})$;

β) $\langle u,v\rangle \in D^n \supset \langle u,v\rangle \in D^{n+1}$;

γ) $\langle p,q\rangle\ L^n\ \langle u,v\rangle \supset \langle p,q\rangle\ L^{n+1}\ \langle u,v\rangle$;

δ_1) $\langle p,q\rangle L^{n+1}\langle u,v\rangle \equiv \langle p,q\rangle L^n \langle u,v\rangle \vee (p \leq n \wedge u=n+1 \wedge p \in F \wedge$
$\wedge\, u \in F \wedge q \in G_p \wedge v \in G_{n+1})$;

δ_2) $\langle n+1,q\rangle\ L^{n+1}\ \langle n+1,v\rangle \equiv q \prec_{n+1} v \wedge n+1 \in F$;

δ_3) $\langle p,q\rangle L^{n+1}\langle n+1,v\rangle \equiv (p \leq n \wedge p \in F \wedge n+1 \in F \wedge q \in G_p \wedge v \in G_{n+1}) \vee$
$\vee\, (p=n+1 \wedge n+1 \in F \wedge q \prec_{n+1} v)$;

ε) $\langle u,v\rangle \in D_n \supset (\langle p,q\rangle\ L^{n+1}\ \langle u,v\rangle \equiv \langle p,q\rangle L^n \langle u,v\rangle)$.

From ε) we get as an immediate consequence $Progr_x(L^{n+1}, A(x)) \supset Progr_x(L^n, A(x))$. Since $Progr_x(L^{n+1}, A(x))$ holds by assumption, it follows that $Progr_x(L^n, A(x))$ holds. From the inductive assumption $TI_x(L^n, A(x))$ we therefore obtain I): $(s,t)(\langle s,t\rangle \in D_n \supset A(\langle s,t\rangle))$. According to α) above, our proof is accomplished if we can show: if $u=n+1$, $n+1 \in F$ and $v \in G_{n+1}$, then $A(\langle u,v\rangle)$. Hence, let us assume $u=n+1$, $n+1 \in F$ and $v \in G_{n+1}$. From $Progr_x(L^{n+1}, A(x))$ and this assumption we infer: II) $(s,t)(\langle s,t\rangle L^{n+1} \langle n+1,v\rangle \supset A(\langle s,t\rangle)) \supset A(\langle n+1,v\rangle)$.

Next we claim: III)

$(p,q)(\langle p,q\rangle L^{n+1} \langle n+1,v\rangle \supset A(\langle p,q\rangle)) \equiv (q)(q \prec_{n+1} v \supset A(\langle n+1,q\rangle))$.

In order to verify the implication from left to right, take $p=n+1$ and use δ_2). In order to prove the implication from right to left, assume $\langle p,q\rangle L^{n+1} \langle n+1,v\rangle$. According to δ_3), this is equivalent to

$(p \leq n \wedge p \in F \wedge n+1 \in F \wedge q \in G_p \wedge v \in G_{n+1} \wedge n+1 \in F) \vee (p=n+1 \wedge n+1 \in F \wedge q \prec_{n+1} v)$. If the first of these alternatives holds, then clearly $\langle p,q\rangle \in D_n$ and therefore $A(\langle p,q\rangle)$ according to I) above. If, however, $p=n+1$, $q \prec_{n+1} v$, then $A(\langle n+1,q\rangle)$ from the assumed righthand side of III). Hence III) is indeed true. This permits us to replace in II) the lefthandside of the implication by the righthandside of III), that is, we get: IV) $(q)(q \prec_{n+1} v \supset A(\langle n+1,q\rangle)) \supset A(\langle n+1,v\rangle)$. In other words we get: V) $Progr_x(\prec_{n+1}, A(\langle n+1,x\rangle))$. Since $n+1 \in F$ by assumption we get $W(\prec_{n+1})$ by assumption A. Now $W(\prec_{n+1}) \supset .Progr_x(\prec_{n+1}, A(\langle n+1,x\rangle)) \supset (z)(z \in G_{n+1} \supset A(\langle n+1,z\rangle))$ is provable in ZTi/I_N^* since $A(\langle n+1,x\rangle)$ belongs to W_N. This together with V) and $W(\prec_{n+1})$ finally implies $(z)(z \in G_{n+1} \supset A(\langle n+1,z\rangle))$ and, in particular, $A(\langle n+1,v\rangle)$, what concludes the induction step. Thus $TI_x(L^z, A(x)) \supset TI_x(L^{z+1}, A(x))$ indeed holds in ZTi/I_N^*. Combining this with the already proved $TI_x(L^o, A(x))$, we obtain $(z)TI_x(L^z, A(x))$ for all $A \in W_N$.

c) It remains to show that $(z)TI_x(L^z, A(x))$ implies $TI_x(L, A(x))$. To this end we list some further consequences of the definitions of L, L_x, D, D_x:

1) $\langle u,v\rangle \in D_x \supset (\langle p,q\rangle L \langle u,v\rangle \equiv \langle p,q\rangle L^x \langle u,v\rangle)$,

2) $\langle x,y\rangle \in D \supset \langle x,y\rangle \in D_x$.

Now assume

$(u,v)(\langle u,v\rangle \in D \supset .(p,q)(\langle p,q\rangle L \langle u,v\rangle \supset A(\langle p,q\rangle)) \supset A(\langle u,v\rangle))$,

that is, $Progr_x(L, A(x))$. Assume in addition $\langle x,y\rangle \in D$. We have to prove $A(\langle x,y\rangle)$. From $Progr_x(L, A(x))$ we easily infer with the aid of 1): $Progr_x(L^z, A(x))$. Namely, let $\langle u,v\rangle$ be in D_z. Then we can replace $\langle p,q\rangle L \langle u,v\rangle$ by $\langle p,q\rangle L^z \langle u,v\rangle$,

according to 1) above. That is, we have
$(u,v)(\langle u,v\rangle \in D_z \supset .(p,q)(\langle p,q \; L^z \langle u,v\rangle \supset A(\langle p,q\rangle))\supset A(\langle u,v\rangle))$,
that is $Progr_x(L^z,A(x))$. Since $TI_x(L^z,A(x))$ is already proved, we can infer from $Progr_x(L^z,A(x))$: $(u,v)(\langle u,v\rangle \in D_z \supset A(\langle u,v\rangle))$. Hence we get, in particular, $\langle x,y\rangle \in D_z \supset A(\langle x,y\rangle)$, or by taking x for z : $\langle x,y\rangle \in D_x \supset A(\langle x,y\rangle)$. However, $\langle x,y\rangle \in D_x$ is true by 2) above and so is $A(\langle x,y\rangle)$, what concludes the proof of lemma 29.

10.5. On extending linear wellorderings

A. In the last section we have considered certain particular relations L which from a classical point of view are wellfounded. We have seen that in virtue of the definition of L, and despite the eventually highly undecidable character of L, one can prove in ZTi/I_N^* transfinite induction with respect to L in the form $TI_x(L,A(x))$, with $A \in W_N$. Such a particular relation L, whose definition will be given later, will serve, roughly speaking, as a substitute for the ordinal Ω in chapter VI, section 6.5. However, not only Ω, but also such ordinals as $\omega_n(\omega_m(\Omega \# 1) \# 1)$ etc. were used. It is the purpose of this and the next section to provide an appropriate constructive substitute for such ordinals and for the functions $\omega_n(\alpha), \alpha \# \beta$.

B. To start with, let $P(z,x,y)$ and $G(z,x)$ be two intuitionistically recursive formulas and $F(z)$ a third formula which satisfies condition A stated at the beginning of part B of the last section. With respect to P,G and F, we use the same abbreviations as in the last section. In addition we assume that P,G,F satisfy also the following additional

Assumption B: 1) P,G,F are in W_N ; 2) for every e, $\prec_e$ is a linear ordering of G_e ; that is: α) if $x,y,z \in G_e$, then $x \prec_e y \vee x=y \vee y \prec_e x$, and $x \prec_e y,\ y \prec_e z \supset x \prec_e z$ and $x \prec_e y \supset \neg y \prec_e x \wedge \neg y=x$ hold ; β) $x \prec_e y \supset x \in G_e \wedge y \in G_e$ holds; 3) there is an e and an $x \in G_e$ such that $F(e)$ holds. In terms of P,G and F, we again introduce a relation L and its domain D by means of clauses 1), 2) in part B of the last section. With respect to L,D, we have the

Lemma 30: a) If $n,m \in D$, then: 1) $nLm \vee n=m \vee mLn$; 2) $nLm, mLp \supset nLp$; 3) $nLm \supset \neg mLn \wedge \neg m=n$; 4) $\neg nLn$; 5) there is an $n \in D$. b) $nLm \supset n \in D \wedge m \in D$.

The proof follows in a straightforward way from assumptions A,B satisfied by P,G,F and from the definition of L,D .

As shown in the last section, we can prove in $ZTi/I_N^* \vdash TI_x(L,A(x))$ for $A \in W_N$. It is clear that in virtue of assumption B,1) the formulas L,D are also in W_N .

Definition 60: The formulas $D(x)$, $L(x,y)$ are said to be an ordering pair if they belong to W_N and if, in addition, clauses a), 1)-5) and b) of lemma 30 are satisfied. They are called a wellordering pair if, in addition, $TI_x(L,A(x))$ is provable in ZTi/I_N^* for every $A \in W_N$.

C. We are now going to extend the relation L . To this end, let + be a new symbol.

Definition 61: α) Let D,L be an ordering pair. Then D^* is the set of strings (words) of the form $n_1 \alpha_1 + \ldots + n_s \alpha_s$ which satisfy the following conditions: 1) $\alpha_1, \ldots, \alpha_s \in D$; 2) $\alpha_{i+1} L \alpha_i$ (in case $s > 1$) ; 3) $n_i > 0$. Thereby we admit $s=1$. β) A relation L^* over D^* is introduced by defining $m_1 \beta_1 + \ldots + m_t \beta_t L^* n_1 \alpha_1 + \ldots + n_s \alpha_s$ if one of the following conditions is satisfied: 1) there is an $i < \min(s,t)$ (possibly 0) such that $m_k = n_k$ and $\alpha_k = \beta_k$ for $k \leqq i$ and either $m_{i+1} < n_{i+1}$ and $\beta_{i+1} = \alpha_{i+1}$, or else $\beta_{i+1} L \alpha_{i+1}$; 2) $t < s$ and $m_k = n_k$, $\alpha_k = \beta_k$ for $k=1,\ldots,t$. By definition, $xL^*y \supset x \in D^* \wedge y \in D^*$. We call L^*,D^* the ordering pair induced by the ordering pair L,D .

Notation: The norm $|\xi|$ of $\xi = n_1 \alpha_1 + \ldots + n_s \alpha_s$ is α_1 .

Remark: It would be an easy matter to represent strings $n_1 \alpha_1 + \ldots + n_s \alpha_s$ by suitably chosen Goedelnumbers; however we omit such an arithmetisation in order to avoid complicated notations.

Concerning L^*,D^* we have

Lemma 31: If L,D is an ordering pair, then L^*,D^* is an ordering pair.

The proof is a straightforward consequence of the definition of L^*, D^* and of the assumption that L, D is an ordering pair. Elements $\alpha \in D$ can be identified with the elements $1\alpha \in D^*$. This identification is justified by the following

<u>Lemma 32:</u> 1) If $\alpha, \beta \in D$ then $\beta L \alpha$ iff $1\alpha L^* 1\beta$.

<u>Notation:</u> For simplicity, we write α instead of 1α for elements $\alpha \in D$. For elements in D^*, we can introduce a natural sum $\#$ which will play about the same role as the natural sum $\#$ usually defined for ordinals. Namely, let ξ and η be $n_1\alpha_1+\dots+n_s\alpha_s$ and $m_1\beta_1+\dots+m_t\beta_t$, respectively. Let S_1 be the set $\{\alpha_1,\dots,\alpha_s\}$, S_2 the set $\{\beta_1,\dots,\beta_t\}$ and $S=S_1 \cup S_2$ the union of both. The elements of S are listed in decreasing order with respect to L: $\gamma_1,\dots,\gamma_a$. Then we define $\xi \# \eta$ to be $p_1\gamma_1+\dots+p_a\gamma_a$, where the coefficients p_i are given as follows: 1) if there is a j and a k such that $\alpha_j=\beta_k=\gamma_i$, then $p_i=(n_j+m_k)$; 2) if there is a j such that $\alpha_j=\gamma_i$, but no k such that $\beta_k=\gamma_i$, then $p_i=n_j$; 3) if there is a k such that $\gamma_i=\beta_k$, but no j such that $\alpha_j=\gamma_i$, then $p_i=m_k$.

<u>Lemma 33:</u> For $\xi, \zeta, \eta \in D^*$ we have 1) $\xi \# \zeta = \zeta \# \xi$; 2) $\xi L^* \xi \# \zeta$; 3) if $\xi \# \zeta = \xi \# \eta$, then $\zeta = \eta$.
This lemma is an easy consequence of the definition of $\#$.
Our principal aim is to prove

<u>Theorem 83:</u> If L, D is a wellordering pair and if $A \in W_N$, then $TI_x(L^*, A(x))$ is provable in ZTi/I_N^*.

Here $TI_x(L^*, A(x))$ is an abbreviation for $Progr_x(L^*, A(x)) \supset (z)(z \in D^* \supset A(z))$, while $Progr_x(L^*, A(x))$ is an abbreviation for $(y)(y \in D^* \supset .(x)(xL^*y \supset A(x)) \supset A(y))$. In order to prove the theorem, it is convenient to introduce a list of further abbreviations. First, we introduce for every $\alpha \in D$ a set D^*_α as follows: $\xi \in D^*_\alpha$ iff $\xi \in D^* \wedge (|\xi| L \alpha \vee |\xi| = \alpha)$. Next we introduce for every $\alpha \in D^*$ the formula $(y)(yL^*\alpha \supset .(x)(xL^*y \supset A(x)) \supset A(y))$ and denote it by $Progr_x(L^*_\alpha, A(x))$; the formula

$Progr_x(L^*_\alpha, A(x)) \supset (z)(zL^*\alpha \supset A(z))$ will be denoted by $TI_x(L^*_\alpha, A(x))$. For $\alpha \in D$ we use $Progr^\alpha_x(L^*, A(x))$ as abbreviation for $(y)(y \in D^*_\alpha \supset .(x)(xL^*y \supset A(x)) \supset A(y))$. Finally, for $\alpha \in D$ we take $TI^\alpha_x(L^*, A(x))$ as abbreviation for $(s)(s \in D^* \supset .Progr^\alpha_x(L^*, A(s \# x) \supset (z)(z \in D^*_\alpha \supset A(s \# z)))$.

D. Instead of **proving** theorem 83 directly, we first prove

Lemma 34: With L,D a wellordering pair, if $A \in W_N$, then the following formula is provable in ZTi/I^*_N:
$(s)(s \in D \supset .(t)(tLs \supset TI^t_x(L^*, A(x))) \supset TI^s_x(L, A(x)))$
(that is, $Progr_s(L, TI^s_x(L^*, A(x))$.

Before coming to the proof of this lemma, we show that theorem 83 is an immediate consequence of it; more precisely, we infer from lemma 34 two corollaries, the second of which is precisely theorem 83.

Corollary 1: $(s)(s \in D \supset TI^s_x(L^*, A(x))$ is provable in ZTi/I^*_N, provided $A \in W_N$.

Proof: According to lemma 29, we have $ZTi/I^*_N \vdash TI_x(L, B(x))$ for all formulas $B \in W_N$. Since $A \in W_N$ it follows that, in particular, $TI_s(L, TI^s_x(L^*, A(x)))$ is provable in ZTi/I^*_N (since $TI^s_x(L^*, A(x))$ is in W_N). That is,
$Progr_s(L, TI^s_x(L^*, A(x))) \supset (z)(z \in D \supset TI^z_x(L^*, A(x)))$ is provable in ZTi/I^*_N. However, according to lemma 34, $Progr_s(L, TI^s_x(L^*, A(x)))$ is provable in ZTi/I^*_N, and so $(s)(s \in D \supset TI^s_x(L^*, A(x)))$ is provable in ZTi/I^*_N, what proves the corollary.

Corollary 2: For $A \in W_N$, the formula $TI_x(L^*, A(x))$ is provable in ZTi/I^*_N.

Proof: a) By definition $TI_x(L^*, B(x))$ is
$Progr_x(L^*, B(x)) \supset (z)(z \in D^* \supset B(x))$. Assume $Progr_x(L^*, B(x))$ and $\xi \in D^*$. Put $|\xi| = \alpha$; by definition, $\alpha \in D$. According to corollary 1, we have $TI^\alpha_x(L^*, B(x))$, that is, I):
$(s)(s \in D^* \supset .Progr^\alpha_x(L^*, B(s \# x)) \supset (z)(z \in D^*_\alpha \supset B(s \# x)))$
provided only that $B \in W_N$. b) Let s_0 be an arbitrary but fixed element from D^*; that there is such an element follows from lemma 31. Take for $B(z)$ the following formula:
$(Ev)(v \# s_0 = z \wedge A(v))$. Clearly, $B(z) \in W_N$. In addition, $B(s_0 \# x)$

is obviously equivalent to $A(x)$, as follows from lemma 33. Hence we conclude from I) above that II) holds:
$Progr^{\alpha}_{x}\ (L^*,A(x)) \supset (z)(z \in D^*_{\alpha} \supset A(z))$. Now it is evident that the following formula III) holds:
$Progr_x(L^*,A(x)) \supset Progr^{\alpha}_{x}\ (L^*,A(x))$. The lefthandside of this implication is $(y)(y \in D^* \supset .(x)(xL^*y \supset A(x)) \supset A(y))$ while the righthandside is by definition:
$(y)(y \in D^*_{\alpha} \supset .(x)(xL^*y \supset A(x)) \supset A(y))$. But $D^*_{\alpha} \subseteq D^*$ by definition of D^*_{α} ; hence III) is clearly provable in ZTi/I^*_N .
Combining III) with II) and using our assumption $Progr_x(L^*,A(x))$, we infer IV): $(z)(z \in D^*_{\alpha} \supset A(z))$. But $|\xi| = \alpha$, that is, $|\xi| \in D^*_{\alpha}$, hence we infer $A(\xi)$ from IV), what concludes the proof in virtue of the arbitrariness of ξ .

That is, theorem 83, which is the same as corollary 2, follows from lemma 34.

E. Prior to the proof of lemma 34, we want to state a remark concerning lemmas 30 and 31 and the use of the law of excluded middle. The relations L,L^* are in general, of course, highly undecidable: given two arbitrary numbers a,b we are in general not able to decide whether aLb, bLa or neither of them holds. Similarly, if we are given two arbitrary expressions $\xi = n_1\alpha_1 + \ldots + n_s\alpha_s$, $\eta = m_1\beta_1 + \ldots + m_t\beta_t$ with the aid of the α_i's and β_k's, which need not necessarily all belong to D , then we are in general not able to decide whether $\xi L^* \eta$, $\eta L^* \xi$ or none of them holds. However, as soon as we are given the information that a,b belong to D, then we know that precisely one of the three relations $a=b$, aLb or bLa holds, and we are able to decide which one of them is true; this is the main content of lemma 30. Similarly, if we are given the information $\alpha_i, \beta_k \in D$, $i=1,\ldots,s$, $k=1,\ldots,t$, then we can decide whether ξ , η belong to D^* and, if so, which of the relations $\xi L^* \eta$, $\eta L^* \xi$, $\xi = \eta$ hold. Finally, if we are told that ξ , $\eta \in D^*$, then we know by definition that $\alpha_i, \beta_k \in D$, and so we are again able to decide which of the relations $\xi L^* \eta$, $\eta L^* \xi$, $\xi = \eta$ holds. In other words, although the statements aLb , $\xi L^* \eta$ are in general highly undecidable, the law of the excluded middle is applicable as soon as we know that the arguments a,b and ξ , η are in D and in D^*, respectively. Keeping this in mind, the reader will verify that no forbidden application of the law of excluded middle occurs in our considerations below.

F. In order to prove lemma 34, we need three properties P_1, P_2, P_3 of D,L and D*,L* which are immediate consequences of definition 61.

P1: If $\alpha \in D$, then $\xi L^* \alpha$, iff $|\xi| L \alpha$.

Proof: Let $\xi \in D^*$ have the form $m_1 \beta_1 + \ldots + m_t \beta_t$; by definition $m_i > 0$, $\beta_i \in D$ and, in case $t > 0$, also $\beta_{i+1} L \beta_i$. Since L,D is an ordering pair, exactly one of the statements $\beta_1 L \alpha$, $\alpha = \beta_1$, $\alpha L \beta_1$ holds. If $\alpha = \beta_1$ or $\alpha L \beta_1$, then $\xi L^* \alpha$ is impossible according to the definition of L^* ; hence, $\beta_1 L \alpha$ has to hold (this argumentation uses the intuitionistic valid formula $A \vee B \wedge \neg A. \supset B$). If, on the other hand, $\beta_1 L \alpha$ then $\xi L^* \alpha$ by definition of L^*.

P2: If $|\xi| L \alpha$, then $|k\alpha \# \xi| = \alpha$.

Proof: Obvious from the definition of norm.

P3: Assume $\xi L^* \alpha$ and $\alpha \in D$. Then $\zeta L^*(n+1) \alpha \# \xi$, if and only if one of the following conditions holds:
1) $\zeta = (n+1)\alpha \# \eta$ and $\eta L^* \xi$; 2) $\zeta = k\alpha \# \eta$ and $0 < k \leqq n$ and $|\eta| L \alpha$; 3) $\zeta = \eta$ and $|\eta| L \alpha$.

Proof: a) Put $\xi = n_1 \alpha_1 + \ldots + n_s \alpha_s$, $\zeta = m_1 \beta_1 + \ldots + m_t \beta_t$ and $\eta = p_1 \gamma_1 + \ldots + p_v \gamma_v$. a_1) Assume 1) to hold. From $\eta L^* \xi$ and lemma 33 we get $(n+1)\alpha \# \eta \; L^*(n+1)\alpha \# \xi$, that is, $\zeta L^*(n+1)\alpha \# \xi$. a_2) Assume 2) to hold. Then $k\alpha \# \eta$ has the form $k\alpha + p_1 \gamma_1 + \ldots + p_v \gamma_v$. Hence, $k\alpha \# \eta \; L^*(n+1) \alpha \# \xi$ according to the definition of L^*. a_3) Assume 3) to hold. Then $\eta L^* \alpha$ according to P1, and hence, $\eta L^*(n+1)\alpha \# \xi$, that is, $\zeta L^*(n+1) \alpha \# \xi$. b) Now assume $\zeta L^*(n+1) \alpha \# \xi$. Since $\xi L^* \alpha$, we infer $|\xi| L \alpha$, and hence, $|(n+1)\alpha \# \xi| = \alpha$. Since $\zeta L^*(n+1) \alpha \# \xi$, there are two possibilities: A) $|\zeta| = \alpha$, B) $|\zeta| L \alpha$. If B) holds, then we take ζ for η and clause 3) of P* is satisfied. Let A) be true. Then $\zeta = k\alpha \# \eta$, $0 < k \leq n+1$, where $|\eta| L \alpha$. Two subcases arise: A_1) : $k \leq n$, A_2) $k = n+1$. If A_1) holds, then clause 2) of P_3 is satisfied. Assume finally $k = n+1$. Then necessarily $\eta L^* \xi$; otherwise

$(n+1)\alpha \# \eta = (n+1)\alpha \# \xi$ or $(n+1)\alpha \# \xi \, L^*(n+1)\alpha \# \eta$ according to whether $\xi = \eta$ or $\xi L^* \eta$, thus giving a contradiction.

G. Now to the proof of lemma 34. We have to show that for $A \in W_N$ we can prove in ZTi/I^*_N the formula $Progr_s(L, TI^s_x(L^*, A(x)))$. That is, we have to prove $TI^\alpha_x(L^*, A(x))$ under the assumption $\alpha \in D$, $(t)(tL\alpha \supset TI^t_x(L^*, A(x)))$ and this finally amounts to prove $(z)(z \in D^*_\alpha \supset A(\gamma \# z))$ under the following assumption AP_o : $a_o)$ $\alpha \in D$; $b_o)$ $\gamma \in D^*$; $c_o)$ $Progr^\alpha_x(L^*, A(\gamma \# x))$; $d_o)$ $(t)(tL\alpha \supset TI^t_x(L^*, A(x)))$. We will do this by proving successively three statements ST1, ST2, ST3, with $(z)(z \in D^*_\alpha \supset A(\gamma \# z))$ an immediate consequence of ST3.

ST1: $(s)(s \in D^* \supset .Progr_x(L^*_\alpha, A(s \# x)) \supset (t)(tL^*\alpha \supset A(s \# t)))$ holds.

Proof: a) In addition to AP_o, we make the following assumptions AP_1 : $a_1)$ $s_o \in D^*$; $b_1)$ $Progr_x(L^*_\alpha, A(s_o \# x))$; $c_1)$ $t_o L^*\alpha$. To prove ST1 amounts to prove $A(s_o \# t_o)$ under the assumptions AP_o and AP_1 . Since $t_o L^*\alpha$, we have $|t_o| \, L \, \alpha$ by P_1, and hence, $TI^{|t_o|}_x (L^*, A(x))$ by AP_o, $d_o)$, that is,

$(s)(s \in D^* \supset .Progr^{|t_o|}_x (L^*, A(s \# x)) \supset (z)(z \in D^*_{|t_o|} \supset A(s_o \# z)))$.

For s_o, in particular, we have I) :

$Progr^{|t_o|}_x(L^*, A(s_o \# x)) \supset (z)(z \in D_{|t_o|} \supset A(s_o \# z))$.

b) Now $Progr_x(L^*_\alpha, A(s_o \# x))$ is

$(y)(yL^*\alpha \supset (x)(xL^*y \supset A(s_o \# x)) \supset A(s_o \# y))$, while

$Progr^{|t_o|}_x (L^*, A(s_o \# x))$ is $(y)(y \in D^*_{|t_o|} \supset .(xL^*y \supset A(s_o \# x)) \supset A(s_o \# y))$.

We claim II) : $Progr_x(L^*_\alpha, A(s_o \# x)) \supset Progr^{|t_o|}_x (L^*, A(s_o \# x))$. To this end, assume $y_o \in D^*_{|t_o|}$ and $Progr_x(L^*_\alpha, A(s_o \# x))$. Then $|y_o| \, L \, |t_o| \vee |y_o| = |t_o|$, and hence, $|y_o| \, L \, \alpha$ in virtue of $|t_o| \, L \, \alpha$. According to P1, this means ${}_oL^*\alpha$, and hence we can infer $(x)(xL^*y \supset A(s_o \# x)) \supset A(s_o \# y_o)$ from $Progr_x(L^*_\alpha, A(s_o \# x))$. This proves II) . Combining I) and II) with $AP_1, b_1)$, we can infer $(z)(z \in D^*_{|t_o|} \supset A(s_o \# z))$. Since $t_o \in D^*_{|t_o|}$, we obtain $A(s_o \# t_o)$ what concludes the proof of ST1 .

ST2: Under the assumption AP_o , if $\xi L^*\alpha$ then $A(\gamma \# \xi)$.

Proof: a) $Progr^\alpha_x (L^*, A(\gamma \# x))$ occurs among the assumptions listed under AP_o . We claim I) :

$Progr_x^{\alpha}$ $(L^*, A(\gamma \# x)) \supset Progr_x^{|\xi|}$ $(L^*, A(\gamma \# x))$. Now
$Progr_x^{\alpha}(L^*, A(\gamma \# x))$ is
$(y)(y \in D_{\alpha}^* \supset .(x)(xL^*y \supset A(\gamma \# x)) \supset A(\gamma \# y))$, while
$Progr_x^{|\xi|}$ $(L^*, A(\gamma \# x))$ is
$(y)(y \in D^* \supset .(x)(xL^*y \supset A(\gamma \# x)) \supset A(\gamma \# y))$. If
$y \in D_{|\xi|}^*$, then $|y|\ L\ |\xi|$ or $|y| = |\xi|$. According to P1 and
$\xi L^* \alpha$, we have $|\xi| L\ \alpha$, hence in any case $|y|\ L\ \alpha$, and so
$y \in D_{\alpha}^*$. This, combined with $Progr_x^{\alpha}$ $(L^*, A(\gamma \# x))$ as assumption, implies $(x)(xL^*y \supset A(\gamma \# x)) \supset A(\gamma \# y)$, what proves I) .
b) As noted, we have $|\xi| L\ \alpha$. From AP_o, d_o) , we can infer
$TI_x^{|\xi|}$ $(L^*, A(x))$, that is, II) :
$(s)(s \in D^* \supset .Progr_x^{|\xi|}$ $(L^*, A(s \# x)) \supset (z)(z \in D_{|\xi|}^* \supset A(s \# z))$.
Since $\gamma \in D^*$ by AP_o, b_o) , we obtain III) :
$Progr_x^{|\xi|}$ $(L^*, A(\gamma \# x)) \supset (z)(z \in D_{|\xi|}^* \supset A(\gamma \# z))$. Combining
I) and AP_o, c_o) , with III), we get IV): $(z)(z \in D_{|\xi|}^* \supset A(\gamma \# z))$.
Since ξ $D_{|\xi|}^*$, we finally obtain $A(\gamma \# \xi)$, proving St.2 .

<u>ST3:</u> If $\xi L^* \alpha$ then $A(\gamma \# n\alpha \# \xi)$
(with $\gamma \# n\alpha \# \xi = \gamma \# \xi$ if $n=0$) .

<u>Proof:</u> The proof is by induction with respect to n . a) If n=0, then the statement is a consequence of St2 . b) Assume that for all k with $0 \leq k \leq n$ we have proved I) : if $\xi L^* \alpha$, then $A(\gamma \# k\alpha \# \xi)$ holds. Since $\gamma \# (n+1)\alpha \in D^*$, it follows from St1 that our statement is proved for n+1 in place of n if $Progr_x(L_{\alpha}^*, A(\gamma \# (n+1)\alpha \# x))$ is provable, that is, if we can prove II) :
$(y)(yL_{\alpha}^* \supset .(x)(xL^*y \supset A(\gamma \# (n+1)\alpha \# x)) \supset A(\gamma \# (n+1)\alpha \# y))$.
According to our assumption AP_o, c_o) , we have at our disposal
$Progr_x^{\alpha}$ $(L^*, A(\gamma \# x))$, that is, III) :
$(y)(y \in D_{\alpha}^* \supset .(x)(xL^*y \supset A(\gamma \# x)) \supset A(\gamma \# y))$. In order to prove II) , assume $\xi L^* \alpha$ and in addition IV) :
$(x)(xL^* \xi \supset A(\gamma \# (n+1)\alpha \# x)$. Put $\lambda = (n+1)\alpha \# \xi$.
For such a λ, we can infer from III) the statement V) :
$(x)(xL^*\lambda \supset A(\gamma \# x)) \supset A(\gamma \# \lambda)$. Now let ζ be such that $\zeta L^* \lambda$. For such ζ we infer from P3 that one of the following three conditions holds: 1) $\zeta = (n+1)\alpha \# \eta$ and $\eta L^* \xi$; 2) $\zeta = k\alpha \# \eta$, $0 < k \leq n$ and $|\eta|\ L\ \alpha$; 3) $\zeta = \eta$ and $|\eta|\ L\ \alpha$. If 1) holds, then $A(\gamma \# (n+1)\alpha \# \eta)$, that is, $A(\gamma \# \zeta)$ holds according to IV) . If 2) holds, then $\zeta = k\alpha \# \eta$, $0 \leq k \leq n$ and $\eta L^* \alpha$ according to P1 ; hence

$A(\gamma \# k\alpha \# \eta)$, that is $A(\gamma \# \zeta)$ holds in virtue of our inductive assumption. If 3) holds, then again $A(\gamma \# \zeta)$ in virtue of ST2 (or also in virtue of our inductive assumption). In any case, whenever $\zeta L^* \lambda$, then $A(\gamma \# \zeta)$. That is, we have proved VI) : $(x)(xL^*\lambda \supset A(\gamma \# x))$. From V) and VI) we infer $A(\gamma \# (n+1)\alpha \# \xi)$, proving thus II) . This concludes the proof of St3 .

Corollary (to St3): $(z)(z \in D^*_\alpha \supset A(\gamma \# z))$.

Proof: If $z \in D^*_\alpha$, then $z = n\alpha \# \xi$ (with $z = \xi$ in case $n=0$) and $\xi L \alpha$; hence, $A(\gamma \# n\alpha \# \xi)$ holds in virtue of St3 . That is, we have proved $(z)(z \in D^*_\alpha \supset A(\gamma \# z))$ under the assumption AP_o , what proves lemma 34.

10.6. Cartesian products of ordering pairs

A. Given two ordering pairs D_1, L_1 and D_2, L_2, we can form a new one, D_+, L_+ , called the cartesian product of D_1, L_1 and D_2, L_2 . The domain D_+ is given as follows: $\langle a,b \rangle \in D_+$ iff $a \in D_1$ and $b \in D_2$. The relation L_+ on D_+ is defined as follows:
$\langle a,b \rangle L_+ \langle u,v \rangle$ iff $\langle a,b \rangle \in D_+ \wedge \langle u,v \rangle \in D_+ \wedge$
$\wedge (aL_1u . \vee . (a=u \wedge bL_2v))$.
Concerning L_+, D_+, we have

Lemma 35: If D_1, L_1 and D_2, L_2 are ordering pairs then L_+, D_+ is an ordering pair.

We omit the completely straightforward proof. In addition, we also have

Lemma 36: If D_1, L_1 and D_2, L_2 are wellordering pairs then D_+, L_+ is a wellordering pair.

Proof: Our aim is to show that $TI_x(L_+, A(x))$ is provable in ZTi/I^*_N if $A \in W_N$. To this end, we assume I) : $Progr_x(L_+, A(x))$. We want to infer $(z)(z \in D_+ \supset A(z))$. a) Instead of proving I) directly we prove II) : $Progr_x(L_1, (s)(s \in D_2 \supset A(\langle x,s \rangle)))$. From II) we infer I) immediately, as follows: from $TI_x(L_1, (s)(s \in D_2 \supset A(\langle x,s \rangle)))$ we infer with the aid of II)

the formula $(t)(t \in D_1 \supset (s)(s \in D_2 \supset A(\langle t,s \rangle)))$ and this is the same as $(z)(z \in D_+ \supset A(z))$. b) In order to prove II), let α be in D_1 and assume III) :
$(t)(tL_1\alpha \supset (s)(s \in D_2 \supset A(\langle t,s \rangle)))$. Our task is accomplished if we can prove IV) : $(s)(s \in D_2 \supset A(\langle \alpha ,s \rangle))$. In virtue of $TI_x(L_2, A(\langle \alpha ,x \rangle))$, this is achieved if we can show V) :
$Progr_x(L_2, A(\langle \alpha ,x \rangle))$. That is, we have to infer $A(\langle \alpha , \xi \rangle)$ from the assumptions VI): 1) $\xi \in D_2$;
2) $(s)(sL_2\xi \supset A(\langle \alpha ,s \rangle))$. Because of I), we have VII) :
$(x,y)(\langle x,y \rangle L_+ \langle \alpha , \xi \rangle \supset A(\langle x,y \rangle)) \supset A(\langle \alpha , \xi \rangle)$.
In virtue of the definition of L_+, there are two cases to be distinguished: 1) $x = \alpha$ and $yL_2 \xi$; 2) $xL_1 \alpha$. In case 1) it follows from assumption VI) , 2) that $A(\langle x,y \rangle)$ holds. In case 2), however, it follows from assumption III) that $A(\langle x,y \rangle)$ holds. Thus the lefthandside of VII) holds, that is, $A(\langle \alpha , \xi \rangle)$, what proves the lemma.

<u>B.</u> Let D,L be an ordering pair and $\circledS$ an element not contained in D . Then we define a new domain and a new relation D^o, L^o, respectively, as follows: 1) $a \in D^o$ iff $a = \circledS \,.\vee.\, (a \in D)$, 2) aL^ob iff $aLb \,.\vee.\, (a = \circledS \wedge b \in D)$. We say that L^o, D^o have been obtained from D,L by addition of a smallest element $\circledS$.

<u>Lemma 37:</u> a) L^o, D^o is an ordering pair. b) If L,D is a wellordering pair, then L^o, D^o is a wellordering pair. c) If $x \in D^o$ and $x \neq \circledS$, then $\circledS L^o x$.

We omit the straightforward proof. Let D,L be a wellordering pair and let e be an arbitrary element of D . Define D_e, L_e as follows: 1) $x \in D_e$ iff eLx ; 2) xL_ey iff $x \in D_e \wedge y \in D_e \wedge xLy$. Concerning D_e, L_e, we have

<u>Lemma 38:</u> a) D_e, L_e is an ordering pair; b) if D,L is a wellordering pair, then D_e, L_e is a wellordering pair.

The proof is rather trivial and hence omitted.

<u>C.</u> Let D^1, L^1 and D^2, L^2 be two wellordering pairs and $\circledS$ an element not contained in D^1 and D^2, respectively. Let D^1_o, L^1_o and D^2_o, L^2_o be obtained from D^1, L^1 and D^2, L^2, respectively, by addition of a smallest element $\circledS$. Let D_+, L_+ be the cartesian product

of D_o^1, L_o^1 and D_o^2, L_o^2 . Then D_+, L_+ is a wellordering pair according to lemmas 36, 37. Finally, put $e = \langle \circledS, \circledS \rangle$. Then $(D_+)_e$, $(L_+)_e$ is a wellordering pair according to lemma 38. We can define $(D_+)_e$, $(L_+)_e$ with $e = \langle \circledS , \circledS \rangle$ also directly as follows:

1) $\langle a,b \rangle \in (D_+)_e$ iff
$(a \in D_1 \wedge b \in D_2) \vee (a \in D_1 \wedge b = \circledS) \vee (a = \circledS \wedge b \in D_2)$;

2) $\langle a,b \rangle (L_+)_e \langle u,v \rangle$ iff
$(\langle a,b \rangle \in (D_+)_e \wedge \langle u,v \rangle \in (D_+)_e) \vee (aL_1U \vee (a = \circledS \wedge u \in D_1) \vee$
$\vee (a=u \wedge bL_2v) \vee (a=u \wedge b=\circledS \wedge v \in D_2))$. For simplicity, we call $(D_+)_e$, $(L_+)_e$, with $e = \langle \circledS , \circledS \rangle$, the extended cartesian product of D^1, L^1 and D^2, L^2 with respect to $\circledS$. With this terminology, we infer from lemmas 36 - 38

<u>Lemma 39:</u> Let D_1, L_1 and D_2, L_2 be wellordering pairs, $\circledS$ an element not in $D_1 \cup D_2$, and $\hat{D}$, $\hat{L}$ the extended cartesian product of D_1, L_1 and D_2, L_2 with respect to $\circledS$. Then $\hat{D}$, $\hat{L}$ is a wellordering pair.

<u>10.7. The ε-construction</u>

<u>A.</u> In what follows we start with a given wellordering pair D, L and construct successively new ones D_o, L_o, D_1, L_1, D_2, L_2, etc. We call this construction ε-construction in view of its similarity with Gentzen's notation for ε_o , used in [1] . Hence, let D, L be a given, fixed wellordering pair and $+$, ω two symbols not contained in D . By definition, D_o, L_o is the wellordering pair induced by D, L according to definition 61; D_o, in particular, is the set of expressions $n_1 \alpha_1 + \dots + n_s \alpha_s$ with $n_i > 0$, $\alpha_i \in D$ and $\alpha_{i+1} L \alpha_i$ (in case $s > 1$) . Now assume that D_n, L_n have already been defined and proved to be a wellordering pair. Then we take for D_{n+1} the set of expressions of the following form:
$n_1 \omega^{\alpha_1} + \dots + n_s \omega^{\alpha_s} + m_1 \beta_1 + \dots + m_t \beta_t$ with $\alpha_i \in D_n$, $\beta_i \in D$, $n_i > 0$, $m_i > 0$, $\alpha_{i+1} L_n \alpha_i$ and $\beta_{i+1} L \beta_i$. Thereby we admit s or t (but not both) to be 0 ; in the first case we obtain an expression of the form $m_1 \beta_1 + \dots + m_t \beta_t$ belonging to D ; in the second case we obtain an expression of the form $n_1 \omega^{\alpha_1} + \dots + n_s \omega^{\alpha_s}$. The relation L_{n+1} is said to hold between $\xi = n_1 \omega^{\alpha_1} + \dots + n_s \omega^{\alpha_s} + m_1 \beta_1 + \dots + m_t \beta_t$ and $\eta = p_1 \omega^{\gamma_1} + \dots + p_a \omega^{\gamma_a} + q_1 \delta_1 + \dots + q_b \delta_b$ (in signs $\xi L_{n+1} \eta$) iff one of the following conditions is satisfied:

1) $s=a$, $n_i=m_i$, $\alpha_i=\gamma_i$ and
$m_1\beta_1+\dots+m_t\beta_t L_o q_1\delta_1+\dots+q_b\delta_b$ ($s=a=0$ admitted);
2) $s<a$ and $n_i=m_i$, $\alpha_i=\gamma_i$ for $i\leq s$ ($s=0$ admitted) ;
3) there is a $j<\min(s,a)$ such that $n_i=m_i$ and $\alpha_i=\gamma_i$ for $i\leq j$, and either $\alpha_{j+1}=\gamma_{j+1}$ and $n_{j+1}=m_{j+1}$ or else $\alpha_{j+1}L\gamma_{j+1}$.

One easily proves by induction with respect to n :

Lemma 40: 1) L_n, D_n are ordering pairs, 2) $D_n\subseteq D_{n+1}$, 3) if $\alpha,\beta\in D_n$ then $\alpha L_{n+1}\beta$ iff $\alpha L_n\beta$.

Finally there is again a natural imbedding of D in D_o and hence in D_n : an $\alpha\in D$ can be identified with $1\alpha\in D_o$. Without danger of confusion, we write simply α in place of 1α for $\alpha\in D$. There is also a notion of natural sum $\#$ whose definition and properties are quite the same as in the previous section and which will be needed later. In order to define $\#$, consider first the case of two elements $\xi=n_1\omega^{\alpha_1}+\dots+n_s\omega^{\alpha_s}$ and $\zeta=m_1\omega^{\gamma_1}+\dots+m_t\omega^{\gamma_t}$ from D_n . Let S_1 and S_2 be $\{\alpha_1,\dots,\alpha_s\}$ and $\{\gamma_1,\dots,\gamma_t\}$, respectively. Put $S=S_1\cup S_2$ and list the elements of S in decreasing order: $\lambda_1>\lambda_2\dots\dots>\lambda_r$. Then we take for $\xi\#\zeta$ the element $\eta=a_1\omega^{\lambda_1}+\dots+a_r\omega^{\lambda_r}$ where the coefficients a_i are defined as follows: 1) if there is a j and a k such that $\alpha_j=\gamma_k=\lambda_i$, then $a_i=n_j+m_k$; 2) if there is a j such that $\alpha_j=\lambda_i$ but no k such that $\gamma_k=\lambda_i$, then $a_i=n_j$; 3) if there is a k such that $\gamma_k=\lambda_i$ but no j such that $\alpha_j=\lambda_i$, then $a_i=m_k$. The direct sum $\xi\#\eta$ of elements $\xi=m_1\beta_1+\dots+m_t\beta_t$, $\zeta=q_1\delta_1+\dots+q_b\delta_b$ ($\beta_i,\delta_i\in D$) is defined in the same way as in part C of the last section. Now we extend the sum $\#$ to arbitrary elements
$\xi=n_1\omega^{\alpha_1}+\dots+n_s\omega^{\alpha_s}+m_1\beta_1+\dots+m_t\beta_t$ ($\beta_i\in D$) and $\zeta=p_1\omega^{\gamma_1}+\dots p_a\omega^{\gamma_a}+q_1\delta_1+\dots+q_b\delta_b$ ($\delta_i\in D$) by taking for $\xi\#\zeta$ the element
$((n_1\omega^{\alpha_1}+\dots n_s\omega^{\alpha_s})\#(p_1\omega^{\gamma_1}+\dots p_a\omega^{\gamma_a}))+$
$+((m_1\beta_1+\dots m_t\beta_t)\#(q_1\delta_1+\dots+q_b\delta_b))$. If, in particular, eg. $s=0$, $a\neq 0$, then the last expression reduces by definition to $(p_1\omega^{\gamma_1}+\dots+p_a\omega^{\gamma_a})+((m_1\beta_1+\dots+m_t\beta_t)\#$
$\#(q_1\delta_1+\dots+q_b\delta_b))$. If $s=a=0$, then we obtain by definition $(m_1\beta_1+\dots+m_t\beta_t)\#(q_1\delta_1+\dots+q_b\delta_b))$; similarly, in other situations such as $a=0$, $s\neq 0$ and $t\neq 0$, $b=0$ etc. Again we have

Lemma 41: If $\xi, \eta, \zeta \in D$ then 1) $\xi \# \zeta = \zeta \# \xi$, 2) $\xi L_n \xi \# \zeta$, 3) if $\xi \# \zeta = \xi \# \eta$ then $\zeta = \eta$.

B. It remains to prove

Theorem 84: If D,L is a wellordering pair, then we can prove in ZTi/I^*_N for every n the formula $TI_x(L_n, A(x))$ for $A \in W_N$.

Proof: The proof is by induction with respect to n. If n=0, then the statement is a consequence of theorem 83. Assume the theorem proved up to n. Hence D_n, L_n is a wellordering pair. In order to form the induced pair of D_n, L_n according to definition 61, we take a new sign $\oplus$ and define $(D_n)^*, (L_n)^*$ as in definition 61 but with $\oplus$ in place of +. Denote by $\hat{D}_{n+1}$ the subset of elements of D_{n+1} having the form $n_1 \omega^{\alpha_1} + \ldots + n_s \omega^{\alpha_s}$ and let $\hat{L}_{n+1}$ be the restriction of L_{n+1} to $\hat{D}_{n+1}$; the pair $\hat{D}_{n+1}, \hat{L}_{n+1}$ can easily be proved to be an ordering pair. The mapping which associates with every element $n_1 \omega^{\alpha_1} + \ldots + n_s \omega^{\alpha_s}$ from $\hat{D}_{n+1}$ the element $n_1 \alpha_1 \oplus \ldots \oplus n_s \omega_s$ is clearly an order isomorphism from $\hat{D}_{n+1}, \hat{L}_{n+1}$ onto $(D_n)^*, (L_n)^*$. From this it follows easily that $\hat{D}_{n+1}, \hat{L}_{n+1}$ is a wellordering pair. But it is not difficult to see that D_{n+1}, L_{n+1} is order-isomorphic with the extended cartesian product of $\hat{D}_{n+1}, \hat{L}_{n+1}$ and D_o, L_o. This, however, implies that D_{n+1}, L_{n+1} is a wellordering pair.

The sequence D_n, L_n, n=0,1,..... thus constructed with the aid of D,L is called the ε-construction based on D,L.

10.8. Direct sums of ordering pairs

A. Consider two ordering pairs D_1, L_1 and D_2, L_2; assume $D_1 \cap D_2 = \phi$. Then we can form a new ordering pair D^+, L^+, called the sum of D_1, L_1 and D_2, L_2. Thereby $D^+ = D_1 \cup D_2$, while xL^+y, iff one of the following conditions is satisfied: 1) $x \in D_1$ and $y \in D_2$; 2) $x, y \in D_1$ and xL_1y; 3) $x, y \in D_2$ and xL_2y. That D^+, L^+ is indeed an ordering pair can easily be proved. We also have

Lemma 42: If D_1, L_1 and D_2, L_2 are wellordering pairs, then D^+, L^+ is a wellordering pair.

Proof: We have to show: $Progr_x(L^+,A(x)) \supset (z)(z \in D^+ \supset A(z))$. That is, we have to prove $(z)(z \in D^+ \supset A(z))$ under the assumption I) : $Progr_x(L^+,A(x))$. The first step consists in proving II) : $Progr_x(L_1,A(x))$, using assumption I). We omit the verification of this in virtue of its simplicity. From II) we can infer III) : $(z)(z \in D_1 \supset A(z))$. We are through if we can prove IV) : $(z)(z \in D_2 \supset A(z))$. This is achieved if we can prove V) : $Progr_x(L_2,A(x))$. To this end, assume VI) : 1) $y \in D_2$, 2) $(x)(xL_2y \supset A(x))$. All we have to do is to prove $A(y)$ and this in turn is achieved if we can prove VII) : $(x)(xL^+y \supset A(x))$. Now $xL^+y \supset x \in D_1 \vee xL_2y$ is an immediate consequence of the definition of L^+ and of $y \in D_2$. But $x \in D_1 \supset A(x)$ holds according to III) and $xL_2y \supset A(x)$ according to VI), 2). Hence, $xL^+y \supset A(x)$, what concludes the proof.

B. There is an obvious generalisation of the above concept. If $D_1,L_1,\ldots\ldots,D_s,L_s$ is a list of ordering pairs such that $D_i \cap D_k = \phi$ for $i \neq k$, then we can form a sum D^+,L^+ by taking for D^+ the union $D_1 \cup \ldots\ldots \cup D_s$, while xL^+y iff one of the following conditions is satisfied: 1) $x,y \in D_i$ and xL_iy; 2) $x \in D_i$, $y \in D_k$ and $i < k$. For D^+,L^+ thus defined we have

Lemma 43: 1) D^+,L^+ is an ordering pair. 2) If D_i,L_i, $i=1,\ldots\ldots s$ are wellordering pairs, then D^+,L^+ is a wellordering pair.

The proof of 1) is straightforward. The proof of 2) can be reduced to the last lemma by an easy induction with respect to s. We call D^+,L^+ the sum of $D_1,L_1,\ldots\ldots,D_s,L_s$.

10.9. One-one mappings of ordering pairs

A. Consider an ordering pair D,L. Let m be a fixed number > 0 and define $\hat{D}$, $\hat{L}$ as follows: 1) $x \in \hat{D}$ iff $(Ey)(my=x \wedge y \in D)$; 2) for $mx, my \in D$ put $mxLmy$ iff xLy. Then we have

Lemma 44: a) $\hat{D}, \hat{L}$ is an ordering pair. b) If D, L is a wellordering pair, then $\hat{D}, \hat{L}$ is a wellordering pair.

We omit the obvious proof.

10.10. A particular ε-construction

A. Our aim is to replace the abstract ordinals used in chapter VI by a suitable ε-construction. To this end, let P be an s.n.s. proof in ZTFi/V whose endsequent has the form $(x)p_1(x)=0,\ldots,(x)p_s(x)=0 \longrightarrow \mathfrak{A}(<_R)$; with every such proof we can associate a certain domain D and a partial ordering $\sqsubset$ of D : namely the domain D and the partial ordering $\sqsubset$ associated with the formula

$(\xi)(Ex)(\neg\xi(x+1) \subset_R \xi(x).\vee.p_1(x)\neq 0.\vee.\ldots.\vee.p_s(x)\neq 0)$

according to the definition given in section 6.1., part A (chapter VI). There we also have associated with D a certain domain D* of sequence numbers and, denoted by $\prec^*$, the Kleene-Brouwer ordering of D* . We call D* the domain associated with P and $\prec^*$ the Kleene-Brouwer ordering associated with P . The statement "z is a (Goedelnumber of an) s.n.s. proof P in ZTFi/V whose endsequent has the form $(x)p_1(x)=0,\ldots,(x)p_s(x)=0 \longrightarrow \mathfrak{A}(<_R)$ and x belongs to the domain D* associated with P " can obviously be formalised and gives rise to an intuitionistically recursive formula $G_o(z,x)$ which expresses precisely this statement. Similarly, we can express the statement "z is (a Goedelnumber of) an s.n.s. proof P in ZTFi/V whose endsequent has the form $(x)p_1(x)=0,\ldots,(x)p_s(x)=0 \longrightarrow \mathfrak{A}(<_R)$ and $x \prec^* y$, where $\prec^*$ is the Kleene-Brouwer ordering associated with P " by means of an intuitionistically recursive formula $P_o(z,x,y)$. Finally, there is a formula F(z) which expresses the statement "z is (a Goedelnumber of) an s.n.s. proof P in ZTFi/V whose endsequent has the form $(x)p_1(x)=0,\ldots,(x)p_s(x)=0 \longrightarrow \mathfrak{A}(<_R)$ and L_P is wellfounded"; it is not difficult to see that there is such an F(z) in W_N . There are two other statements which can be formalized by means of intuitionistic recursive formulas, namely, "z is not (a Goedelnumber of) an s.n.s. proof P in ZTFi/V with endsequent $(x)p_1(x)=0,\ldots,(x)p_s(x)=0 \longrightarrow \mathfrak{A}(<_R)$ ", and "z is not (a Goedelnumber of) an s.n.s. proof P in ZTFi/V with endsequent $(x)p_1(x)=0,\ldots,(x)p_s(x)=0 \longrightarrow \mathfrak{A}(<_R)$, and $x<y$ " . The two intuitionistically recursive formulas which formalize the first and second statement, respectively, are denoted by $G_1(z)$ and $P_1(z,x,y)$, respectively; by definition, $P_1(z,x,y)$ is just $x<y \wedge G_1(z)$. Now let G(z,x) and P(z,x,y) be two intuitionistically recursive formulas for which the following holds: 1) $G(z,x) \equiv G_o(z,x) \vee G_1(z,x)$, 2) $P(z,x,y) \equiv P_o(z,x,y) \vee P_1(z,x,y)$. It is not difficult to find

such formulas G,P . With respect to the triple G,P,F we retain the notation used in the last section; in particular, we write $x \prec_z y$ in place of $P(z,x,y)$. The properties of G,P,F are summarized by the following lemma:

Lemma 45: 1) $x \prec_z y \supset G(z,x) \wedge G(z,y)$; 2) $F(z) \supset W(\prec_z)$; 3) P,G,F are in W_N ; 4) for every e , $\prec_e$ is a linear ordering of $G_e = \{x/G(e,x)\}$; 5) there is an e and a z such that $z \in G$ and $F(z)$ holds.

Clause 1) of this lemma is an obvious consequence of the definition of P,G . Clause 2) is nothing else than a restatement of theorem 40, which, as noted in section 10.3., is provable in ZTi/I_N^* . Clause 3) is obvious for P,G . As noted above, it is always possible to take F from the set W_N , and in virtue of this choice, clause 3) is true. Clause 4) is satisfied in virtue of the definition of P,G . In order to verify clause 5), it is sufficient to take for e the Goedelnumber of a proof P in ZTi whose endsequent has the form $\longrightarrow \mathfrak{W}(<_R)$ with $\{x/R(x)\}$ nonempty.

B. In terms of G,P,F we now introduce the wellordering pair D',L' by means of clauses 1),2) in part B of section 10.4. (with D',L' in place of D,L) . With the aid of D',L', we form a new wellordering pair $\hat{D},\hat{L}$ as follows: 1) $x \in \hat{D}$ iff $(Ey)(x=3y \wedge y \in D')$; 2) $3x\hat{L}3y$ iff $xL'y$ (see section 10.9.) . There are two further wellordering pairs which will be used: D_o', L_o' and D_1', L_1' . As D_o', we take the set of numbers congruent two modulo three (that is 2,5,8,....) and, as D_1', the set of numbers congruent one modulo three (1,4,7,....) . As L_o' and L_1', we take the restriction of $<$ to D_o' and D_1' , respectively. Now we form the sum of D_o', L_o' , D',L' and D_1', L_1' in this order, according to the definition in part B of section 10.8., and denote it by D,L .

In order to describe briefly the behaviour of D,L, let e,f be Goedelnumbers of s.n.s. proofs P_1, P_2 in ZTFi/V , both having an endsequent of the form
$(x)p_1(x)=0,\ldots,(x)p_s(x)=0 \longrightarrow \mathfrak{W}(<_R)$. Let D_e^* and D_f^* be the domains of sequence numbers associated with P_1 and P_2, respectively; let $\prec_e^*$ and $\prec_f^*$ be the Kleene-Brouwer orderings associated with P_1 and P_2, respectively. Assume in addition $e < f$ and let x_1, x_2, y_1, y_2 be four numbers such that $x_1 \prec_e^* x_2$ and $y_1 \prec_f^* y_2$

hold. Then we have: 1) $3n+2 L 3\langle e,x_i\rangle$, $i=1,2$ for all n ; 2) $3n+2 L 3\langle f,y_i\rangle$, $i=1,2$ for all n ; 3) $3\langle e,x_i\rangle L 3n+1$, $i=1,2$ for all n ; 4) $3\langle e,y_i\rangle L 3n+1$, $i=1,2$ for all n ; 5) $3\langle e,x_i\rangle L 3\langle f,y_k\rangle$, $i,k=1,2$; 6) $3\langle e,x_1\rangle L 3\langle e,x_2\rangle$; 7) $3\langle f,y_1\rangle L 3\langle f,y_2\rangle$. In particular, $2L3\langle e,x_1\rangle$, $3\langle e,x_1\rangle L1$, $3\langle e,x_1\rangle L4$, and similarly with $\langle e,x_2\rangle$, $\langle f,y_1\rangle$ and $\langle f,y_2\rangle$ in place of $\langle e,x_1\rangle$. In addition, we note $1L4$.

C. Now we form the ε-construction based on D,L . With respect to D_n,L_n , $n=1,2...$ we use the following notation: 1) $\omega_0(\alpha)=\alpha$; 2) $\omega_{n+1}(\alpha)=\omega^{\omega_n(\alpha)}$. This particular ε-construction will serve as a substitute for the abstract ordinals used in chapter VI. We note that elements $\alpha\in D$ can be identified with the elements $1\alpha\in D_0$, and that for $\alpha,\beta\in D$ we have $\alpha L\beta$ iff $1\alpha L 1\beta$ $(n=0,1,2,....)$. As before, we write without danger of confusion α in place of 1α for elements $\in D$. We remind that for elements $\alpha,\beta \in \bigcup_n D_n$ we have defined a natural sum $\alpha \# \beta$ which has the properties described by lemma 41.

10.11. An ordinal assignment

A. An s.n.s. proof P in ZTFi/V with endsequent $(x)p_1(x)=0,....,(x)p_s(x)=0 \longrightarrow \mathfrak{W}(<_R)$ is called "good" according to definitions 41 and 43 if and only if L_P is wellfounded. This means that $F(e)$ is true if and only if e is the Goedelnumber of such a good proof P . Graded proofs on the other hand are s.n.s. proofs in ZTF/V all whose side proofs are good. Now we are going to define an ordinal assignement for graded proofs with the aid of that particular ε-construction described in the last section. More precisely, we associate with each sequent S in a graded proof P a certain element $\in \bigcup_n D_n$, to be denoted by $o(S)$. The definition of $o(S)$ is by induction according to the clauses listed below.

1. S is an axiom. Then $o(S)=2$.

2. S is conclusion of a conversion or a one-premiss structural inference S_1/S . Then $o(S)=o(S_1)$.

3. S is the conclusion of a one-premiss logical inference S_1/S . Then $o(S)=o(S_1)\# 2$.

4. S is the conclusion of a two-premiss logical inference $S_1, S_2/S$. Then $o(S)=o(S_1) \# o(S_2)$.

5. S is the conclusion of a cut $S_1, S_2/S$. Then $o(S)=\omega_d(o(S_1) \# o(S_2))$, where $d=h(S_1)-h(S)$, and with $h(S_1), h(S)$ the height of S_1 and S respectively.

6. S is the conclusion of an induction S_1/S. We distinguish two cases.

Case 1: $o(S_1)=n_1\,\omega^{\alpha_1}+\dots\dots$. Then we put $o(S)=\omega_d(\omega^{\alpha_1 \# 2})$ where $d=h(S_1)-h(S)$.

Case 2: $o(S_1)=n_1\,\alpha_1+\dots\dots$ with $\alpha_i \in D$. Then we put $o(S)=\omega_d(\omega^2)$ where $d=h(S_1)-h(S)$.

7. S is the conclusion of a V-inference S_1/S.

Case 1: $o(S_1)=n_1\,\omega^{\alpha_1}+\dots\dots$. Then we put $o(S)=\omega_d(\omega^{\alpha_1 \# 4})$ where $d=h(S_1)-h(S)$.

Case 2: $o(S_1)=n_1\ {}_1+\dots\dots$ with $\alpha_i \in D$. Then we put $o(S)=\omega_d(\omega^4)$ where $d=h(S_1)-h(S)$.

8. S is the conclusion of a $T(P_1)$-inference S_1/S.

Case 1: $o(S_1)=n_1\,\omega^{\alpha_1}+\dots\dots$. Then we put $o(S)=\omega_d(\omega^{\alpha_1 \# 1})$ with $d=h(S_1)-h(S)$.

Case 2: $o(S_1)=n_1\,\alpha_1+\dots\dots$ with $\alpha_i \in D$. Then we put $o(S)=\omega_d(\omega^1)$.

9. S is the conclusion of a $T(P_1,a)$-inference S_1/S. Let e be the Goedelnumber of P_1. Since P is a graded proof, P_1 is good and $F(e)$ holds. By definition, a is an unsecured element of D^*, with D^* the domain associated with P. Hence $3\langle e,a\rangle \in D$, that is, $3\langle e,a\rangle \in \bigcup_n D_n$.

Case 1: $o(S_1)=n_1\,\omega^{\alpha_1}+\dots\dots$. Then we put $o(S)=\omega_d(\omega^{\alpha_1 \# 3\langle e,a\rangle})$ with $d=h(S_1)-h(S)$.

Case 2: $o(S_1)=n_1\,\alpha_1+\dots\dots$, $\alpha_i \in D$. Then we put $o(S)=\omega_d(\omega^{3\langle e,a\rangle})$, where $d=h(S_1)-h(S)$.

As ordinal of P, we take as usual the ordinal of its endsequent; we denote it by $o(P)$.

B. Our next task is to prove that the above ordinal assignement has the same properties as the ordinal assignements introduced in chapters II, IV, etc. More precisely one has to prove

Theorem 41*: Let P be a graded s.n.s. proof in ZTF/V and let any of the following reduction steps be applied to P. a) The operation "omission of a cut" lowers the ordinal of P . b) A preliminary reduction step does not increase the ordinal of P . c) A fork elimination (intuitionistic or classical) lowers the ordinal of P . d) An induction reduction lowers the ordinal. e) A T_1-reduction step lowers the ordinal. f) A T_2-reduction step lowers the ordinal. g) A subformula reduction step (as defined in part E of section 6.4.) lowers the ordinal of P .

This is the counterpart of theorem 41. We also need the counterparts of theorem 42 and of basic lemma III_1 , which are word by word the same with the only proviso that the word "ordinal" refers to the ordinal assignement defined here with the aid of the ε-construction. We denote these counterparts by theorem 42* and basic lemma III_1^* . The corollary of basic lemma III_1 is evidently true in the present case, provided basic lemma III_1^* is true. We refer to this corollary, interpreted in the present sense, as corollary * . Basic lemma III* and theorem 42*, in turn, are straightforward consequences of theorem 41* . The proof of theorem 41* consists in a step by step verification of a)-g). This verification, performed in detail, is quite lengthy, but entirely routine. We therefore content ourself with some indications.

Consider a) of theorem 41*: in order to prove a), it is essentially sufficient to prove a counterpart of lemma 8 (call it lemma 8*) (sect. 2.6., chapter II). To this end one introduces again all the notions listed under definitions 13, 14 and 15; the $T(P_1)$- , $T(P_1, a)$- and V-inferences are thereby included among the strong inferences. The proof of theorem 8* in turn essentially reduces to the proof of the counterpart of a statement A) which appears in the proof of lemma 8. This counterpart (call it A*) is the following statement: if S is a good sequent, if ξ is the ordinal of S with respect to f, and $\bar{\xi}$ the ordinal of S with respect to g, then $\bar{\xi} L_n \xi$ (for suitably large n) . This verification splits up into several cases, whose discussion is straightforward and which we omit.

Consider b) of theorem 41* : once part a) of theorem 41* is verified, part b) is an immediate consequence.

Consider c) of theorem 41* : in order to prove c) it is sufficient to show that classical fork elimination lowers the ordinal of the proof P to which it is applied. For intuitionistic fork elimination, the statement then follows immediately with the aid of parts a) and b). The case of classical fork elimination, however, leads to the verification of the following inequality: $\omega_b(\omega_a^{\alpha_1} \# \omega_a^{\alpha_2}) L_n \omega_{a+b}^{\alpha}$ (for sufficiently large n), where $\alpha_1 L_n \alpha$, $\alpha_2 L_n \alpha$ and $a \neq 0$ are assumed. From the definition of L_n and $\#$, we immediately infer $\omega_a^{\alpha_1} L_n \omega_a^{\alpha}$, $\omega_a^{\alpha_2} L_n \omega_a^{\alpha}$, $\omega_a^{\alpha_1} \# \omega_a^{\alpha_2} L_n \omega_a^{\alpha}$ and hence $\omega_b(\omega_a^{\alpha_1} \# \omega_a^{\alpha_2}) L_n \omega_{a+b}^{\alpha}$.

Consider d) of theorem 41* : a verification of d) essentially amounts to a proof of the following inequalities: 1) if $\xi =_{n_1} \omega^{\alpha_1} + \ldots$, then $(\xi \# \ldots \# \xi) L_n \omega^{\alpha_1 \# 2}$ (n sufficiently large) ; 2) if $\xi =_{n_1} \alpha_1 + \ldots$ $(\alpha_i \in D)$, then $(\xi \# \ldots \# \xi) L_n \omega^2$. Both inequalities are immediate consequences of the definition of L_n and of $\#$.

Consider e) of theorem 41* : consider the case of a critical $T(P_1)$-inference S_1/S , and assume that e is the Goedelnumber of P_1 . Application of a T_1-reduction step to the $T(P_1)$-inference S_1/S transforms this inference into a series of new inferences; among these, there occurs a particular $T(P_1,a)$-inference, where a is an element of D* , with D* the domain associated with P_1 . Assume $o(S_1)= \xi$. In order to prove that the T_1-reduction step in question lowers the ordinal of the proof one is finally led to the verification of the following inequalities: 1) if $\xi =_{n_1} \omega^{\alpha_1} + \ldots$, then $\omega_d(\omega^{\alpha_1 \# 3 \langle e,a \rangle} \# 2 \# 2 \# \xi) L_n \omega_d(\omega^{\alpha_1 \# 1})$; 2) $\omega_d(\omega^{3 \langle e,a \rangle} \# 2 \# 2 \# \xi) L_n \omega_d(\omega^1)$ (n sufficiently large in both cases). We leave it to the reader to verify that these inequalities are straightforward consequences of our definitions of L_n and $\#$.

Consider f) of theorem 41* : consider the case of a critical $T(P_1,a)$-inference S_1/S ; let e be the Goedelnumber of P_1 . Application of a T_2-reduction step transforms the inference S_1/S into a series of new inferences; among these there occurs a particular $T(P_1,b)$-inference such that $b \prec^* a$ holds, where $\prec^*$ is the Kleene-Brouwer ordering associated with P_1 . Put $o(S_1)= \xi$. The proof of f) leads to the verification of the following inequalities: 1) if $\xi =_{n_1} \omega^{\alpha_1} + \ldots$ then $\omega_d(\omega^{\alpha_1 \# 3 \langle e,b \rangle} \# 2 \# 2 \# \xi) L_n \omega_d(\omega^{\alpha_1 \# 3 \langle e,a \rangle})$;

2) if $\xi = n_1 \alpha_1 + \ldots$, with $\alpha_i \in D$, then $\omega_d(\omega^{3<e,b>^1} \# 2 \# 2 \# \xi) L_n \omega_d(\omega^{3<e,a>})$ for n sufficiently large. As before, these inequalities are straightforward consequences of the definitions of L_n and $\#$.

Consider g) of theorem 41* : a verification of g) essentially reduces to the verification of the following inequalities:
1) $\alpha L_n \alpha \# 2$, 2) $\alpha L_n \alpha \# \beta$. Both are contained in lemma 41.

Finally, consider theorem 42* : its proof essentially reduces to the verification of the following two inequalities:
1) $\omega_d(\omega^{\alpha \# 1}) L_n \omega_d(\omega^{\alpha \# 4})$,
2) $\omega_d(\omega^1) L_n \omega_d(\omega^4)$.

10.12. The wellfoundedness proof

A. We now come to our final task, namely, to the proof of an appropriate counterpart of theorem 43. To begin with, we have to convince ourself that if we restrict our attention to graded s.n.s. proofs P not containing formulas with more than n logical symbols, then one has to use only ordinals belonging to a certain D_m , with m depending on n . In order to do this, we associate with every $\alpha \in \bigcup_n D_n$ a natural number $\lambda(\alpha)$ inductively as follows:
1) if $\alpha \in D_o$, then $\lambda(\alpha)=0$; 2) if λ has already been defined on D_n , and if $\xi = n_1 \omega^{\alpha_1} + \ldots$ is in D_{n+1} , then we put $\lambda(\xi) = \lambda(\alpha_1)+1$. Concerning λ , we can prove several simple properties by induction with respect to n , whose proof we omit for simplicity. These properties are summarized by

Lemma 46: 1) if $\alpha \in D$, then $\lambda(\alpha) \leq n$; 2) if $\alpha L_n \beta$ then $\lambda(\alpha) \leq \lambda(\beta)$; 3) $\lambda(\alpha \# \beta) = \max(\lambda(\alpha), \lambda(\beta))$; 4) $\lambda(n_1 \omega^{\alpha_1} + \ldots) = \lambda(\alpha_1)+1$; 5) $\lambda(\omega_d(\alpha)) = \lambda(\alpha)+d$; 6) if $\lambda(\alpha)=n$, then $\alpha \in D_n$.

Definition 62: Let P be a proof in ZTF/V and $S_o, \ldots, S_m$ a path in P , that is a list of sequents having the properties: 1) S is an axiom; 2) S_{i+1} is the successor of S_i . We do not require that S_m is the endsequent of P . With any such path (denote it by C) we associate the number

$D(C)=(h(S_o)-h(S_1))+....+(h(S_{m-1})-h(S_m))$ (that is, $h(S_o)-h(S_m)$) ; $D(C)$ is 0 if $m=0$. Now let S be any sequent in P and F_S the set of paths which contain S as last element. Then we put $d(S)=\max D(C)$, $C \in F_S$, ($h(S)$ the heigth of S) .

A relation between λ and $d(S)$ is given by

Lemma 47: Let P be a graded s.n.s. proof in ZTF/V which does not contain formulas with more than n logical symbols. If S is a sequent in P, then $\lambda(o(S)) \leqq d(S)+1$.

Proof: a) We begin by listing two properties of d . First, if S is conclusion of a two-premiss inference, say, $S_1,S_2/S$, then $d(S)=\max(d(S_1),d(S_2))+h(S_1)-h(S)$. If, on the other hand, S is the conclusion of a one-premiss inference S_1/S , then $d(S)=d(S_1)+h(S_1)-h(S)$. b) Now we claim: $\lambda(o(S)) \leq d(S)+1$. The proof is by induction over P . We proceed in steps. 1) We omit the discussion of the following cases which are trivial to handle: α) S is an axiom; β) S is the conclusion of a one-premiss structural inference; γ) S is the conclusion of a logical inference. 2) S is the conclusion of a cut $S_1,S_2/S$, with $\lambda(S_i) \leq d(S_i)+1$, $i=1,2$ by induction. Put $d=h(S_1)-h(S)$. Then $o(S)=\omega_d(o(S_1) \# o(S_2))$, and hence $\lambda(o(S))=d+\max(\lambda(o(S_1)), \lambda(o(S_2)))$. From this we get $\lambda(o(S)) \leqq d+\max(d(S_1)+1,d(S_2)+1) \leqq d+\max(d(S_1),d(S_2))+1 = d(S)+1$. 3) S is the conclusion of an induction S_1/S . If $\lambda(o(S_1))=0$, then $o(S)=\omega_d(\omega^2)$, $\lambda(o(S))=d+1$ and $d(S)=d(S_1)+d$, hence $\lambda(o(S)) \leqq d(S)+1$ (with $d=h(S_1)-h(S)$) . If $o(S_1)=n_1\omega^{\alpha_1}+\dots$, then $\lambda(o(S_1))=\lambda(\alpha_1)+1$ and $o(S)=\omega_d(\omega^{\alpha_1 \# 2})$. From this we get $\lambda(o(S))=d+1+\lambda(\alpha_1)=d+\lambda(o(S_1))$. Since $\lambda(o(S_1)) \leqq d(S_1)+1$, we have $\lambda(o(S)) \leq d+d(S_1)+1=d(S)+1$, what proves the statement also in this case. 4) The cases where S is the conclusion of a $T(P_1)$-, $T(P_1,a)$- or V-inference S_1/S are treated in the same way as the case of induction. We omit their discussion.

In connection with this lemma, we say that a proof has bound n if no formula with more than n logical symbols occurs in this proof. From the last lemma, we infer

Lemma 48: If P is a graded proof in ZTF/V with bound n , then $o(P) \in D_{n+1}$. We also have the evident

Lemma 49: Reduction steps of any kind do not increase the bound of a proof.

B. Now we come to the main task of this section, namely, the proof of

Theorem 43*: For every fixed n, we can prove in ZTi/I^*_N the Goedelized version of the following statement: "if P is a graded s.n.s. proof in $ZTFi/V$ with bound n, then L_P is wellfounded".

Proof: Let n be fixed. By $A(x,P)$ we denote a formula which says: P is a graded s.n.s. proof in $ZTFi/V$ with bound n and x is the ordinal associated with P. In virtue of lemma 48 we have: $A(x,P) \supset x \in D_{n+1}$. Let $W(L_P)$ be a formula which says that L_P is wellfounded. By a suitable choice, both $A(x,P)$ and $W(L_P)$ are in W_N. By $B(x)$ we denote the statement: $(P)(A(x,P) \supset W(L_P))$. Obviously, $B(x)$ is in W_N. In virtue of theorem 84, we have: $TI_x(L_{n+1},B(x))$. The theorem is proved if we can show I) : $Progr_x(L_{n+1},B(x))$. Hence assume II) : a) $y \in D_{n+1}$; b) $(x)(xL_{n+1}y \supset B(x))$. We are through if we have proved III) : $B(y)$. Let P be any graded s.n.s. proof in $ZTFi/V$ with bound n and ordinal y. According to its definition, L_P is wellfounded if $L_{P'}$ is wellfounded for all P' with $P'LP$. Hence, we are through if, on the basis of II) a),b), we can prove IV) : $(P')(P'LP \supset W(L_{P'}))$. As in the proof of theorem 43, we distinguish three cases. Case 1: P is strongly saturated and does not admit preliminary reduction steps. Then $P'LP$ holds iff P' follows from P by means of an essential reduction step. Subcase 1: P' follows from P by means of a reduction step other than a V-reduction step. In virtue of theorem 41*, we have $o(P')L_m o(P)$ for sufficiently large m. It follows from lemmas 48, 49 that we can chose $n+1$ for m. Hence $A(x,P')$ holds, where $x=o(P')$. From II),b) and the form of $B(x)$ we infer: $W(L_{P'})$. Subcase 2: P' follows from P by means of a V-reduction step. Let

$$V \quad \frac{t_R(y)=0\ ,\ (x)\,<_R y\, A(x)\ ,\ \vdash\!\longrightarrow A(y)}{\overset{o}{W}(<_R)\ ,\ t_R(q)=0\ ,\ \vdash\!\longrightarrow A(q)}$$

be the critical V-inference in P, to which the V-reduction step in question is applied. Let P_1 be the side proof determined by $\overset{o}{W}(<_R)$. According to basic lemma III* and lemmas 48, 49, P is a graded s.n.s. proof in ZTFi/V whose ordinal $o(P_1)$ belongs to D_{n+1} and for which $o(P_1)L_{n+1}o(P)$ holds. Denote $o(P_1)$ by x_1. From II),b) and the form of $B(x)$ we infer $W(L_P)$. Hence, P' is a graded s.n.s. proof in ZTFi/V whose ordinal $z=o(P')$ is smaller than y (that is, $zL_{n+1}y$) according to theorem 42* . But then it follows again from II),b) and the form of $B(x)$ that $W(L_{P'})$ holds. Therefore $W(L_P)$ holds, proving thus the theorem under the assumptions of case 1. There remains the discussion of the following two cases: 2) P is strongly saturated but admits preliminary reduction steps; 3) P is not strongly saturated and admits preliminary reduction steps. Case 2) is handled in exactly the same way as in the proof of theorem 35, while case 3) is reduced to case 2), as in the proof of theorem 35 or 43.

Let us draw a few corollaries from theorem 43*.

Corollary 1: For fixed n, we can prove in ZTi/I_N^* the Goedelized version of the following statement: "If P is an s.n.s. proof in ZTi/V with bound n, then L_P is wellfounded" .

Proof: Since P has no side proofs at all, it is by definition a graded s.n.s. proof in ZTFi/V and hence subject to theorem 43* .

Corollary 2: For fixed n, we can prove in ZTi/I_N^* the Goedelized version of the following statement: "Let P be an s.n.s. proof in ZTi/V with bound n of $\longrightarrow (\alpha)(Ex)\neg\, \alpha(x+1) <_R \alpha(x)$ and assume that no special function constants occur in its endsequent. Then there is a continuity function τ with the property: if $\tau(u)\neq 0$, then there is an m and a proof P_m of $\longrightarrow \neg\, \alpha_u(m+1) <_R \alpha_u(m)$".

We omit the proof of this corollary, which is an easy consequence of corollary 1, and which proceeds along the same lines as similar proofs in earlier cases, eg. the proof of theorem 24 (chapter IV). Another straightforward consequence of corollary 1 is

Corollary 3: For fixed n we can prove in ZTi/I_N^* the Goedelized version of the following statement: "Let P be an s.n.s. proof in ZTi/V with bound n whose endsequent has the form $\longrightarrow t=q$, with t,q saturated. Then $|t| = |q|$."

A combination of corollaries 2 and 3 finally yields

Corollary 4: Let n be fixed. In ZTi/I_N^* we can prove the Goedelized version of the following statement: "Let P be an s.n.s. proof in ZTi/V which does not contain special function constants and whose bound is n. Let the endsequent of P have the form $\longrightarrow W(<_R)$, (with $R(x)$ by definition a prime formula). Then $W(<_R)$ is true".

We have omitted the proofs of corollaries 2 - 4 since they do not present the slightest difficulties and are completely analogous to the proofs of similar statements, presented earlier.

10.13. Applications

A. In order to mention two applications, we note the

Lemma 50: For every n, we find an N with the property: if P is a proof in ZT/V with bound n of $A_1,\dots,A_s \longrightarrow B$, then there is a proof P' in ZTi/V with bound N of $A_1^o,\dots,A_s^o \longrightarrow B^o$.

We omit the routine proof of this lemma. From this lemma and corollary 3 we obtain

Theorem 85: For every n the following statement is provable in ZTi/I_N^* : "If P is a proof in ZT/V with bound n of $\longrightarrow p=q$ (with p,q numerals), then $p=q$ is true".

As corollary we obtain

Corollary: If ZTi/I_N^* is consistent, then ZT/V is consistent.

B. According to a corollary stated at the end of section 4.7., chapter IV, we can prove in ZTi/V the following form of Markov's principle: $\bar{W}(<_R) \longrightarrow W(<_R)$. Combining this with corollary 4 to theorem 43* and lemma 50, we obtain

Theorem 86: For every n the following statement is provable in ZTi/I_N^* : "Let P be a proof in ZT/V with bound n of $\longrightarrow W(<_R)$ with $W(<_R)$, not containing special function constants or free variables. Then $W(<_R)$ is true".

In other words, if we can prove in ZT/V that a certain primitive recursive linear ordering is a wellordering, then we can prove this also in ZTi/I_N^* . A similar situation is described by our last

<u>Theorem 87:</u> For every n the following statement is provable in ZTi/I_N^* : "Let P be a proof in ZT/V of $\longrightarrow (x)(Ey)R(x,y)$, with $R(x,y)$ a quantifierfree formula not containing special function constants and with x,y as its only free variables, then $(x)(Ey)R(x,y)$ is true".

The proof is an immediate consequence of the corollary stated at the end of section 4.7., of lemma 50 and of corollary 4 to theorem 43*.

This concludes our investigations about the constructive character of the reasoning presented in chapter VI, in particular, and our investigations about the proof theoretic treatment of intuitionistic systems of analysis in general.

REFERENCES

1. G. Gentzen: Neue Fassung des Widerspruchfreiheitsbeweises für die reine Zahlentheorie. Forschungen zur Logik und Grundlegung der exakten Wissenschaften, Neue Folge (1938).

2. R. Harrop: Concerning formulas of types $A \longrightarrow B \quad C$, $A \longrightarrow (Ex)A(x)$ in intuitionistic formal systems. Journal of Symbolic Logic 25, 27-32 (1960).

3. W. Howard, G. Kreisel: Transfinite induction and bar induction of type zero and the role of continuity in intuitionistic analysis. Journal of Symbolic Logic 31, no. 3 (1966).

4. S.C. Kleene: Introduction to Metamathematics. North Holland 1962.

5. S.C. Kleene and R.E. Vesley: The foundations of intuitionistic mathematics. North Holland 1965.

6. S.C. Kleene: Formalized recursive functionals. Memoirs of the American Math. Soc. (1969).

7. G. Kreisel and A.S. Troelstra: Formal systems for some branches of intuitionistic systems of analysis. Annals of Mathematical Logic, vol. 1, no. 3 (1970).

8. B. Scarpellini: Some applications of Gentzen's second consistency proof. Mathematische Annalen 181, 325-344 (1969).

9. B. Scarpellini: On cut elimination in intuitionistic systems of analysis. Buffalo Conference on intuitionism and proof theory, summer 1968.

10. K. Schütte: Beweistheorie. Berlin-Göttingen-Heidelberg, Springer 1960.

11. J. Shoenfield: Mathematical Logic. Addison Wesley 1967.

12. Stanford Report: vol. I, II. Winter 1963-64, mimeographed notes.